TRENDS IN CHEMICAL PHYSICS RESEARCH

TRENDS IN CHEMICAL PHYSICS RESEARCH

A. N. LINKE
EDITOR

Nova Science Publishers, Inc.
New York

For permission to use material from this book please contact us:
Telephone 631-231-7269; Fax 631-231-8175
Web Site: http://www.novapublishers.com

LIBRARY OF CONGRESS CATALOGING-IN-PUBLICATION DATA

Trends in chemical physics research / A.N. Linke (editor).
p. cm.
Includes bibliographical references and index.
ISBN 1-59454-483-2
1. Chemistry, Physical and Theoretical--Research. I. Linke, A. N.
QD453.3.T74 2005
541'.072--dc22 2005015501

Published by Nova Science Publishers, Inc. ✦ New York

CONTENTS

Preface

Chemical physics and physical chemistry are closely related fields of study. Together they are distinguished from other disciplines by the incredible range of problems addressed by their practitioners. An effective physical chemist or chemical physicist is a "jack-of-all-trades," able to apply the principles and techniques of the field to everything from high-tech materials to biology. Just as the fields of chemistry and physics have expanded, so have chemical physics subject areas, which include polymers, materials, surfaces/interfaces, and biological macromolecules, along with the traditional small molecule and condensed phase systems. This new book gathers research from around the world presenting important new developments.

Chapter 1 reviews the analytical methods of solving the stochastic equations for barrier-type dynamical behavior in plasma systems. The path-integral approach is examined as a particularly efficient method of determination of the statistical properties.

Anodic polarization behavior of iron in phosphoric acid displays oscillatory phenomena at different electrolyte concentrations. However, oscillatory behavior at low values of phosphoric acid concentration is different, as the current oscillations do not occur at the negative branch of the polarization curve (where faradaic impedance is negative to fulfill the necessary condition for the occurrence of electrochemical oscillations). For investigating the sign of faradaic impedance, electrochemical impedance spectroscopy was employed in chapter 2. The frequency response or impedance spectra showed the existence of negative faradaic impedance in hidden form. This is the first report on the observation of so-called "hidden negative faradaic impedance" during electrochemical oscillators involving metal dissolution under potentiostatic condition.

Means are presented for incorporating metric contributions into transition rates in the attempt to relate more specific properties of the potential energy surface at stationary points with the relaxation towards the global minimum. Metric contributions are corrections due to reducing the reaction coordinate from ordinary coordinates to internal degree of freedom. These mentain the signature of corresponding geometric arrangements at stationary points on the canonical behavior of the system. As an application, the authors work out the *four*-level system of the H_2CO surface and finger out the level at which the metric contributions may affect the accuracy of characterizing the ability of the system to find the global minimum. The output of chapter 3 can lead to identify a way to enrich our knowledge about the ability of the complex systems to relax only to preferential structures.

Chapter 4 reviews recent attempts aimed at the recursive calculations of radial matrix elements in relativistic quantum mechanics. The authors first discuss how to obtain sum rules relating matrix elements of f and of βf with matrix elements of their first and second derivatives —f is an arbitrary radial function and β is the standard Dirac matrix. Such elements are assumed to be taken between radial energy eigenfunctions corresponding to two different radial potentials $V_1(r)$ and $V_2(r)$ in the unshifted case. That is, the validity of our obtained relations requires that both potentials attain a minimum at the same radial position. To obtain directly usable results we then insert in our general formulas specific expressions for the two radial potentials and a definite form of the radial function to obtain recurrence relations between the relativistic matrix elements of the mentioned function. They give specific examples of such procedure obtaining both new and previously reported recurrence relations between matrix elements of powers of the radial coordinate, r^λ and βr^λ, where λ can in general be a complex number, between radial hydrogenic states. They also obtain recurrence relations for two center matrix elements of r^λ in potential energy functions of the form $V_i(r) = a_i r^{\alpha_i}, i = 1, 2$, where a_i and α_i are constants.

Chapter 5 shows that the difference between the heat of chemical reaction derived by the Van't-Hoff equation, ΔH^0, and measured by calorimetry, ΔQ^0, is $P\Delta V^0$. An explanation is given.

According to the first law of thermodynamics, $(\partial S/\partial V)_U$ is always positive. According to other equations of thermodynamics, for heat exchange it is positive if the thermal expansion coefficient is positive, and negative if the thermal expansion is negative. A contradiction in thermodynamic equations is shown in chapter 6. It is also shown that this result is successfully used for calculation of phase diagrams for substances with negative thermal expansion (ice, iodine) and for calculation of anomalous dependence of the heat capacity of ZrW_2O_8, HfW_2O_8, water and other substances on pressure. This contradiction is explained.

Ab initio density functional theory is increasingly applied to accurately estimate parameters for the Marcus theory of electron transfer. In particular, the conceptually simple outer-sphere self-exchange of transition-metal ions in aqueous solution has been a frequent choice. Although transition metal ions only exist in trace quantities in natural aquatic environments, their redox cycling in the oceans and atmosphere is geo- and biochemically important. The experimental determination of their reaction rates is furthermore extremely tedious, partly because different transition-metal ions can have rates that vary by several orders of magnitude. A convenient theoretical approach consists of separating the various contributions to the electron transfer rate into a pre-exponential and an exponential, Arrhenius-type factor, exp(-$\Delta G^*/kT$), with ΔG^* being the important activation free energy. Chapter 7 reviews contributions from density functional theory to the study of outer-sphere self-exchange reactions. The calculation of activation energies (and of related reorganization energies) is our main focus. A variety of approaches will be discussed, such as the George-Griffith-Marcus method, the Klimkāns and Larsson method, the polynomial curve-fitting method, as well as the Hessian matrix method.

Recently, optical beams with a new and distinctive type of helicity have become the subject of considerable interest. While circularly polarised light is associated with electromagnetic fields whose locus takes a helical form, for these optically engineered phase-bearing or 'twisted laser beams it is the wave-front surface that is helical. Consequently, a new and intriguing manifestation of optical handedness is available for characterising optical interactions with matter. Many of the interactions that have been studied with such beams utilise or exhibit their intrinsic 'orbital angular momentum. To date, however, such studies have mostly focused on interactions with achiral (generally mirror-symmetric) matter. It is timely to assess what new features, if any, can be expected when such beams are used to interrogate a chiral (molecularly right- or left-handed) system. In chapter 8 the authors introduce and analyse the concepts that determine whether optically twisted beams should offer a controllable specificity in their interactions with chiral matter, associated with orbital angular momentum transfer. By explicit analysis of the grounds for chiroptical behaviour, and with a specific focus on circular dichroism, they establish the principle that the helicity of optical vortices cannot directly engage through any parametric or non-parametric optical process with the chirality of a molecular system, up to the level of electric quadrupole interactions. These conclusions have been corroborated by recent experimental results.

Recent interest is focused on establishing the necessary conditions that guarantee a long-term stable evolution of both natural and artificial systems as described in chapter 9. Two necessary conditions, called global and local boundedness, are that a system stays stable if and only if the amount and the rate of exchange of energy and/or matter currently involved in any transition do not exceed the thresholds of stability of the system. The relationship between the local and global boundedness and the stability of the system introduces two new general properties of the state space and the motion in it, namely: the state space is always bounded, the successive steps of motion are always finite and involve only nearest neighbors. An immediate consequence of the boundedness is that the invariant measure of the state space is the normal distribution. The necessary condition for the asymptotic stability of the invariant measure is derived. It is found out that the state space exhibits strong chaotic properties regardless to the particularities of the system considered. An example of kinetics, called chaotic one, that is compatible with both global and local boundedness is considered.

Large scale computer simulations (those involving explicit consideration of a large number of atoms and/or long simulation times) are needed in order to get a proper understanding of many materials properties. Traditional *ab initio* electronic structure methods, which provide a fairly accurate description of atomic interactions, can not yet be routinely employed in large scale simulations, due to their prohibitive computational expense. The advent of so-called linear-scaling *ab initio* methods, which take advantage of the exponential decay of the one-body density matrix in real space, has much improved this situation for insulators and semiconductors. However, these methods can not be applied to metallic systems, where the absence of a finite band gap imposes a power-like decay of one-body density matrix and inhibits electron localization. In chapter 10, the authors describe a natural scheme to obtain *ab initio* linear scaling for metallic systems. It involves the use of the density functional theory (DFT) within its Hohenberg-Kohn (HK) representation where the valence electron density stands as the basic variable, thus avoiding employment of auxiliary one-particle orbitals as in its Kohn-Sham (KS) representation. The electronic kinetic energy is expressed by an explicit functional of the density, and the interaction of valence electrons with ionic cores is represented by local pseudopotentials; the rest of the energy functional is the same as in traditional KS approaches. In order to perform molecular dynamics simulations, forces on atoms are obtained by means of Hellmann-Feynman theorem. The authors demonstrate applicability of the present methodology to the efficient simulation of problems which otherwise would be very difficult to simulate with the usual KS representation. Specifically, they provide new results for the melting-like transition in medium-size heterogeneous alkali clusters, as well as for the static and dynamic properties of bulk simple (*sp bonded*) liquid metals and the liquid-vapor interfaces.

Generation of femtosecond and subfemtosecond pulses is a key to the direct observation of electron and nuclear rearrangements in molecules in real time. The feasibility of investigations relies on the concept of coherence. Using ultrafast laser pulses in time-resolved inner-shell spectroscopy allows one to probe dynamics of core-excited atoms and molecules. Understanding of dynamical changes in molecules, as well as advances in laser technology and molecular spectroscopy stimulate the study of laser control of molecular motion. Modern adaptive control techniques use feed-back to correct the shape of the pulse. Selective excitation of molecular vibrations is achieved in stimulated Raman scattering experiments using adaptive control based on genetic type algorithm. To build up a connection between the pulse shapes and mode selective excitation, a semiclassical theory is developed that investigates the possibility of control of excitations of closely spaced Raman transitions in molecules using broadband pulse shaping. In chapter 11, a method is proposed for achieving either enhancement or suppression of particular vibrational excitations using broadband pulses. It implies pulse shapes possessing a broad spectral dip at the frequency to be suppressed and a suitably chosen intensity of the field at the frequency to be efficiently excited. The role of coupling between vibrational modes is elucidated in the controllability of selective excitation in weak and strong fields. Phase control of the initially populated vibrational states gives new possibilities for the coherence manipulation in molecules

The interaction energies between the different types of bases of a single strand of DNA molecule have been calculated. Using these original values of energies the harmonic behavior of a number of base patterns of DNA has been studied in chapter 12. In view of the great interest aroused by the discovery of the role of the telomere segment of the DNA in the replication process and its possible link with the aging process, we have investigated, with simple models, the harmonic behavior of the telomeric pattern of bases as well as the thermodynamic response in the biological system. With these results a conclusion on the probable role of the telomeric pattern on aging has also been drawn. Here the calculated values of harmonic frequencies of the telomeric pattern of bases and of other possible patterns show that the telomeric pattern is associated with the highest vibrational frequency among all patterns of base combinations at the replication end of DNA. This seems to suggest that due to the existing telomeric pattern being closest to the frequencies of the electromagnetic radiation coming from sunlight, resonance of the telomeric frequency with such radiation may be responsible for damage to the reproductive ability of the cells and consequent aging and other problems. On the other hand in the last part of this work we have calculated the thermal vibrational amplitudes of the telomeric pattern and other possible patterns which show that the amplitude for the telomeric pattern is the least, and this suggests that the telomeric pattern is more mechanically and thermally stable than other possible patterns in the biological environment against damage from ordinary heat and mechanical effects.

In: Trends in Chemical Physics Research
Editor: A.N. Linke, pp. 1-69

ISBN 1-59454-483-2

Chapter 1

REVIEW OF ANALYTICAL TREATMENTS OF BARRIER-TYPE PROBLEMS IN PLASMA THEORY

***F. Spineanu*[1,2], *M. Vlad*[1,2], *K. Itoh*[1], *S.-I. Itoh*[3]**
[1]National Institute for Fusion Science
322-6 Oroshi-cho, Toki-shi, Gifu-ken 509-5292, Japan
[2]Association EURATOM-MECT Romania
NILPRP, P.O.Box MG-36 Magurele, Bucharest, Romania
[3]Research Institute for Applied Mechanics, Kyushu University,
Kasuga 816-8580, Japan

Abstract

We review the analytical methods of solving the stochastic equations for barrier-type dynamical behavior in plasma systems. The path-integral approach is examined as a particularly efficient method of determination of the statistical properties.

1 Introduction

A large number of physical systems, in particular in plasma physics, are described in terms of a variable evolving in a deterministic velocity field under the effect of a random perturbation. This is described by a stochastic differential equation of the form

$$\dot{x} = V(x) + \sqrt{2D}\xi(t) \tag{1}$$

where $V(x)$ is the deterministic velocity field and the perturbation is the white noise

$$\langle \xi(t)\xi(t')\rangle = \delta(t-t') \tag{2}$$

This general form has been invoked in several applications. More recently, in a series of works devoted to the explanation of the intermittent behavior of the statistical characteristics of the turbulence in magnetically confined plasma it has been developed a formalism based on barrier crossing. Previous works that have discussed subcritical excitation of plasma instabilities are Refs. [1], [2], [3], [4], [5], [6], [7], [8], [9]. Various other applications of the generic model of transition in plasma have been investigated [10], [11], [12], [13], [14]. Some models are related with the presence of two distinct regimes in plasma transport [15], [16] or with structures [17]. See also Ref.[18].

In the present work we provide a comparative analysis of two functional integral approaches to the determination of the statistical properties of the system's variable $x(t)$ for the case where the space dependence of $V(x)$ is characterized by the presence of three equilibrium points, $V(x) = 0$. For all explicit applications we will take

$$V(x) = ax - bx^3 \text{ , with } a, b > 0 \tag{3}$$

2 The Functional Approach

2.1 Overview of the Functional Methods

In studying the stochastic processes the functional methods can be very useful and obtain systematic results otherwise less accessible to alternative methods. The method has been developed initially in quantum theory [19] and it is now a basic instrument in condensed matter, field theory, statistical physics, etc. In general it is based on the formulation of the problem in terms of an action functional. There are two distinct advantages from this formulation: (1) the system's behavior appears to be determined by all classes of trajectories that extremize the action and their contributions are summed after appropriate weights are applied; (2) the method naturally includes the contributions from states close to the extrema, so that uctuations can be accounted for.

There are technical limitations to the applicability of this method. In the statistical problems (including barrier-type problems) it is simpler to treat cases with white noise, while colored noise can be treated perturbatively. In the latter case, the procedure is however still useful since the diagrammatic series can be formulated systematically. Alternatively, the colored noise can be treated by extending the space of variables: the stochastic variable with finite correlation is generated by integration of a new, white noise variable. For the barrier-type problem [20] the functional method has been developed in numerous publications [22], [23], [24], [25], etc.

The one dimensional version can be developed up to final explicit result. Since however the barrier type problem is frequently formulated in two-dimensions, one has to look for extrema of the action and ennumerate all possible trajectories. It is however known that, in these cases, the behavior of the system is dominated by a particular path, "the optimum escape path", and a reasonnable approximation is to reduce the problem to a one-dimensional one along this system's trajectory.

2.2 The Path Integral with the Martin-Siggia-Rose (MSR) Action

We will brie y mention the steps of constructing the MSR action functional [26], in the Jensen path integral reformulation [27]. We begin by choosing a particular realisation of the noise $\xi(t)$. All the functions and derivatives can be discretised on a lattice of points in the time interval $[-T, T]$ (at the limit we will take $\pm\infty$). The solution of the equation (1) is a "configuration" of the field $x(t)$ which can be seen as a point in a space of functions. We extend the space of configurations $x(t)$ to this space of functions, including all possible forms of $x(t)$, not necessarily solutions. In this space the solution itself will be individualised by a functional Dirac δ function.

$$\delta\left(\dot{x} - V(x) - \sqrt{2D}\xi(t)\right)$$

Any functional of the system's real configuration (*i.e.* solutions of the equations) can be formally expressed by taking as argument an arbitrary functional variable, multiplying by this δ functional and integrating over the space of all functions.

We will skip the discretization and the Fourier representation of the δ functions, followed by reverting to the continuous functions. The result is the following functional (see [28], [29], [30])

$$Z_\xi = \int D[x(t)]\, D[k(t)] \exp\left[i \int_{-T}^{T} dt \left(-k\dot{x} + kV(x) + \sqrt{2D}k\xi\right)\right]$$

The index ξ means that the functional is still defined by a choice of a particular realization of the noise. The generating functional is obtained by averaging over ξ.

$$Z = \langle Z_\xi \rangle = \int D[x(t)]\, D[k(t)] \exp\left[i \int dt \left(-k\dot{x} + kV(x) + iDk^2\right)\right] \tag{4}$$

We add a formal interaction with two currents

$$Z_J = \int D[x(t)]\, D[k(t)] \exp\left[i \int dt \left(-k\dot{x} + kV(x) + iDk^2 + J_1 x + J_2 k\right)\right] \tag{5}$$

in view of future use to the determination of correlations. This functional integral must be calculated explicitely. The standard way to proceed to the calculation of Z_J is to find the saddle point in the function space and then expand the action around this point to include the uctuating trajectories. This requires first to solve the Euler-Lagrange equations

$$\begin{cases} \dot{k} = -ak + 3bkx^2 - J_1 \\ \dot{x} = ax - bx^3 + 2iDk + J_2 \end{cases} \tag{6}$$

The simplest case should be examined first. We assume there is no deterministic velocity ($a = b = 0$) in order to see how the purely diffusive behavior is obtained in this framework

$$\begin{aligned} \dot{x} - 2iDk &= J_2 \\ \dot{k} &= -J_1 \end{aligned}$$

The equations can be trivially integrated [29], [30]

$$
\begin{aligned}
x^{(0)}(t) &= \int_{-T}^{T} dt \Delta_{11}(t,t') J_1(t') + \int_{-T}^{T} dt \Delta_{12}(t,t') J_2(t') \\
k^{(0)}(t) &= \int_{-T}^{T} dt \Delta_{21}(t,t') J_1(t') + \int_{-T}^{T} dt \Delta_{22}(t,t') J_2(t')
\end{aligned}
$$

where

$$
\begin{aligned}
\Delta_{11}(t,t') &= 2iD\left[t\Theta(t'-t) + t'\Theta(t-t')\right] \\
\Delta_{12}(t,t') &= \Theta(t-t') \\
\Delta_{21}(t,t') &= \Theta(t'-t) \\
\Delta_{22}(t,t') &= 0
\end{aligned}
$$

with the symmetry

$$\Delta_{ij}(t,t') = \Delta_{ji}(t',t)$$

The symbol $\Theta(x)$ is the Heaviside funstion.

The lowest approximation to the functional integral Z_J is obtained form this saddle point solution, by calculating the action along this system's trajectory. We insert this solutions in the expression of the generating functional, for $V(x) \equiv 0$

$$
\begin{aligned}
Z_J^{(0)} &= \int D[x(t)] D[k(t)] \exp\left[i \int_{-T}^{T} dt \left(-k\dot{x} + iDk^2 + J_1 x + J_2 k\right)\right]\Bigg|_{x^{(0)},k^{(0)}} \\
&= \exp\left[\frac{1}{2} i \int_{-T}^{T} dt \int_{-T}^{T} dt' J_i(t) \Delta_{ij}(t,t') J_j(t)\right]
\end{aligned}
$$

The dispersion of the stochastic variable $x(t)$ can be obtained by a double functional derivative followed by taking $J_i \equiv 0$. We obtain

$$
\begin{aligned}
\langle x(t) x(t') \rangle &= \frac{1}{Z_J^{(0)}} \frac{\delta^2}{\delta J_1(t) \delta J_1(t')} \exp\left[\frac{1}{2} i \int_{-T}^{T} dt \int_{-T}^{T} dt' J_i(t) \Delta_{ij}(t,t') J_j(t)\right]\Bigg|_{J_{1,2}=0} \\
&= D \min(t,t')
\end{aligned}
$$

which is the diffusion. The same mechanism will be used in the following, with the difference that the equations cannot be solved in explicit form due to the nonlinearity.

In general the nonlinearity can be treated by perturbation expansion, if the amplitude can be considered small. This is a procedure analogous to that used in field theory and leads to a series of terms represented by Feynman diagrams. We can separate in the Lagrangian the part that can be explicitely integrated (quadratic, or Gaussian) and make a perturbative treatment for the non-quadratic term. This is possible when we assume a particular (polynomial) form of the deterministic velocity, $V(x)$. Obviously, in Eq.(5) this term is $k(t)\,x(t)^3$. The functional integral can be written, taking account of this separation

$$Z_J = \exp\left[i \int_0^T dt\,(-b) \frac{\delta}{i\delta J_2(t)} \frac{\delta}{i\delta J_1(t)} \frac{\delta}{i\delta J_1(t)} \frac{\delta}{i\delta J_1(t)}\right] Z_J^{(q)} \qquad (7)$$

where the remaining part in the Lagrangian is quadratic

$$Z_J^{(q)} = \int D\left[x\left(t\right)\right] D\left[k\left(t\right)\right] \exp\left[i \int dt \left(-k\dot{x} + akx + iDk^2 + J_1 x + J_2 k\right)\right] \quad (8)$$

The Euler-Lagrange equations are

$$\begin{aligned} \dot{k} &= -ak - J_1 \\ \dot{x} &= ax + 2iDk + J_2 \end{aligned}$$

The solutions can be expressed as follows

$$\begin{aligned} x\left(t\right) &= x_0 \exp\left(at\right) + \int_0^T dt' \Delta_{1j}\left(t,t'\right) J_j\left(t'\right) \\ k\left(t\right) &= \int_0^T dt' \Delta_{2j}\left(t,t'\right) J_j\left(t'\right) \end{aligned} \quad (9)$$

with

$$\begin{aligned} \Delta_{11}\left(t,t'\right) &= \frac{iD}{(-a)} \exp\left(at\right) \\ &\quad \times \left\{\left[\exp\left(-2at\right) - 1\right] \Theta\left(t'-t\right) + \left[\exp\left(-2at'\right) - 1\right] \Theta\left(t-t'\right)\right\} \\ &\quad \times \exp\left(at'\right) \\ \Delta_{21}\left(t,t'\right) &= \exp\left(-at\right) \Theta\left(t'-t\right) \exp\left(at'\right) \\ \Delta_{12}\left(t,t'\right) &= \exp\left(at\right) \Theta\left(t-t'\right) \exp\left(-at'\right) \\ \Delta_{22}\left(t,t'\right) &= 0 \end{aligned} \quad (10)$$

The form of the generating functional derived from the quadratic part is

$$Z_J^{(q)} = \exp\left\{i \int_0^T dt x_0 \exp\left(at\right) J_1\left(t\right) + \frac{i}{2} \int_0^T dt \int_0^T dt' J_i\left(t\right) \Delta_{ij}\left(t,t'\right) J_j\left(t'\right)\right\} \quad (11)$$

The occurence of the first term in the exponent is the price to pay for not making the expansion around x_0. However, such an expansion would have produced two non-quadratic terms in the Lagrangian density: $(-3bx_0)\,\varepsilon^2 k$ and $(-b)\,\varepsilon^3 k$ where ε is the small parameter. This would render the perturbative expansion extremly complicated since we would have to introduce two vertices : one, of order four, is that shown in Eq.(7) and another, of order three, related to the first of the nonlinearities mentioned above.

Even in the present case, the calculation appears very tedious. We have to expand the vertex part of Z_J as an exponential, in series of powers of the vertex operator. In the same time we have to expand the exponential in Eq.(11) as a formal series. Then we have to apply term by term the first series on the second series. The individual terms can be represented by diagrams. In this particular case we have a finite contribution even at the zero-loop order (the "tree" graph). It is however much more difficult to extract the statistics since we will need at least the diagrams leaving two free ends with currents J_1.

In the case we examine here, the perturbative treatment is not particularly useful since the form of the potential (from which the velocity field is obtained) supports topologically distinct classes of saddle point solutions and this cannot be represented by a series expansion.

2.3 The Onsager-Machlup Functional

To compare with other approaches, we take $J_{1,2} = 0$ and integrate over the functional variable k.

$$\begin{aligned} Z &= \int D[x(t)] D[k(t)] \exp\left\{ \int_0^T dt \left[-Dk^2 + i\left(-k\dot{x} + kV \right) \right] \right\} \\ &= \int D[x(t)] D[k(t)] \exp\left\{ -\int_0^T dt \left[\sqrt{D}k + \frac{i\dot{x} - iV}{2\sqrt{D}} \right]^2 + \int_0^T dt \frac{\left(i\dot{x} - iV \right)^2}{4D} \right\} \end{aligned}$$

$$Z = N_1 \int D[x(t)] \exp\left\{ -\frac{1}{4D} \int_0^T dt \left(\dot{x} - V \right)^2 \right\}$$

In other notations

$$Z = N_1 \int D[x(t)] \exp\left[-\frac{S}{D} \right]$$

where

$$S = \int_0^T dt \frac{1}{4} \left(\dot{x} - V \right)^2$$

In Ref. [31] this expression is called the Onsager-Machlup action functional and the analysis is based on this formula.

If we go further with our comparison we will find inconsistencies. We take account of the fact that the velocity is derived form a potential

$$\begin{aligned} V[x(t)] &= -\frac{dU[x(t)]}{dx} \\ &\equiv -U'[x(t)] \end{aligned}$$

$$\begin{aligned} S &= \int_0^T dt \frac{1}{4} \left(\dot{x}^2 + 2\dot{x}U' + U'^2 \right) \\ &= \int_0^T dt \left[\frac{1}{4} \left(\dot{x}^2 + U'^2 \right) + \frac{1}{2} \frac{dU}{dt} \right] \\ &= \frac{1}{2} \left[U(T) - U(0) \right] + \int_0^T dt \left[\frac{1}{4} \left(\dot{x}^2 + U'^2 \right) \right] \end{aligned}$$

This leads to the form of the generating functional

$$Z = \frac{\exp\left[-\frac{U(T)}{2D}\right]}{\exp\left[-\frac{U(0)}{2D}\right]} K(x,t;x_i,t_i)$$

with

$$K(x,t;x_i,t_i) = \int D[x(t)] \exp\left(-\frac{S}{D}\right)$$

$$S = \int_0^T dt \left[\frac{1}{4}\left(\dot{x}^2 + U'^2\right)\right] \tag{12}$$

These are almost identical to the formulas (2-5) of the reference [23] (except that $\varepsilon \rightarrow 2D$). Also, it is quite close of the Eqs.(7a-7c) of the ref.[34].

However there is an important difference.

There is a term missing in Eq.(12) which however is present in the two above references. The full form of the action S, instead of Eq.(12) is

$$S = \int_0^T dt \left(\frac{1}{4}\dot{x}^2 + W\right)$$

$$W = \frac{U'^2}{4} - \frac{D}{2}U''$$

The last term comes from the Jacobian that is hidden in the discretization of the functional δ integration.

2.4 Connection between the MSR Formalism and Onsager-Machlup

In our approach the most natural way of proceeding with a stochastic differential equation is to use the MSR type reasonning in the Jensen reformulation. The equation is discretized in space and time and then selected with δ functions in an ensemble of functions (actually in sets of arbitrary numbers at every point of discretization). The result is a functional integral. There is however a particular aspect that needs careful analysis, as mentioned in the previous Subsection. It is the problem of the Jacobian associated with the fact that the argument of the δ functional is the expression representing the original equation for the unkown function $x(t)$, and not the unknown function directly. This problem is discussed in Ref.[39].

The equation they analyse is presented in most general form as

$$\frac{\partial \phi_j(t)}{\partial t} = -(\Gamma_0)_{jk} \frac{\delta H}{\delta \phi_k(t)} + V_j[\phi(t)] + \theta_j$$

where the number of stochastic equations is N , H is functional of the fields, V_j is the streaming term which obeys a current-conserving type relation

$$\frac{\delta}{\delta \phi_j} V_j[\phi] \exp\{-H[\phi]\} = 0$$

The noise is θ_j.

The following generating functional can be written

$$Z_\theta = \int D\left[\phi_j(t)\right] \exp \int dt \left[l_j \phi_j(t)\right] \prod_{j,t} \delta\left(\frac{\partial \phi_j(t)}{\partial t} + K_j\left[\phi(t)\right] - \theta_j\right) J\left[\phi\right]$$

the functions $l_j(t)$ are currents,

$$K_j\left[\phi(t)\right] \equiv -\left(\Gamma_0\right)_{jk} \frac{\delta H}{\delta \phi_k(t)} + V_j\left[\phi(t)\right]$$

and $J[\phi]$ is the Jacobian associated to the Dirac δ functions in each point of discretization.

The Jacobian can be written

$$J = \det\left[\left(\delta_{jk}\frac{\partial}{\partial t} + \frac{\delta K_j\left[\phi\right]}{\delta \phi_k}\right)\delta\left(t - t'\right)\right]$$

Up to a multiplicative constant

$$J = \exp\left(Tr \ln\left[\left(\frac{\partial}{\partial t} + \frac{\delta K}{\delta \phi}\right)\frac{\delta\left(t-t'\right)}{\frac{\partial}{\partial t}\delta\left(t-t'\right)}\right]\right)$$

or

$$J = \exp\left(Tr \ln\left[1 + \left(\frac{\partial}{\partial t}\right)^{-1}\frac{\delta K(t)}{\delta \phi(t')}\right]\right)$$

Since the operator $\left(\frac{\partial}{\partial t}\right)^{-1}$ is retarded, only the lowest order term survives after taking the trace

$$J = \exp\left[-\frac{1}{2}\int dt \frac{\delta K_j\left[\phi(t)\right]}{\delta \phi_j(t)}\right]$$

The factor $1/2$ comes from value of the Θ function at zero.

In the treatment which preserves the dual function $\widehat{\phi}$ associated to ϕ in the functional, there is a part of the action

$$\widehat{\phi} K\left[\phi\right]$$

Then a $\widehat{\phi}$ and a ϕ of the same coupling term from $\widehat{\phi}K[\phi]$ close onto a loop.Since $G_{\widehat{\phi}\phi}$ is retarded, all these contributions vanish except the one with a single propagator line. This cancels exactly, in all orders, the part coming from the Jacobian.

Then it is used to ignore all such loops and together with the Jacobian.

We can now see that in our notation this is precisely the term needed in the expression of the action.

$$\begin{aligned}
\phi(t) &\rightarrow x(t) \\
K_j\left[\phi(t)\right] &\rightarrow U'\left[x(t)\right] \\
\frac{\delta K_j\left[\phi(t)\right]}{\delta \phi_j(t)} &\rightarrow -U''\left[x(t)\right]
\end{aligned}$$

and the action (12) is completed with the new term

$$\int_0^T dt \left(-\frac{D}{2}U''\right)$$

Now the generating functional is

$$Z = \frac{\exp\left[-\frac{U(T)}{2D}\right]}{\exp\left[-\frac{U(0)}{2D}\right]} K\left(x, t; x_i, t_i\right)$$

with

$$K\left(x, t; x_i, t_i\right) = \int D\left[x\left(t\right)\right] \exp\left(-\frac{S}{D}\right)$$

and

$$S = \int_0^T dt \left(\frac{1}{4}\dot{x}^2 + W\right) \tag{13}$$

$$W = \frac{U'^2}{4} - \frac{D}{2}U'' \tag{14}$$

Now the two expressions are identical with those in the references cited. This will be the starting point of our analysis.

In conclusion we have compared the two starting points in a functional approach: The one that uses dual functions $x(t)$ and $k(t)$, closer in spirit to MSR. And the approach based on Onsager-Machlup functional, traditionally employed for the determination of the probabilities [34], [23]. Either we keep $k(t)$ and ignore the Jacobian (the first approach) or integrate over $k(t)$ and include the Jacobian. The approaches are equivalent and, as we will show below, lead to the same results.

A final observation concerning the choice of one or another method: in the MSR method, the trajectories include the diffusion from the direct solution of the Euler-Lagrange equations. In the Onsager-Machlup method the paths extremizing the action are deterministic and the diffusion is introduced by integrating on a neighborhood in the space of function, around the deterministic motion.

3 The Transition Solutions (Instantons)

3.1 Numerical Trajectories

The equations for the saddle point trajectory are expressed in complex numbers so we must also extend the variable in complex space

$$\begin{aligned} x &\rightarrow (x_R, x_I) \\ k &\rightarrow (k_R, k_I) \end{aligned}$$

We obtain a system of four nonlinear ordinary differential equations which can be integrated numerically.

A typical form of the solution $x(t)$ is similar to the kink instanton (i.e. the $\tanh$ function). The function $x(t)$ spends very much time in the region close to the equilibrium point; then it performs a fast transition to the neighbour equilibrium point, where it remains for the rest of the time interval.

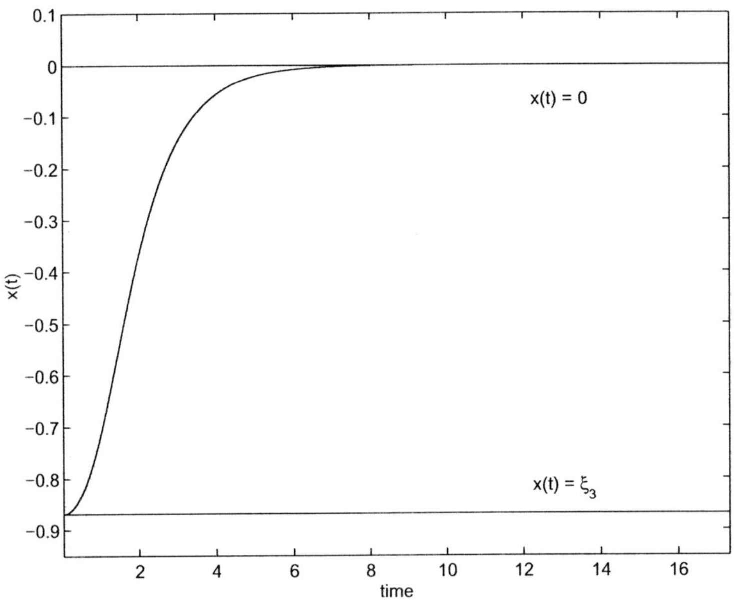

Figure 1: The kink-like instanton connecting a position close to the left (stable) equilibrium point to the middle $x = 0$ (unstable) equilibrium point in the effective potential $-W(x)$

3.2 Elliptic Functions Instantons (from Onsager-Machlup Action)

The action functional Eqs.(13) and (14) leads to the following differential equation (which replaces Eqs.(6))

$$\frac{1}{2}\ddot{x} = \left(\frac{3b^2}{2}\right)x^5 + (-2ab)\,x^3 + \left(\frac{a^2}{2} - 3Db\right)x$$

Multiplying by $\dot{x}$ and integrating we have

$$\dot{x} = \pm\left[b^2 x^6 + (-2ab)\,x^4 + \left(a^2 - 6Db\right)x^2 + c_1\right]^{1/2} \tag{15}$$

We are interested in the functions $x(t)$ that has the following physical property: they stay for very long time stuck to the equilibrium points and perform a fast jump between them at

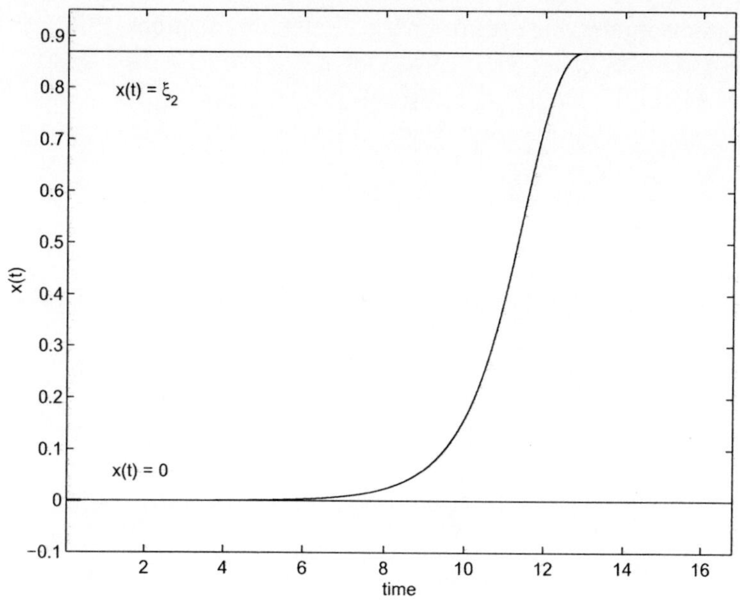

Figure 2: The kink-like instanton connecting the middle $x = 0$ (unstable) equilibrium point with a position close to the right (stable) equilibrium point to in the effective potential $-W(x)$

a certain moment of time. Then we can take $c_1 = 0$. The solution can be obtained form the integration

$$\int_x^{x_q} \frac{d\xi}{\xi\sqrt{\xi^4 + (-2a/b)\,\xi^2 + a^2/b^2 - 6D/b}} = \pm bt + c_2$$

The upper limit x_q will be specified later. For the next calculation it will be taken as the smallest of the roots of the polynomial under the square root. The details of the calculations in terms of elliptic functions can be found in Byrd and Friedman [40].

The roots of the forth degree polynomial are introduced by the factorization

$$\xi^4 + (-2a/b)\,\xi^2 + a^2/b^2 - 6D/b = (\xi_1 - \xi)\,(\xi_2 - \xi)\,(\xi_3 - \xi)\,(\xi_4 - \xi)$$

where

$$\begin{array}{ll} \xi_1 = \sqrt{\alpha_1 + \alpha_2} & \xi_3 = -\sqrt{\alpha_1 - \alpha_2} \\ \xi_2 = \sqrt{\alpha_1 - \alpha_2} & \xi_4 = -\sqrt{\alpha_1 + \alpha_2} \end{array}$$

such as to have $\xi_1 > \xi_2 > \xi_3 > \xi_4$; then we will use $x_q \equiv \xi_4$. The notations are

$$\alpha_1 \equiv a/b\;, \;\; \alpha_2 \equiv \sqrt{6D/b}$$

The following substitutions are useful

$$\begin{aligned} sn^2 u &= \frac{(\xi_1 - \xi_3)(\xi_4 - \xi)}{(\xi_1 - \xi_4)(\xi_3 - \xi)} \\ k^2 &= \frac{(\xi_2 - \xi_3)(\xi_1 - \xi_4)}{(\xi_1 - \xi_3)(\xi_2 - \xi_4)} \\ g &= \frac{2}{\sqrt{(\xi_1 - \xi_3)(\xi_2 - \xi_4)}} \end{aligned}$$

and a new variable is introduced identifying the lower limit of the integral, $x \to u_1$

$$\begin{aligned} \alpha^2 &= \frac{\xi_1 - \xi_4}{\xi_1 - \xi_3} > 1 \\ \varphi &= am\, u_1 = \arcsin \sqrt{\frac{(\xi_1 - \xi_3)(\xi_4 - x)}{(\xi_1 - \xi_4)(\xi_3 - x)}} \\ sn\, u_1 &= \sin \varphi \end{aligned}$$

The integral can be written

$$\begin{aligned} &\int_x^{\xi_4} \frac{d\xi}{\xi \sqrt{(\xi_1 - \xi)(\xi_2 - \xi)(\xi_3 - \xi)(\xi_4 - \xi)}} \\ &= \frac{g}{\xi_4} \int_0^{u_1} \frac{1 - \alpha^2 sn^2(u)}{1 - \frac{\xi_3 \alpha^2}{\xi_4} sn^2(u)} du \end{aligned}$$

This integral can be expressed in terms of elliptic functions. We take $p \equiv \frac{\xi_3 \alpha^2}{\xi_4}$

$$\begin{aligned} &\int_0^{u_1} \frac{1 - \alpha^2 sn^2(u)}{1 - p\, sn^2(u)} du \\ &= \frac{1}{p^6} \left[\alpha^6 u + 3\alpha^4 \left(p^2 - \alpha^2\right) V_1 + 3\alpha^2 \left(p^2 - \alpha^2\right)^2 V_2 + \left(p^2 - \alpha^2\right)^3 V_3 \right] \end{aligned}$$

Here the notations are

$$V_1 = \int \frac{du}{1 - p^2 sn^2(u)} = \Pi\left(\varphi, p^2, k\right)$$

$$\begin{aligned} V_2 &= \int \frac{du}{\left(1 - p^2 sn^2(u)\right)^2} \\ &= \frac{1}{2\left(p^2 - 1\right)\left(k^2 - p^2\right)} \left[p^2 E(u) + \left(k^2 - p^2\right) u + \right. \\ &\quad \left. + \left(2p^2 k^2 + 2p^2 - p^4 - 3k^2\right) \Pi\left(\varphi, p^2, k\right) - \frac{p^4 sn(u)\, cn(u)\, dn(u)}{1 - p^2 sn^2(u)} \right] \end{aligned}$$

$$\begin{aligned} V_3 &= \frac{1}{4\left(1-p^2\right)\left(k^2-p^2\right)}\left[k^2 V_0 + \right. \\ & +2\left(p^2k^2+p^2-3k^2\right)V_1+ \\ & +3\left(p^4-2p^2k^2-2p^2+3k^2\right)V_2+ \\ & \left. -\frac{p^4 sn\left(u\right) cn\left(u\right) dn\left(u\right)}{\left(1-p^2 sn^2\left(u\right)\right)^2}\right] \end{aligned}$$

and

$$V_0 = \int du = u = F\left(\varphi, k\right) = \int \frac{d\varphi}{\sqrt{1-k^2\sin^2\varphi}}$$

The symbol $\Pi\left(\varphi, p^2, k\right) \equiv \Pi\left(u, p^2\right)$ represents the Legendre 's incomplete elliptic integral of the third kind and $\varphi = am\left(u\right)$ is the amplitude of u. The symbols sn, cn, dn represent the Jacobi elliptic functions.

This calculation may serve to get an idea of the general behavior

3.3 Typical Instanton Solutions

There are several well known examples of instantons. They appear in physical systems whose lowest energy state is degenerate and the minima of the action functional (or the energy, for stationary solutions) are separated by energy barriers. Instantons connect these minima by performing transitions which are classically forbidden and are only possible in imaginary time (the theory is expressed in Euclidean space, with uniform positive metric). It is only by including these instantons that the action functional is correctly calculated and classical, experimental physical quantities can be determined.

From this calculation we can obtain the explicit trajectories that extremize the action functional and in the same time reproduce the jump of the system between the two distant equilibrium positions. These trajectories will be necessary in the calculation of the functional integral. However, since in the Onsager-Machlup approach it has been eliminated the external current and integrated over the dual functional variable $k\left(t\right)$ we cannot derive the statistical properties of $x\left(t\right)$ from a generating functional.

3.4 Approximations of the Instanton Form

In the approach based on the Onsager-Machlup action the instanton is not used in its explicit form (elliptic functions) in the calculation of the action. The reason is that the result can be proved to depend essentially on local properties of the potential $V\left(x\right)$. This will be shown later.

In the approach with dual functions (i.e. the function $k\left(t\right)$), one can reduce the instanton to its simplest form, an instantaneous transition between two states, a jump appearing at an arbitrary moment of time. Using this form as a first approximation we will calculate the solutions of the Euler-Lagrange equations and then the action.

4 Fluctuations Around the Saddle Point (Instanton) Solution

4.1 The Expansion of the Action Functional Around the Extrema

Using the Onsager-Machlup action we express the probability of finding the particle at (x,t) if it was initially in (x_i,t_i).

$$P(x,t;x_i,t_i)=\exp\left[-\frac{U(x)-U(x_i)}{2D}\right]K(x,t;x_i,t_i)$$

The new function K has the expression

$$K(x,t;x_i,t_i)=\int_{x_i}^{x}D[x(\tau)]\exp\left(-\frac{1}{D}\int_{t_i}^{t}d\tau\left[\frac{\dot{x}^2}{4}+W(x)\right]\right) \quad (16)$$

The integrand at the exponent can be considered as the Lagrangean density for a particle of mass $1/2$ moving in a potential given by

$$\text{potential}=-W(x)$$

In a semiclassical treatment (similar to the quantum problem, where $\hbar$ is the small diffusion coefficient D of the present problem), the most important contribution comes from the neighborhood of the classical trajectories, $x_c(\tau)$ that extremalizes the action S.

The "classical" equation of motion is

$$\begin{aligned}\frac{1}{2}\ddot{x}_c &= \left.\frac{dW(x)}{dx}\right|_{x=x_c} \\ x_c(t_i) &= x_i\ ,\ x_c(t)=x\end{aligned} \quad (17)$$

To take into account the trajectories in a functional neighborhood around $x_c(\tau)$ we expand the action to second order introducing the new variables

$$y(\tau)=x(\tau)-x_c(\tau)$$

This gives

$$\begin{aligned}K(x,t;x_i,t_i) &= \exp\left[-\frac{1}{D}S_c(x,t;x_i,t_i)\right] \\ &\times\int_{y(t_i)=0}^{y(t)=0}D[y(\tau)]\exp\left\{-\frac{1}{D}\int_{t_i}^{t}d\tau\left[\frac{1}{4}\dot{y}^2+\frac{1}{2}y^2W''(x_c(\tau))\right]\right\}\end{aligned}$$

The deviation of the action from that obtained at the extremum $x_c(\tau)$, can be rewritten

$$\delta S=\frac{1}{2}\int_{t_i}^{t}d\tau y(\tau)\left[-\frac{1}{2}\frac{d^2}{d\tau^2}+W''(x_c(\tau))\right]y(\tau)$$

The functional integration can be done since it is Gaussian and the result is

$$\sim \frac{1}{\left[\det\left(-\frac{1}{2}\frac{d^2}{d\tau^2} + W''\left(x_c\left(\tau\right)\right)\right)\right]^{1/2}}$$

In order to calculate the determinant, one needs to solve the eigenvalue problem for this operator

$$\left[-\frac{1}{2}\frac{d^2}{d\tau^2} + W''\left(x_c\left(\tau\right)\right)\right] y_n\left(\tau\right) = \lambda_n y_n\left(\tau\right)$$

with the eigenfunctions verifying the conditions

$$y_n\left(t_i\right) = y_n\left(t\right) = 0$$

$$\int_{t_i}^{t} d\tau y_n\left(\tau\right) y_m\left(\tau\right) = \delta_{nm}$$

The formal result for K is

$$K\left(x,t;x_i,t_i\right) = N\frac{1}{\left(\prod_n \lambda_n\right)^{1/2}} \exp\left[-\frac{S_c}{D}\right]$$

where N is a constant that will be calculated by normalizing P. Another way to calculate N is to fit this result to the known harmonic oscillator problem.

It has been shown (Coleman [32]) that the factor arising from the determinant can be written in the form

$$N\frac{1}{\left(\prod_n \lambda_n\right)^{1/2}} \simeq \frac{1}{\left[4\pi D\psi\left(t\right)\right]^{1/2}}$$

where the function ψ is the solution of

$$-\frac{1}{2}\frac{d^2\psi}{d\tau^2} + W''\left(x_c\left(\tau\right)\right)\psi = 0$$

with the boundary conditions

$$\begin{aligned} \psi\left(t_i\right) &= 0 \\ \left.\frac{d\psi}{d\tau}\right|_{t=t_i} &= 1 \end{aligned}$$

A more detailed discussion on particular (technical) aspects of this approach is presented in the Appendix.

In the case where there are degenerate minima in $W\left(x\right)$ the particle can travel from one minimum to another and these solutions are called, as said before, instantons. Consider for example the potential with two degenerate maxima of $-V\left(x\right)$ at $\pm a$ and with a minimum at $x=0$. We want to calculate the probability $P\left(a,t/2;-a,-t/2\right)$.

The classical solution connecting the point $-a$ to the point a is a kink-like instanton. The energy of this solution is exponentially small

$$\begin{aligned} E &= \frac{1}{4}\dot{x}^2 - W \\ &\simeq 2W_a''a^2 \exp\left[-t\left(2W_a''\right)^{1/2}\right] \end{aligned}$$

This solution spends quasi-infinite time in both harmonic regions around $\pm a$ where it has very small velocity; and travels very fast, in a short time Δt between these points (this is the time-width of the instanton).

The special effect of the translational symmetry in time is seen in the presence of the parameter representing the center of the instanton. It can be any moment of time between $-t/2$ and $t/2$. This case must be treated separately and we note that this corresponds to the lowest eigenvalue in the spectrum, since the range of variation of the coefficient in the expansion of any solution in terms of eigenfunction is the inverse of the eigenvalue

$$\delta c_n = \left(\frac{D}{\lambda_n}\right)^{1/2}$$

The widest interval, for the variation of the center of the instanton, must be associated with the smallest eigenvalue and this and its eigenfunction must be known explicitely. Instead of a precise knowledge of the lowest eigenvalue and its corresponding eigenfunction we will use an approximation, exploiting the fact that the function $\dot{x}_c(t)$ is very close of what we need.

We start by noting that $\dot{x}_c(t)$ is a solution of the eigenvalue problem for the operator of second order functional expansion around the instanton. This eigenfunction corresponds to the eigenvalue 0

$$-\frac{1}{2}\frac{d^2}{d\tau^2}\dot{x}_c(t) + W''(x_c(\tau))\,\dot{x}_c(t) = 0$$

and has boundary conditions

$$\dot{x}_c\left(\pm\frac{t}{2}\right) \sim \text{exponential small}$$

very close to 0, which is the exact boundary condition we require from the eigenfunctions of the operator. So the difference between $\dot{x}_c(t)$ and the true eigenfunction are very small. Since $\dot{x}_c(t)$ corresponds to eigenvalue 0 we conclude that, by continuity, the true eigenfunction will have an eigenvalue $\lambda_0(t)$ very small, exponentially small. Then the range of important values of the coefficient c_0 is very large and the Gaussian expansion is invalid since the departure of such a solution from the classical one (the instanton) cannot be considered small. This is known as the zero-mode problem in functional integration.

The degeneracy in the moment of time where the center of the instanton is placed (i.e. the moment of transition) can be solved by treating this parameter as a colective coordinate. The result is (see, for example, [33])

$$K(a, t/2; -a, -t/2) = \left[\frac{\lambda_0(t)}{4\pi D\psi(t/2)}\right]^{1/2} \int_{-t/2}^{t/2} d\theta \left\{\frac{S[x_I(\tau-\theta)]}{4\pi D}\right\}^{1/2} \times \exp\left\{-\frac{1}{DS[x_I(\tau-\theta)]}\right\}$$

the parameter τ in the expression of the instanton solution is the current time variable along the solution that is used to calculate the action. The integartion is performed on the intermediate transition moment θ. We also note that

$$S[x_I(\tau-\theta)] = \lim_{t\to\infty} S[x_I(\tau)] = S_0$$

since, except for the very small intervals (approx. the width of the instanton) at the begining and the end of the interval $\left(-\frac{t}{2}, \frac{t}{2}\right)$, the value of the action S is not sensitive to the position of the transition moment.

$$K(a, t/2; -a, -t/2) = \left[\frac{\lambda_0(t)}{4\pi D\psi(t/2)}\right]^{1/2} \left(\frac{S_0}{4\pi D}\right)^{1/2} t \exp\left(-\frac{S_0}{D}\right)$$

We note that the instanton degeneracy introduces a linear time dependence of the probability.

In general, for a function $U(x)$ that has two minima separated by a barrier (a maximum) the potential $W(x)$ calculated form the action will have three minima and these are not degenerate. The inverse of this potential, $-W(x)$, which appears in the equation of motion, will have three maxima in general nondegenerate and the differences in the values of $-W(x)$ at these maxima is connected with the presence of the term containing D. Since we assume that D is small, the non-degeneracy is also small. The previous discussion in which the notion of instanton was introduced and K was calculated, takes into consideration the degenerate maxima and the instanton transition at equal initial and final W.

Let us consider the general shape for $-W(x)$ with three maxima, at $x = b, 0$ and a. The heigths of these maximas are

$$\frac{DU''_\alpha}{2}, \quad \alpha = b, 0, a \qquad (18)$$
$$U''_b > 0,\ U''_a > 0,\ U''_0 < 0$$

This is because the extrema of $-W(x)$

$$\frac{d[-W(x)]}{dx} = 0$$

coincides according to the equation of motion with the pointes where

$$\ddot{x} = 0 \text{ , or } \dot{x}(t) = \text{const}$$

and the constant cannot be taken other value but zero

$$\dot{x}(t) = \text{const} = 0$$

Then, since we have approximately that $\dot{x} \approx V(x)$ (for small D), at these extrema of $-W(x)$ we have $V(x) = 0$ and only the second term in the expression of $W(x)$ remains. This justifies Eq.(18).

It will also be assumed that $U_b'' > U_a''$.

4.2 The Harmonic Region (Diffusion Around the Equilibrium Points)

4.2.1 Contribution from the Trivial Fixed Point Solutions

We want to calculate, on a Kramers time scale, $\tau_K \sim \exp(\Delta U/D)$ the probability

$$P(b,t;b,0) = P\left(b,\frac{t}{2};b,-\frac{t}{2}\right) = K\left(b,\frac{t}{2};b,-\frac{t}{2}\right)$$

We have to find the classical solution of the equation of motion connecting $\left(b,-\frac{t}{2}\right)$ with $\left(b,\frac{t}{2}\right)$. This is the trivial solution, particle sitting at b

$$x_c(t) = b$$

We have to calculate explicitely the form of the propagator in this case

$$K(x,t;x_i,t_i) = N\frac{1}{\left(\prod_n \lambda_n\right)^{1/2}}\exp\left[-\frac{S_c}{D}\right]$$

We use the formulas given before

$$N\frac{1}{\left(\prod_n \lambda_n\right)^{1/2}} \simeq \frac{1}{\left[4\pi D\psi(t)\right]^{1/2}} \tag{19}$$

where the function ψ is the solution of

$$-\frac{1}{2}\frac{d^2\psi}{d\tau^2} + W''(x_c(\tau))\psi = 0 \tag{20}$$

with the boundary conditions

$$\begin{aligned} \psi(t_i) &= 0 \\ \left.\frac{d\psi}{d\tau}\right|_{t=t_i} &= 1 \end{aligned} \tag{21}$$

We use simply U' and U'' for the respective functions calculated at the fixed point b.

We have

$$\begin{aligned} W'' &= \frac{d^2}{dx^2}\left[\frac{1}{4}(U')^2 - \frac{D}{2}U''\right] \\ &\approx \frac{1}{2}(U'')^2 + \text{ terms of order } D \end{aligned}$$

The equation for the eigenvalues becomes

$$\psi'' - (U'')^2 \psi = 0$$

$$\psi(t) = a_1 \exp(|U''| t) + a_2 \exp(-|U''| t)$$

and it results form the boundary conditions

$$\begin{aligned} a_1 &= \frac{1}{2|U''|} \exp[-|U''| t_i] \\ a_2 &= -\frac{1}{2|U''|} \exp[|U''| t_i] \end{aligned}$$

Then

$$\psi(t) = \frac{1}{2|U''|}\{\exp[|U''|(t-t_i)] - \exp[-|U''|(t-t_i)]\}$$

NOTE. This is the sinh which is obtained in the calculation of the ground level splitting by quantum tunneling for a particle in two-well potential.

Now we can calculate

$$\frac{1}{[4\pi D\psi(t)]^{1/2}} = \frac{1}{(4\pi D)^{1/2}}[2|U''|]^{1/2} \frac{\exp\left[-\frac{1}{2}|U''|(t-t_i)\right]}{(1-\exp[-2|U''|(t-t_i)])^{1/2}}$$

It remains to calculate the action for this trivial trajectory $x_c(t) = b$

$$\begin{aligned} S_c &= \int_{t_i}^{t} dt \left[\frac{1}{4}\dot{x}_c^2 + W(x_c)\right] \\ &= \int_{t_i}^{t} dt W(x_c) \\ &= \int_{t_i}^{t} dt \left[\frac{1}{4}U'^2(x_c) - \frac{D}{2}U''(x_c)\right] \end{aligned}$$

Since the position of the extremum of W is very close (to order D) to the position where $V(x)$ is zero, and since $V = -U'$, we can take with good approximation the first term in the integrand zero. Then

$$S_c = -\frac{D}{2}\int_{t_i}^{t} dt U''(x_c) = -\frac{D}{2}U''(t-t_i)$$

We have to put together the two factors of the propagator and take into account that at b we have $U'' > 0$

$$\begin{aligned} K(x,t;x_i,t_i) &= \frac{1}{[4\pi D\psi(t)]^{1/2}}\exp\left[-\frac{S_c}{D}\right] \\ &= \left(\frac{U''}{2\pi D}\right)^{1/2}\frac{\exp\left[-\frac{1}{2}U''(t-t_i)\right]}{(1-\exp\left[-2U''(t-t_i)\right])^{1/2}}\exp\left(\frac{1}{2}U''(t-t_i)\right) \end{aligned}$$

$$K(x,t;x_i,t_i) = \left(\frac{U''}{2\pi D}\right)^{1/2}\frac{1}{(1-\exp\left[-2U''(t-t_i)\right])^{1/2}}$$

The contribution to the action is

$$\begin{aligned} K^0\left(b,\frac{t}{2};b,-\frac{t}{2}\right) &= \left(\frac{U''_b}{2\pi D}\right)^{1/2}\frac{1}{\left[1-\exp\left(-2U''_b t\right)\right]^{1/2}} \\ &\simeq \left(\frac{U''_b}{2\pi D}\right)^{1/2} \quad \text{(at large } t\text{)} \end{aligned}$$

It should be noticed that no other classical solution exists since there are no turning points permitting the solution to come back to b.

4.2.2 Contribution from the Trivial Fixed Point Solution in the Dual Function Approach

We apply the procedure described for the purely diffusive case to the case $V(x) \neq 0$ and for this we need the solution of the Euler-Lagrange equations. An approximation is possible if the diffusion coefficient is small. In this case the diffusion will take place around the equilibrium positions $-\sqrt{a/b}$ and $+\sqrt{a/b}$. Taking the equilibrium $x(t) = x_0 = \pm\sqrt{a/b}$ in the equation for $k(t)$ we have

$$k(t) = \int_0^T dt'\Delta_{21}(t,t')\,J_1(t') \tag{22}$$

where

$$\Delta_{21}(t,t') = \exp\left[\left(-a+3bx_0^2\right)t\right]\Theta\left(t'-t\right)\exp\left[-\left(-a+3bx_0^2\right)t'\right] \tag{23}$$

The symbol Θ stands for the Heaviside function. In the equation for $x(t)$ we expand around the equilibrium position

$$x = x_0 + \varepsilon(t) \tag{24}$$

and solve the equation

$$\dot{\varepsilon} = \left(a - 3bx_0^2\right)\varepsilon + 2iDk + J_2 \tag{25}$$

taking account of Eq.(22)

$$x(t) = x_0 + \int_0^T dt' \Delta_{11}(t,t') J_1(t') + \int_0^T dt' \Delta_{12}(t,t') J_2(t') \tag{26}$$

where

$$\begin{aligned} \Delta_{11}(t,t') &= (2iD)\exp\left[\left(a-3bx_0^2\right)t\right] \\ &\times\left\{\frac{1}{2\left(-a+3bx_0^2\right)}\left(\exp\left[2\left(-a+3bx_0^2\right)t\right]-1\right)\Theta\left(t'-t\right)\right. \\ &\left.+\frac{1}{2\left(-a+3bx_0^2\right)}\left(\exp\left[2\left(-a+3bx_0^2\right)t'\right]-1\right)\Theta\left(t-t'\right)\right\} \\ &\times\exp\left[\left(a-3bx_0^2\right)t'\right] \end{aligned} \tag{27}$$

and

$$\Delta_{12}(t,t') = \exp\left[\left(a-3bx_0^2\right)t\right]\Theta\left(t-t'\right)\exp\left[\left(-a+3bx_0^2\right)t'\right] \tag{28}$$

We note that at the limit where no potential would be present, $\left(-a+3bx_0^2\right) \to 0$, the propagators Δ become

$$\Delta_{11}(t,t') \to 2iD\left[t\Theta\left(t'-t\right)+t'\Theta\left(t-t'\right)\right]$$

$$\Delta_{12}(t,t') \to \Theta\left(t-t'\right)$$

i.e. the propagators of a purely diffusive process (see [29]).

Using the solutions Eqs.(22) and (26) we can calculate the action along this path.

$$S_J = \int_0^T dt\left[-k\dot{x}+kV(x)+iDk^2+J_1x+J_2k\right]$$

We will insert the expansion Eq.(24), perform an integration by parts over the first term and take into account the equations, i.e. the first line of Eq.(6) and Eq.(25)

$$\begin{aligned} S_J &= \int_0^T dt\left\{-\frac{1}{2}k\dot{\varepsilon}+kV'(x_0)\varepsilon+iDk^2+J_1x_0+J_1\varepsilon\right. \\ &\left.+\frac{1}{2}k\left[-V'(x_0)-J_1\right]\varepsilon+J_2k\right\} \end{aligned}$$

$$S_J = x_0\int_0^T dtJ_1 + \frac{1}{2}\int_0^T dt\left(\varepsilon J_1+kJ_2\right)$$

Now we express the two solutions, for $k(t)$ and $\varepsilon(t)$ in terms of the propagators Δ

$$S_J = x_0\int_0^T dtJ_1(t) + \frac{1}{2}\int_0^T dt\int_0^T dt' J_i(t)\Delta_{ij}(t,t')J_j(t') \tag{29}$$

where summation over $i, j = 1, 2$ is assumed, and $\Delta_{22}(t, t') \equiv 0$.

We now dispose of the generating functional of the system when this is in a region around $x_0 = \pm\sqrt{a/b}$, the fixed equilibrium points. To see what is the effect of the diffusion in this case we calculate for the variable $x(t)$ the average and the dispersion.

$$\langle x(t) \rangle = \frac{\delta}{i\delta J_1(t)} Z_J \bigg|_{J_{1,2}=0} = x_0$$

which was to be expected. And

$$\left\langle x(t)\, x(t') \right\rangle = \frac{\delta}{i\delta J_1(t)} \frac{\delta}{i\delta J_1(t')} Z_J \bigg|_{J_{1,2}=0}$$

$$\begin{aligned}
&\frac{\delta}{i\delta J_1(t)} \frac{\delta}{i\delta J_1(t')} Z_J \\
= &\ \frac{1}{2}\frac{1}{i}\Delta_{11}(t,t')\exp(iS_J) \\
+ &\left[x_0 + \frac{1}{2}\int_0^T dt'\Delta_{11}(t,t')\, J_1(t') + \frac{1}{2}\int_0^T dt'\Delta_{12}(t,t')\, J_2(t') + \frac{1}{2}\int_0^T dt'' J_2(t'')\,\Delta_{21}(t'',t)\right] \\
\times &\left[x_0 + \frac{1}{2}\int_0^T dt''\Delta_{11}(t',t'')\, J_1(t'') + \frac{1}{2}\int_0^T dt''\Delta_{12}(t',t'')\, J_2(t'') + \frac{1}{2}\int_0^T dt J_2(t)\,\Delta_{21}(t,t')\right] \\
&\times \exp(iS_J)
\end{aligned}$$

This gives the result

$$\begin{aligned}
\left\langle x(t)\, x(t') \right\rangle &= x_0^2 + \exp\left[\left(a - 3bx_0^2\right)t\right] D \\
&\times \left\{ \frac{1}{2\left(-a + 3bx_0^2\right)} \left(\exp\left[2\left(-a + 3bx_0^2\right)t\right] - 1\right)\Theta(t' - t) \right. \\
&\left. + \frac{1}{2\left(-a + 3bx_0^2\right)} \left(\exp\left[2\left(-a + 3bx_0^2\right)t'\right] - 1\right)\Theta(t - t') \right\} \\
&\times \exp\left[\left(a - 3bx_0^2\right)t'\right]
\end{aligned} \tag{30}$$

In the absence of the potential $a \to 0$ and $b \to 0$, this is simply

$$\left\langle x(t)\, x(t') \right\rangle = x_0^2 + D\min(t, t')$$

i.e. the diffusion around the position x_0. In the present case, we note that $a - 3bx_0^2 \equiv V'(x_0) = -U''(x_0)$ with $U_0'' \equiv U''(x_0) > 0$. Fixing the parameters, we take $t > t'$ and obtain

$$\left\langle x(t)\, x(t') \right\rangle = x_0^2 + D\exp\left(-U_0''t\right)\frac{1}{2U_0''}\left[\exp\left(2U_0''t'\right) - 1\right]\exp\left(-U_0''t'\right)$$

For $t = t'$ (the dispersion) we obtain (with $\Theta \to 1/2$)

$$\left\langle x(t)^2 \right\rangle = x_0^2 + \left(\frac{D}{2U_0''} \right) \left[1 - \exp\left(-2U_0'' t \right) \right] \tag{31}$$

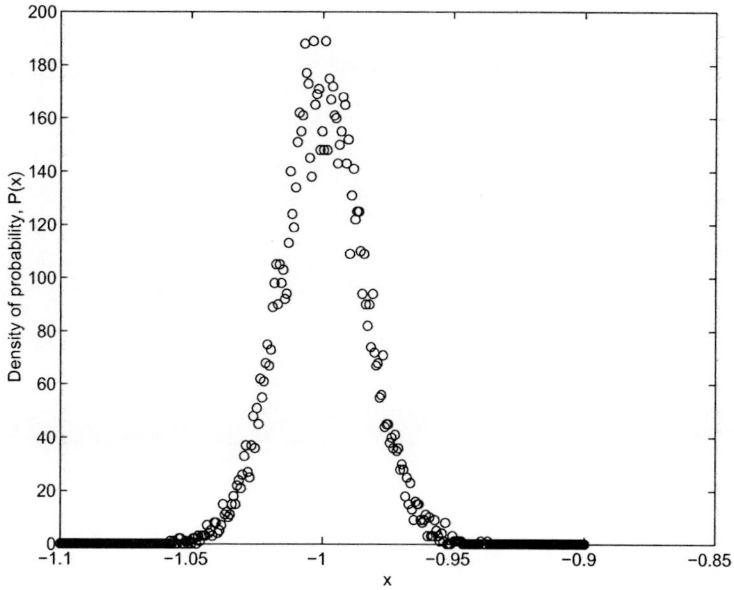

Figure 3: The probability distribution in the harmonic region around the left equilibrium stable point.

It is straightforward to calculate the higher order statistics for this process, since the functional derivatives can easily be done. We have to remember that this derivation was based on the approximation consisting in taking the equilibrium position in the Euler-Lagrange equation for $k(t)$.

4.3 Assembling the Instanton-Anti-instanton Solutions

4.3.1 The Formulation Based on Onsager-Machlup Action

Since $W(0) - W(b)$ and $W(a) - W(b)$ are small quantities (for small D) we can suppose that the instantons, even if they are not exact solutions of the equations of motion, give an essential contribution to the action. We will calculate this contribution following the approach of Caroli, Caroli and Roulet [34]. The instantons connects the points of maximum not of $-W(x)$ (because they are not equal) but of a different potential, a modified $-W(x)$ to order one in D that has degenerate maxima. This potential will be called $W^{(0)}(x)$ and it

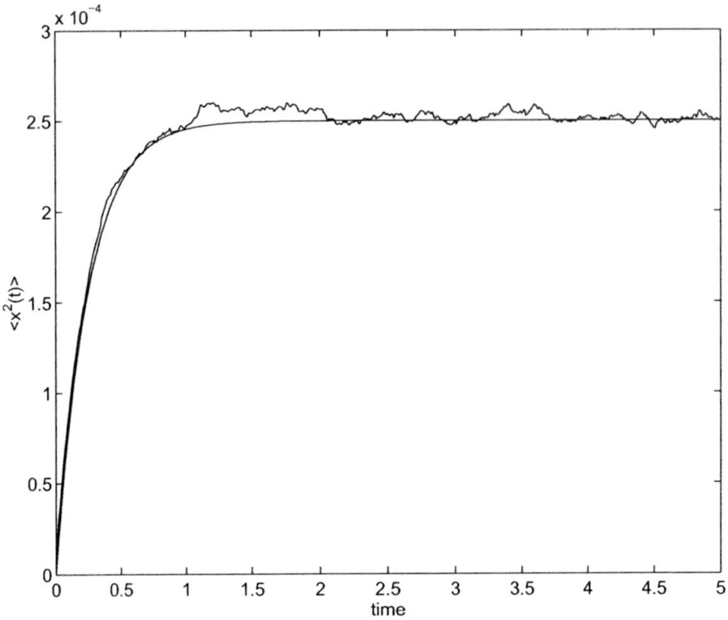

Figure 4: Comparison between the analytical formula Eq.(31) (continuous line) and numerical integration of the stochastic equation.

will be considered in the calculation of the contribution of the instantons and antiinstantons.

$$\begin{aligned} W^{(0)}(x) &= W(x) - \delta W(x) \\ \delta W(x) &= 0\ ,\ x < x_m \\ &= W_0 - W_b\ ,\ x > x_m \end{aligned}$$

where x_m is the point corresponding to the minimum situated between the two maximas.

It is introduced the family of trajectories $x_I(\tau - t_0)$ which leaves $x = b$ at time $-\infty$ and reach $x = 0$ at time ∞. They have all the same energy

$$E = -W_b$$

and the classical action S_{b0}.

We now consider the travel from $x = b$ to $x = 0$ made by an instanton $x_I(\tau - t_0)$ with the center located at time t_0 ; next the return made by an antiinstanton which is actually an instanton $x_I(t_1 - \tau)$ starting from $x = 0$ and going to $x = b$ with the center located at time t_1. With these two instantons we create a single classical solution

$$x_{IA}(\tau; t_0, t_1) = \begin{cases} x_I(\tau - t_0) & \tau < \frac{1}{2}(t_0 + t_1) \\ x_I(t_1 - \tau) & \tau > \frac{1}{2}(t_0 + t_1) \end{cases}$$

The contribution to the action of this assambled solution is

$$\begin{aligned} S_{IA}(t;t_0,t_1) &\simeq (W_0 - W_b)(t_1 - t_0) + W_b t + 2S_{b0} \\ &\quad -2\int_{(t_1+t_0)/2}^{\infty} d\tau \frac{\left[\dot{x}_I(\tau - t_0)\right]^2}{2} \end{aligned}$$

The contributions in this formula comes from the potential energy and the kinetic energy along the trajectory. We should remember that the action is the integral on time of the density of Lagrangian, where there is the kinetic energy term and minus the potential energy.

$$S = \int dt \left\{ \frac{1}{2}\dot{x}^2 - [-W(x)] \right\}$$

If the particle would have remained in $x = b$ immobile for all times $(-t/2, t/2)$ then the contribution from the potential would have been

$$W(b)\,t$$

The instanton spends $(t_1 - t_0)$ time in the point 0 before returning to b. Then it accumulates the action equal with the difference in potential between b and 0 multiplied with this time interval.

$$[(-W_b) - (-W_0)](t_1 - t_0) = (W_0 - W_b)(t_1 - t_0)$$

(since $(-W_b) > (-W_0)$).

Define the kinetic energy and the energy

$$\begin{aligned} K &= \int_{t_0}^{t} \frac{1}{2}\dot{x}^2(t)\,dt \\ E &= \frac{1}{2}\dot{x}^2(t) + [-W(x)] \\ \dot{x}(t) &= \sqrt{2(E + W(x))} \end{aligned}$$

then

$$\begin{aligned} K &= \int_{t_0}^{t} (E + W(x))\,dt = \int_{t_0}^{t} \frac{E + W(x)}{\dot{x}(t)} dx \\ &= \frac{1}{2}\int_{x_0}^{x} \sqrt{2(E + W(x))} dx \end{aligned}$$

In this formula we have to replace the expression of the trajectory $x_{IA}(t)$ and carry out the integration.

In our case the energy has the same value as in the initial position. There the velocity is zero and the potential is $-W(b)$

$$E|_{x=b} = -W(b)$$

and the potential is actually the current value of the zeroth -order potential

$$\begin{aligned} E + W(x) &= [-W(b)] - \left[-W^{(0)}(x)\right] \\ &= W^{(0)}(x) - W(b) \end{aligned}$$

If we want to calculate the integral of the kinetic energy along the trajectory, we have to consider separately the intervals where the kinetic energy is strongly determined by the velocity, i.e. the region where the instanton transition occurs, from the rest of the trajectory where, the velocity being practically zero, the potential is a better description and can be easily approximated. The approximation will relay on the fact that the particle is practically immobile in b or in 0, after the transition has been made. So we will use both expressions for the kinetic energy

$$K|_{x=b}^{x=0} = \int_0^{\frac{t_1+t_0}{2}} dt \frac{1}{2} \dot{x}^2(t) + \int_b^0 dx \left[W^{(0)}(x) - W_b\right]^{1/2}$$

The terms must be considered twice and the interval of integration can be extended for the region of transition, since in any case it is very small

$$2\int_b^0 dx \left[W^{(0)}(x) - W_b\right]^{1/2} - 2\int_{\frac{t_1+t_0}{2}}^{\infty} d\tau \frac{1}{2} \dot{x}^2(\tau - t_0)$$

The term S_{b0} is

$$S_{b0} = \int_b^0 dx \left[W^{(0)}(x) - W_b\right]^{1/2}$$

Using a step-like approximation for the form of the instanton, it results

$$\begin{aligned} S_{IA}(t; t_0, t_1) &= (W_0 - W_b)(t_1 - t_0) + W_b t + 2S_{b0} \\ &\quad - \frac{x_m^2 \left(W_0''\right)^{1/2}}{\sqrt{2}} \exp\left[\left(2W_0''\right)^{1/2}(2\Delta_{m0} - t_1 + t_0)\right] \end{aligned}$$

where

$$W_0'' = \left.\frac{d^2 W_0}{dx^2}\right|_{x=0}$$

and

$$\Delta_{m0} = \frac{1}{2}\int_{x_m}^0 dx \left\{ \frac{1}{\left[W^{(0)}(x) - W_b\right]^{1/2}} - \frac{1}{\left[W_h^{(0)}(x) - W_b\right]^{1/2}} \right\}$$

The local harmonic approximation to $W^{(0)}$ is $W_h^{(0)}$

$$W_h^{(0)} = \begin{cases} W_b + \frac{W_0''}{2} x^2 & x > x_m \\ W_b + \frac{W_b''}{2} (x-b)^2 & x < x_m \end{cases}$$

NOTE Since the trajectory $x_{IA}(\tau; t_0, t_1)$ is not an exact solution of the equation of motion in the potential W, the expansion of the action S will not be limitted to the zeroth and the second order terms. It will also contain a first order term, $(\delta S/\delta x)|_{x=x_c(\tau)}$. It can be shown that this contribution is negligible in the order $O(D)$.

The calculation of the contribution to the functional integral from the second order expansion around x_{IA} is done as usual by finding the eigenvalues of the determinant of the corresponding operator. As before, the product of the eigenvalues should not include the first eigenvalue since this is connected with the translational symmetry (in time) of the instanton solution. Now there will be two eigenvalues, one for t_0 (the transition performed by the $b \to 0$ instanton) and the second for t_1 (the transition prtformed by the antiinstanton, or the transition $0 \to b$). Another way of expressing this invariance to the two time translations is to say that the pair of instantons has not a determined central moment and, in addition, there is an internal degree of freedom of the breathing solution (the breathing solution is the traditional name for the pair instanton-antiinstanton in the sine-Gordon theory).

To take into account these modes, whose eigenvalues are zero, we need to integrate in the functional integral, over the two times, t_0 and t_1.

The measure of integration for the two translational symmetries t_0 and t_1 is

$$\frac{S_{b0}}{2\pi D}$$

This quantity is the Jacobian of the change of variables in the functional integration over the uctuations around the instanton solution. The uctuation that corresponds to the lowest (almost zero) eigenvalue is replaced in the measure of integration with the differential of the time variable representing the moment of transition. Then it results this Jacobian.

It can be shown that in the approximation given by exponentially small terms, the contribution to the path integral of the small uctuations around the classical instanton-antiinstanton x_{IA} solution is the product of the uctuation terms around the instanton and antiinstanton separately.

$$\begin{aligned} K^{(1)}\left(b, \frac{t}{2}; b, -\frac{t}{2}\right) &= \int_{-t/2}^{t/2} dt_0 \int_{t_0}^{t/2} dt_1 \left(\frac{S_{bo}}{2\pi D}\right) \\ &\times \left(\frac{\lambda_0}{4\pi D\psi}\right)_I^{1/2} \left(\frac{\lambda_0}{4\pi D\psi}\right)_A^{1/2} \\ &\times \left[\frac{2\pi D}{(2W_0'')^{1/2}}\right]^{1/2} \exp\left[-\frac{S_{IA}(t; t_0, t_1)}{D}\right] \end{aligned}$$

where $(\lambda_0)_{I,A}$ represent the lowest eigenvalue of the operator arising from the second order expansion of the action, defined on the time intervals

$$\begin{aligned} x_c(\tau) &\equiv x_I(\tau - t_0) \ , \ -\frac{t}{2} < \tau < \frac{t_0+t_1}{2} \\ x_c(\tau) &\equiv x_I(t_1 - \tau) \ , \ \frac{t_0+t_1}{2} < \tau < \frac{t}{2} \end{aligned}$$

The notation ψ_I and respectively ψ_A represents

$$\psi_I \text{ value at } \frac{t_0+t_1}{2} \text{ of the eigenfunction } \psi \text{ that starts at } t_i = -\frac{t}{2}$$

$$\psi_A \text{ value at } \frac{t}{2} \text{ of the eigenfunction } \psi \text{ that starts at } t_i = \frac{t_0+t_1}{2}$$

This corresponds to the formula of Coleman which replaces the infinite product of eigenvalues with $\left(\frac{\lambda_0}{4\pi D\psi}\right)^{1/2}$ where ψ is calculated at the end of the interval of time, where ψ verifies the boundary conditions (21).

The result is

$$\begin{aligned} &K^{(1)}\left(b,\frac{t}{2};b,-\frac{t}{2}\right) \qquad (32) \\ &= \left[\frac{2\pi D}{\left(2W_0''\right)^{1/2}}\right]^{1/2} \frac{(x_m-b)\left|x_m\right| W_0'' W_b''}{2\pi^2 D^2} \exp\left[-\frac{\varepsilon_0^{(b)} t}{D}\right] \\ &\times \exp\left[-\frac{2S_{b0}}{D} + \Delta_{bm}\left(2W_b''\right)^{1/2} + \Delta_{bm}\left(2W_0''\right)^{1/2}\right] \\ &\times \int_{-t/2}^{t/2} dt_0 \int_{t_0}^{t/2} dt_1 \exp\left\{-\frac{\varepsilon_0^{(0)}-\varepsilon_0^{(b)}}{D}(t_1-t_0) + \frac{C}{D}\exp\left[-\left(2W_0''\right)^{1/2}(t_1-t_0)\right]\right\} \end{aligned}$$

where

$$\varepsilon_0^{(b)} = W_i + D\left(\frac{W_i''}{2}\right)^{1/2}$$

is the lowest eigenvalue of the Schrodinger equation associated with the Fokker-Planck diffusion equation in the local harmonic approximation of the potential W in the well i.

$$C = x_m^2\left(\frac{W_0''}{2}\right)^{1/2} \exp\left[2\left(2W_0''\right)^{1/2}\Delta_{m0}\right] \qquad (33)$$

and

$$\Delta_{ij} = \frac{1}{2}\int_{x_i}^{x_j} dx \left\{\frac{1}{\left[W^{(0)}(x) - W_b\right]^{1/2}} - \frac{1}{\left[W_h^{(0)}(x) - W_b\right]^{1/2}}\right\}$$

There is a very important problem with this formula: the coefficient C is positive and the contribution of this part in the t_1 integration comes from time intervals

$$t_1 - t_0 \ll \frac{1}{\left(2W_0''\right)^{1/2}} \sim \Delta t$$

Since this is a very small time interval, it results that the contributions are due to states where the instanton and the antiinstanton are very close one to the other, which is unphysical. It will be necessary to calculate in a particular way this part, introducing a contour in the complex $t_1 - t_0$ plane, with an excursion on the imaginary axis.

The calculation of the propagator $K^{(1)}\left(b, \frac{t}{2}; b, -\frac{t}{2}\right)$. Here (1) means that only one pair of instanton and anti-instanton is considered.

We particularize the formula above using the expression for $W(x)$ given in terms of $U(x)$. Now we have

$$\begin{aligned} W_b &= -\frac{DU_b''}{2} \\ W_0 &= -\frac{DU_0''}{2} \\ W_b &= -\frac{U_b''^2}{2} \\ W_0 &= -\frac{U_0''^2}{2} \end{aligned}$$

The expression of S_{b0} is

$$S_{b0} = \frac{U_0 - U_b}{2} + \frac{D}{2} \ln \left| \frac{U_b''(x_m - b)}{U_0'' x_m} \right| + \frac{D}{2}\left(U_b'' \Delta_{bm} + U_0'' \Delta_{m0}\right)$$

$$\begin{aligned} K^{(1)}\left(b, \frac{t}{2}; b, -\frac{t}{2}\right) &= \frac{U_b'' \left|U_0''\right| x_m^2}{2} \left(\frac{\left|U_0''\right|}{2\pi D}\right)^{3/2} I(t) \\ &\times \exp\left(-\frac{U_0 - U_b}{D} + 2\Delta_{m0}\left|U_0''\right|\right) \end{aligned}$$

where

$$I(t) = \int_{-t/2}^{t/2} dt_0 \int_C dz \exp\left[-\left|U_0''\right| z + \frac{C}{D} \exp\left(-\left|U_0''\right| z\right)\right]$$

4.3.2 The Formulation Using Dual Functions

We will now use the alternative functional approach (in the setting that has been developed by us in Subsection 2.2) to the calculation of the probability of transition from one minimum to the same minimum with an intermediate stay at the unstable maximum point (symmetric potential).

The initial equation is

$$\dot{x}(t) = V(x) + \xi(t)$$

The action functional with external current added is

$$S = i \int_{-T}^{T} dtL$$

$$L = -k(t)\dot{x}(t) + k(t)V(x) + iDk^2(t) + J_1(t)x(t)$$

The equations of motion are

$$\begin{aligned} \dot{x} - V(x) - 2iDk &= 0 \\ \dot{k} - k\left(\frac{dV}{dx}\right) &= -J_1 \end{aligned}$$

We have to solve these equations, and replace the solutions $x(t)$ and $k(t)$ in the action functional. Then the functional derivatives to the external current $J_1(t)$ will give us the correlations for the stochastic variable $x(t)$. We can also calculate the probability that the particle, starting from one point at a certain time will be found at another point at other time. This will be done in the following.

The first step is to obtain an analytical solution to the Euler-Lagrange equation. The method to solve these equation is essentially a successive approximation, as we have done above, for the diffusive motion around a stable position in the potential (harmonic region). We know that the classical trajectory must be of the type of a transition between the initial point, taken here as the left minimum of the potential and the final point, the unstable maximum of the potential. From there, the particle will return to the left minimum by an inverse transition. We have to calculate simultaneously $x(t)$ and $k(t)$, but we have sufficient information to find an approximation for $x(t)$, by neglecting the effect of diffusion (the last term in the differential equation for $x(t)$).

$$\dot{x} = ax - bx^3$$

This gives

$$x(\tau) = \pm\sqrt{\frac{a}{b}} \frac{1}{\{1 + \exp[-2a(\tau - t_0)]\}^{1/2}}$$

This solution is a transition between either of the minima $\pm\sqrt{a/b}$ and 0. It shows the same characteristics as found numerically or by integrating the elliptic form of the equation, in the case of Onasger-Machlup action. The particle spends long time in the initial and final points and makes a fast transition between them at an arbitrary time t_0. The width of transition is small compared to the rest of quasi-imobile stays in the two points, especially if a is large. Then we will make an approximation, taking the solution as

$$x_\eta^{(1)}(\tau) = -\sqrt{\frac{a}{b}}\Theta(t_0 - \tau) + 0 \times \Theta(\tau - t_0)$$

where $-\sqrt{a/b}$ and 0 are the initial and final positions, the indice η means that this is the calssical solution (extremum of the action) and (1) means the first part of the full trajectory, which will also include the inverse transition, from 0 to $-\sqrt{a/b}$. The following structure of the total trajectory is examined: The total time interval is between $-T$ and T (later the parameter T will be identified with $-t/2$ for comparison with the results from the literature). The current time variable is τ and in the present notations, t is any moment of time in the interval $(-T, T)$. At time t_0 the particle makes a jump to the position $x = 0$ and remains there until $\tau = t_1$. At t_1 it performs a jump to the position $-\sqrt{a/b}$, where it remains for the rest of time, until T.

With this approximative solution of the Euler-Lagrange equations (since we have neglected the term with D in the equation of $x(\tau)$) we return to the equation for $k(\tau)$. We first calculate

$$\left.\frac{dV}{dx}\right|_{x_\eta^{(1)}(\tau)} = (-2a)\Theta(t_0 - \tau) + a\Theta(\tau - t_0)$$

We need the integration of this quantity in the inverse direction starting from the end of the motion toward the initial time

$$\int_T^t d\tau \left(\frac{dV}{dx}\right)_{x_\eta^{(1)}(\tau)} = [-a(T-t)]\Theta(t - t_0) + [-aT + 3at_0 - 2at]\Theta(t_0 - t)$$

We now introduce the second part of the motion: at time $\tau = t_1$ the particle makes the inverse transition

$$x_\eta^{(2)}(\tau) = 0 \times \Theta(t_1 - \tau) + \left(-\sqrt{\frac{a}{b}}\right)\Theta(\tau - t_1)$$

with the similar quantities.

After explaining the steps of the calculation, we change to work with the full process, assembling the two transitions and the static parts into a single trajectory

$$\begin{aligned} x_\eta(\tau) &= x_\eta^{(1)}(\tau) + x_\eta^{(2)}(\tau) \\ &= \left(-\sqrt{\frac{a}{b}}\right)\Theta(t_0 - \tau) + \left(-\sqrt{\frac{a}{b}}\right)\Theta(\tau - t_1) \end{aligned}$$

$$\left.\frac{dV}{dx}\right|_{x_\eta(\tau)} = (-2a)\Theta(t_0 - \tau) + a\Theta(t_1 - \tau)\Theta(\tau - t_0) + (-2a)\Theta(\tau - t_1)$$

and

$$\int_T^t d\tau \left(\frac{dV}{dx}\right)_{x_\eta^{(1)}(\tau)} = W(t)$$

We have introduced the notation

$$\begin{aligned} W(t) &\equiv (2a)(T-t)\Theta(t-t_1) \\ &+ (2aT - 3at_1 + at)\Theta(t-t_0)\Theta(t_1-t) \\ &+ (2aT - 3at_1 + 3at_0 - 2at)\Theta(t_0-t) \end{aligned} \tag{34}$$

Then the solution of the equation for the dual variable is

$$k_\eta(\tau) = \exp[-W(\tau)]\left\{k_T + \int_T^T dt' \left[-J_1(t')\right]\exp\left[W(t')\right]\right\}$$

According to the procedure explained before we will need to express the solutions as bilinear combinations of currents, so we identify

$$\Delta_{21}(t,t') = \Theta(t'-t)\exp[-W(t)]\exp\left[W(t')\right] \tag{35}$$

and the solution can be rewritten

$$k_\eta(\tau) = k_T\exp[-W(\tau)] + \int_{-T}^T dt'\Delta_{21}(t,t')J_1(t') \tag{36}$$

Using these first approximations for the extremizing path $(x_\eta(\tau), k_\eta(\tau))$ we return to the Euler Lagrange equations and expand the variable x as

$$x(\tau) = x_\eta(\tau) + \delta x(\tau) \tag{37}$$

whose equation is

$$\dot{\delta x}(\tau) = 2iDk_\eta(\tau) + \left(\frac{dV}{dx}\right)_{x_\eta(\tau)}\delta x(\tau)$$

and the solution

$$\delta x(t) = \exp\left[\int_0^t d\tau\left(\frac{dV}{dx}\right)_{x_\eta(\tau)}\right]\left\{B_0 + \int_0^t dt' 2iDk_\eta(t')\exp\left[-\int_0^{t'} dt''\left(\frac{dV}{dx}\right)_{x_\eta(t'')}\right]\right\}$$

Here B_0 is a constant to be determined by the condition that δx vanishes at the final point. We note that here all integrations are performed forward in time. We also notice that this expression will contain the current J_1 and we will introduce the propagator Δ_{11}. Performing the detailed calculations we obtain

$$\begin{aligned} \delta x(t) &= B_0\exp[W(t)] + \exp[W(t)]\int_{-T}^T dt'(2iDk_T)\Theta(t-t')\exp\left[-W(t')\right] \\ &+ \int_{-T}^T dt'\Delta_{11}(t,t')J_1(t') \end{aligned} \tag{38}$$

The first two terms depend on constants of integrations and the propagator is

$$\Delta_{11}\left(t,t'\right)=\exp\left[W\left(t\right)\right]\int_{-T}^{T}dt''2iD\Theta\left(t-t''\right)\Delta_{21}\left(t'',t'\right)\exp\left[-W\left(t''\right)\right] \tag{39}$$

Using the solutions (37) and (36) we can calculate the action along this trajectory.

$$\begin{aligned}
S &= i\int_{-T}^{T}dt\left(-k\dot{x}+kV+iDk^2+J_1x\right)\\
&= i\int_{-T}^{T}dt\left[\frac{1}{2}\left(-k\dot{x}+kV+2iDk^2\right)+\frac{1}{2}\left(-k\dot{x}+kV\right)+J_1x\right]\\
&= i\int_{-T}^{T}dt\left(\frac{1}{2}x\dot{k}+\frac{1}{2}kV+J_1x\right)\\
&= i\int_{-T}^{T}dt\left[\frac{1}{2}x\left(-J_1+k\frac{dV}{dx}\right)+\frac{1}{2}kV+J_1x\right]\\
&= i\int_{-T}^{T}dt\left\{\frac{1}{2}J_1x+\frac{1}{2}k\left[V-\left(\frac{dV}{dx}\right)x\right]\right\}
\end{aligned}$$

[It becomes obvious that when the potential is linear (which means that the Lagrangean is quadratic the two functional variables $x\left(t\right)$ and $k\left(t\right)$) the potential does not contribute to the action along the extremal path].

Using the solutions we have

$$\begin{aligned}
S &= i\int_{-T}^{T}dt\frac{1}{2}J_1\left(t\right)\left[x_\eta\left(t\right)+\delta x\left(t\right)\right]\\
&\quad +i\int_{-T}^{T}dt\frac{1}{2}\left[V\left(x_\eta\right)-\left(\frac{dV}{dx}\right)_{x_\eta}x_\eta\right]i\int_{-T}^{T}dt'\Delta_{21}\left(t,t'\right)J_1\left(t'\right)
\end{aligned}$$

It will become clear later that the parts containing constants are not signifiant for the final answer, the probability. We will focus on the terms containing explicitely the current J_1 since the statistical properties are determined by functional derivatives to this parameter. The action is

$$\begin{aligned}
S &= i\int_{-T}^{T}dt\left\{\frac{1}{2}J_1\left(t\right)x_\eta\left(t\right)\right.\\
&\quad +\frac{1}{2}J_1\left(t\right)\int_{-T}^{T}dt'\Delta_{11}\left(t,t'\right)J_1\left(t'\right)\\
&\quad \left.+\frac{1}{2}\left[V\left(x_\eta\left(t\right)\right)-\left(\frac{dV}{dx}\right)_{x_\eta(t)}x_\eta\left(t\right)\right]\int_{-T}^{T}dt'\Delta_{21}\left(t,t'\right)J_1\left(t'\right)\right\}
\end{aligned}$$

The generating functional of the correlations (at any order) is

$$Z\left[J_1\right]=\exp\left\{S\left[J_1\right]\right\} \tag{40}$$

and we can calculate any quantity by simply parforming functional derivatives and finally taking $J_1 = 0$.

Instead of that and in order to validate our procedure, we will calculate the probability for the process: a particle in the initial position $x = -\sqrt{a/b}$ at time $-T$ can be found at the final position $x = x_c$ at the time $t = t_c$. (Later we will particularize to $x_c = -\sqrt{a/b}$ and $t_c = T$). The calculation of this probability P can be done in the functional approach in the following way.

We have defined the statistical ensemble of possible particle trajectories starting at $x = -\sqrt{a/b}$ at $t = -T$ and reaching an arbitrary point x at time t. In the Martin-Siggia-Rose-Jensen approach it is derived the generating functional as a functional integration over this statistical ensemble of a weight measure expressed as the exponential of the classical action. Now we restrict the statistical ensemble by imposing that the particle is found at time $t = t_c$ in the point $x = x_c$. By integration this will give us precisely the probability required.

The condition can be introduced in the functional integration by a Dirac δ function which modifies the functional measure

$$D\left[x\left(\tau\right)\right] \rightarrow D\left[x\left(\tau\right)\right]\delta\left[x\left(t_c\right) - x_c\right]$$

and we use the Fourier transform of the δ function

$$\begin{aligned} P\left(x_c, t_c; -\sqrt{\frac{a}{b}}, -T\right) &= \int D\left[x\left(\tau\right)\right]\delta\left[x\left(t_c\right) - x_c\right]\exp\left(S\right) \\ &= \frac{1}{2\pi}\int d\lambda \exp\left(-i\lambda x_c\right)\int D\left[x\left(\tau\right)\right]\exp\left[S + i\lambda x\left(t_c\right)\right] \end{aligned}$$

It is convenient to write

$$\begin{aligned} S + i\lambda x\left(t_c\right) &= i\int_{-T}^{T} d\tau L\left[x\left(t\right), k\left(t\right)\right] + i\int_{-T}^{T} d\tau \lambda x\left(\tau\right)\delta\left(\tau - t_c\right) \\ &= i\int_{-T}^{T} d\tau\left[L + \lambda x\left(\tau\right)\delta\left(\tau - t_c\right)\right] \end{aligned}$$

We can see that the new term plays the same rôle as the external current $J_1\left(\tau\right)$ and this sugests to return to the Eq.(40) and to perform the modification

$$J_1\left(\tau\right) \rightarrow J_1'\left(\tau\right) = J_1\left(\tau\right) + \lambda\delta\left(\tau - t_c\right)$$

obtaining

$$P\left(x_c, t_c; -\sqrt{\frac{a}{b}}, -T\right) = \frac{1}{2\pi}\int d\lambda \exp\left(-i\lambda x_c\right) Z^{(\lambda)}\left[J_1\right]\Big|_{J_1=0}$$

Using the previous results we have

$$
\begin{aligned}
Z^{(\lambda)}[J_1]\Big|_{J_1=0} &= \exp\left\{ i\frac{1}{2}\lambda x_\eta(t_c) \right. \\
&\quad + i\frac{1}{2}\lambda^2\Delta_{11}(t_c,t_c) \\
&\quad \left. + i\frac{1}{2}\left[V - \left(\frac{dV}{dx}\right)x\right]_{t=t_c}\lambda\Delta_{21}(t_c,t_c)\right\}\Bigg|_{J_1=0}
\end{aligned}
\tag{41}
$$

This expression will have to be calculated at $t_c = T$ and $x_c = -\sqrt{a/b}$.

$$
\begin{aligned}
V\left(-\sqrt{\frac{a}{b}}\right) &= 0 \\
\frac{dV}{dx}\bigg|_{x=-\sqrt{\frac{a}{b}}} &= 0
\end{aligned}
$$

so the last term is zero.

$$
Z^{(\lambda)}[J_1]\Big|_{J_1=0} = \exp\left[-i\frac{1}{2}\lambda\sqrt{\frac{a}{b}} + i\frac{1}{2}\lambda^2\Delta_{11}(T,T)\right]\Bigg|_{J_1=0}
\tag{42}
$$

We need the propagator Δ_{21} in the expression for Δ_{11},

$$
\Delta_{21}(t',T) = \Theta(T-t')\exp\left[-W(t')\right]\exp\left[W(T)\right]
$$

$$
W(T) = 0
$$

Then the integration giving the propagator Δ_{11} is

$$
\begin{aligned}
&\int_{-T}^{T} dt' 2iD\Theta(T-t')\,\Theta(T-t')\exp\left[-W(t')\right]\exp\left[-W(t')\right] \\
&= 2iD\int_{-T}^{T} dt'\exp\left[-2W(t')\right] \\
&= 2iD\left\{\exp\left[-4aT + 6at_1 - 6at_0\right]\int_{-T}^{t_0} dt\exp(4at)\right. \\
&\quad + \exp\left[-4aT + 6at_1\right]\int_{t_0}^{t_1} dt\exp(-2at) + \\
&\quad \left. + \exp\left[-4aT\right]\int_{t_1}^{T} dt\exp(4at)\right\}
\end{aligned}
$$

We introduce the notation

$$
\begin{aligned}
Y(t_0,t_1) \equiv\; & 1 + \\
& + 3\exp\left[-4a(T-t_1)\right]\left\{\exp\left[2a(t_1-t_0)\right] - 1\right\} \\
& - \exp\left[-8aT\right]\exp\left[6a(t_1-t_0)\right]
\end{aligned}
$$

and we obtain the expression

$$\Delta_{11}(T,T) = \frac{iD}{2a} Y(t_0, t_1)$$

Returning to (42)

$$Z^{(\lambda)}[J_1 = 0] = \exp\left[-i\frac{\lambda}{2}\sqrt{\frac{a}{b}} - \frac{\lambda^2 D}{4a} Y\right]$$

and the probability is

$$\begin{aligned} P(t_0, t_1) &= \frac{1}{2\pi}\int_{-\infty}^{\infty} d\lambda \exp\left[i\lambda\sqrt{\frac{a}{b}}\right] Z^{(\lambda)}[J_1 = 0] \\ &= \frac{1}{2\pi}\int_{-\infty}^{\infty} d\lambda \exp\left[\frac{1}{2}i\lambda\sqrt{\frac{a}{b}} - \frac{\lambda^2 D}{4a} Y(t_0, t_1)\right] \end{aligned}$$

We have emphasized the dependence of the probability on the arbitrary times of jump t_0, t_1. Integrating on λ

$$P(t_0, t_1) = \sqrt{\frac{a}{\pi D Y}} \exp\left(-\frac{a^2}{4bDY}\right)$$

This expression contains an arbitrary parameter, which is the duration of stay in the intermediate position, i.e. at the unstable extremum of the potential, $x = 0$. Integrating over this duration, $t_0 - t_1$, adimensionalized with the unit $1/a$, for all possible values between 0 and the entire time interval, $2T$, will give the probability. In doing this time integration we will make a simplification to consider that T is a large quantity, such that the exponential terms in Y will vanish. Then Y reduces to 1 and the integration is trivial

$$\begin{aligned} P &= \int_0^{2T} d\left[a(t_1 - t_0)\right] P(t_0, t_1) \\ &= 2Ta\sqrt{\frac{a}{\pi D}} \exp\left(-\frac{a^2}{4bD}\right) \end{aligned}$$

This is our answer. To compare with the result of Caroli et al. [34], we first adopt their notation $T \equiv t/2$ and we write the potential as

$$V(x) = -U'(x)$$

with

$$U(x) = -\frac{1}{2}ax^2 + \frac{1}{4}bx^4$$

We note that

$$U(x=0) - U\left(-\sqrt{\frac{a}{b}}\right) = 0 + \frac{a^2}{4b}$$

which allows to write

$$P = t\frac{a^{3/2}}{\sqrt{\pi D}} \exp\left[-\frac{U(0) - U\left(-\sqrt{\frac{a}{b}}\right)}{D}\right]$$

which corresponds to the result of Caroli et al. [34] for a single instanton-anti-instanton pair, where

$$a^{3/2} \sim \left|U_0''\right|^{1/2} U''_{-\sqrt{a/b}}$$

The only problem is the need to consider the reverse of sign due to the virtual re ection at the position $x = 0$. In the tratment of Caroli, it is included by an integration in complex time plane, after the second jump, t_1.

4.4 The Trajectory Reflection and the Integration along the Contour in the Complex Time Plane

The problem arises at the examination of the Eq.(32) where it is noticed that the constant C multiplying the exponential term in the exponential in the integrations over t_0 and t_1 is positive (see Eq.(33)). Then the dependence of the integrand on $t_1 - t_0$ makes that the most important contribution to the integration over the variable t_1 to come from

$$t_1 - t_0 \ll \frac{1}{(2W_0'')^{1/2}} = \omega_0^{-1} \sim \Delta t$$

Here the integrand is of the order

$$\sim \exp\left(\frac{C}{D}\right)$$

This would imply, in physical terms, that the most important configurations contributing to the path integral would be pairs of instanton anti-instanton very closely separated, i.e. transitions with very short time of stay in the point 0. This is not correct since we expect that in large time regime the trajectories assembled from instantons and anti-instantons with arbitrary large separations (duration of residence in 0 before returning to b) should contribute to K.

This problem can be solved by extending the integration over the variable representing the time separation between the transitions

$$z = t_1 - t_0$$

in complex (see [35]).

NOTE. The quantity

$$\left(2W_0''\right)^{1/2} = \omega_0$$

is the frequency of the harmonic oscillations in the parabolic approximation of the potential.

The contour C is defined as

$$\begin{aligned} \mathrm{Im}z &= \frac{\pi}{|U_0''|} \\ 0 &\leqslant \mathrm{Re}z \leqslant \frac{t}{2} - t_0 \end{aligned}$$

The form of the trajectory x_{IA} connecting the point b to itself, b, assembled from one instanton x_I and one anti-instanton x_A can now be described as follows.

The trajectory starts in the point b and follows the usual form of the instanton solution, which means that for times

$$-\frac{t}{2} < \tau \lesssim t_0$$

it practically conicides with the fixed position b. At t_0 it makes the transition to the point 0, the duration of the transition (width of the instanton) being Δt. Then it remains almost fixed at the point 0. The problem is that there is no turning point at $x = 0$ since there the potential $-W(x)$ is smaller than at $x = b$. Then we do not have the classical instanton which takes an infinite time to reach exactly $x = 0$, etc. Also there is no exact instanton solution which starts from $x = 0$ to go back to $x = b$. In the assambled solution $x_I + x_A$, there is a discontinuity at $x = 0$ in the derivative, the velocity simply is forced (assumed) to change from almost zero positive (directed toward $x = 0$) to almost zero negative (directed toward $x = b$). What we do is to introduce a turning point, which limits effectively the transition solution, but for this we have to go to complex time, to compensate for the negative value of the potential. Making time imaginary changes the sign of the velocity squared (kinetic energy) and correspondingly can be seen as a change of sign of the potential which now becomes positive and presents a wall from where the solution can re ect. This is actually a local Euclidean-ization of the theory, around the point $x = 0$.

At time $(t_0 + t_1)/2$ the trajectory starts moving along the imaginary z axis. It reaches an imaginary turning point at $\frac{i\pi}{|U_0''|}$ and returns to the point $Rez = x_{IA}\left(\frac{t_0+t_1}{2}\right)$, $Imz = 0$.

This trajectory can be considered the limiting form of the trajectory of a fictitious particle that traverses a part of potential (forbidden classically) when this part goes to zero.

Then the expression is

$$K^{(1)}\left(b, \frac{t}{2}; b, -\frac{t}{2}\right) = -t\left(\frac{|U_0'''|}{2\pi D}\right)^{1/2} \frac{\left(U_b''\,|U_0''|\right)^{1/2}}{2\pi} \exp\left(-\frac{U_0 - U_b}{D}\right)$$

where the minus sign is due to the re ection at the turning point.

4.5 Summation of Terms from Multiple Transitions

Calculation of the transitions implying the points b and a

This is the term

$$K_{ba}^{(2)}\left(b,-\frac{t}{2};a,\frac{t}{2}\right)$$

and corresponds to the transitions

$$b\rightarrow 0\rightarrow a\rightarrow 0\rightarrow b$$

Again we have to introduce a new potential replacing W. The real potential $W(x)$ as it results from the Langevin equation, depende on D which is supposed to be small. Then the "expansion" in this small parameter D has as zerth order the new potential $W^{(0)}(x)$ that will be used in the computations below. This is

$$W^{(0)}=W-\delta W$$

$$\delta W(x)=\begin{cases}0 & x<x_m\\ W_0-W_b & x_m<x<x_p\\ W_a-W_b & x>x_p\end{cases}$$

where the new point x_p is in the region of the right-hand minimum of $-W(x)$. This point will correspond to the center of the instanton connecting 0 to a.

We introduce the new family of instanton solution

$$\widetilde{x}_I(\tau-t_1)$$

in the potential $-W^{(0)}$ that leave $x=0$ at $-\infty$ and reach $x=a$ at time ∞.

With this instanton and the previously defines ones, we assamble a function that provides approximately the transitions connecting b to b via a. This, obviously, is not a classical solution of the equation of motion, although it is very close to a solution.

$$x_{IIAA}(\tau;t_0,t_1,t_2,t_3)=\begin{cases}x_I(\tau-t_0) & \tau<\frac{1}{2}(t_0+t_1)\\ \widetilde{x}_I(\tau-t_1) & \frac{1}{2}(t_0+t_1)<\tau<\frac{1}{2}(t_1+t_2)\\ \widetilde{x}_I(t_2-\tau) & \frac{1}{2}(t_1+t_2)<\tau<\frac{1}{2}(t_2+t_3)\\ x_I(t_3-\tau) & \tau>\frac{t_2-t_3}{2}\end{cases}$$

These trajectories are rather artificially, being assambled form pieces of true solutions. They have singularities: in the middle, at $(t_1+t_2)/2$ the function is continuous but the derivative is discontinuous; in the points where one instanton arrives and another must continue (i.e. $(t_0+t_1)/2$ and $(t_2+t_3)/2$, both the function and the derivatives are discontinuous.

This is corrected by extending the trajectories in complex time plane. One defines the variables

$$\begin{aligned}z_1 &= t_1-t_0\\ z_2 &= t_2-t_1\\ z_3 &= t_3-t_2\end{aligned}$$

and the three integration contours

$$\begin{array}{lll} C_1 & \mathrm{Im} z_1 = \frac{\pi}{|U_0''|} & 0 \leqslant \mathrm{Re} z_1 \leqslant \frac{t}{2} - t_0 \\ C_2 & \mathrm{Im} z_2 = \frac{\pi}{U_a''} & 0 \leqslant \mathrm{Re} z_2 \leqslant \mathrm{Re}\left(\frac{t}{2} - t_1\right) \\ C_3 & \mathrm{Im} z_3 = \frac{\pi}{|U_0''|} & 0 \leqslant \mathrm{Re} z_3 \leqslant \frac{t}{2} - t_2 \end{array}$$

Together these contours must ensure the continuity of the function x_{IA} in the region 0, from which the condition arises

$$x_I \left[\mathrm{Re}\left(\frac{t_1 - t_0}{2}\right)\right] = -\widetilde{x}_I \left[\mathrm{Re}\left(\frac{t_0 - t_1}{2}\right)\right]$$

which gives for large separations $t_1 - t_0$

$$x_p \exp\left(\left|U_0''\right| \Delta_{0p}\right) = |x_m| \exp\left(\left|U_0''\right| \Delta_{m0}\right)$$

The exprsssion of K at this moment is

$$\begin{aligned} K_{ba}^{(2)}\left(b, \frac{t}{2}; b, -\frac{t}{2}\right) &= \left(\frac{2\pi D}{U_a''}\right)^{1/2} \frac{U_b'' U_a'' U_0''^2}{4} x_m^2 x_p^2 \left(\frac{|U_0''|}{2\pi D}\right)^3 \\ &\times \exp\left[-\frac{2U_0 - U_a - U_b}{D} + 2\left|U_0''\right|\left(\Delta_{m0} + \Delta_{0p}\right)\right] \\ &\times \int_{-t/2}^{t/2} dt_0 \int_{C_1} dz_1 \int_{C_2} dz_2 \int_{C_3} dz_3 \exp\left[F(z_1) + G(z_2) + F(z_3)\right] \end{aligned}$$

where

$$\begin{aligned} F(z) &= \frac{C_0}{D} \exp\left(-\left|U_0''\right| z\right) \\ &- \left|U_0''\right| \mathrm{Re} z \end{aligned} \tag{43}$$

$$G(z) = \frac{C_a}{D} \exp\left(-U_a'' z\right) \tag{44}$$

and

$$C_0 = \frac{x_m^2 \left|U_0''\right|}{2} \exp\left(2\left|U_0''\right| \Delta_{m0}\right) \tag{45}$$

$$C_a = \frac{\left(a - x_p\right)^2 U_a''}{2} \exp\left(2U_a'' \Delta_{pa}\right) \tag{46}$$

It is necessary to carry out the complex time integrals.

The z_3 integral. Here we have

$$\int_{C_3} dz_3 \exp\left\{\frac{C_0}{D} \exp\left(-\left|U_0''\right| z\right) - \left|U_0''\right| \mathrm{Re} z\right\}$$

along the real axis of z, we can substitute

$$u = \exp\left(-\left|U_0''\right| z\right)$$

and obtain

$$\int_{\mathrm{Re}C_3'} \frac{-1}{|U_0''|} du \exp\left(-\frac{C_0}{D}u\right) = -\frac{D}{|U_0''| C_0} \int dw \exp\left(-w\right)$$

This simply results in the constant

$$\frac{D}{C_0 |U_0''|} \tag{47}$$

The z_2 integral. The z_2 integral involves only the exponential of $G(z_2)$ and the latter function does not contain a term proportional with Rez_2 (comapre with (43)), but only the exponential term. The function $G(z_2)$ is very close to 1 as soon as

$$\mathrm{Re}z_2 \gg t_S^{(a)} \simeq \frac{1}{U_a''} \ln\left(\frac{C_a}{D}\right)$$

where $t_S^{(a)}$ is the Suzuki time for the region a

$$t_S^{(a)} \ll t \sim \tau_K$$

Then, with very good approximation

$$\int_{C_2} dz_2 \exp\left[G(z_2)\right] = \int_0^{\frac{t}{2}-t_1} dz_2 \exp\left[G(z_2)\right] \simeq \frac{t}{2}$$

The z_1 integral. This is similar to the integral over z_3 and will imply again $\frac{D}{C_0|U_0''|}$ as in (47).

Finally we note that only the constant C_0 relative to the intermediate point 0 appears, and the one for the farthest point of the trajectory, a, is absent.

There will be then a time integration over the moment of the first transition, t_0, of an integrand that contains the time t arising from the integration on the contour C_2, in the complex plane of the variable $z_2 = t_2 - t_1$. This variable represents the duration between the last transition to a and the first transition from a, i.e. is the duration of stay in a. This time integration will produce a term proportional with t^2.

Replacing in the expressions resulting from the integrations on z_1 and z_3 the constant C_0 from (45) the factors x_m and x_p disappear.

The result is

$$\begin{aligned} K_{ba}^{(2)}\left(b, \frac{t}{2}; b, -\frac{t}{2}\right) &= \frac{t^2}{2!}\left(\frac{|U_b''|}{2\pi D}\right)^{1/2} \frac{\left(|U_0''| U_a''\right)^{1/2}}{2\pi} \frac{\left(|U_0''| U_b''\right)^{1/2}}{2\pi} \\ &\times \exp\left(-\frac{2U_0 - U_a - U_b}{D}\right) \end{aligned}$$

5 The Probability

In the following we develop calculations according to the approach of Caroli et al. [34] but the same results can be obtained, as we have shown, by the path integral integration. The objective is to calculate $P\left(b,\frac{t}{2};b,-\frac{t}{2}\right)$ on the basis of the above results. We have to sum over any number of independent pseudomolecules of the four species

$$\begin{array}{ccc} (b0b) & \longleftrightarrow & -\alpha_b \\ (b0a) & \longleftrightarrow & \alpha_b \\ (a0b) & \longleftrightarrow & \alpha_a \\ (b0b) & \longleftrightarrow & -\alpha_a \end{array}$$

where the factor which is associated to each pseudomolecule is

$$\alpha_i = \frac{\left(|U_0''|\,U_a''\right)^{1/2}}{2\pi}\exp\left(-\frac{U_0-U_i}{D}\right)\ ,\ i=a,b$$

(we use the term pseudomolecule as in Caroli [34] to designate various combinations of instanton-antiinstanton transitions).

A particular path containing n $(b0b)$ and m $(a0a)$ has the general form

$$(-1)^{n+m}\frac{t^{n+m+2p}}{(n+m+2p)!}\alpha_b^{n+p}\alpha_a^{p+m}\left(\frac{|U_b''|}{2\pi D}\right)^{1/2}$$

It was taken into account that a path $(b\to b)$ contains the same number p of $(b0a)$ and of $(a0b)$.

5.0.1 Transition and Self-transition Probabilities at the Equilibrium Points

The results from the previous calculations are listed below.

The probability that a particle initially at $\left(b,-\frac{t}{2}\right)$ will be found at $\left(b,\frac{t}{2}\right)$

$$P\left(b,\frac{t}{2};b,-\frac{t}{2}\right)=\left(\frac{|U_b''|}{2\pi D}\right)^{1/2}\frac{1}{\alpha_a+\alpha_b}\left\{\alpha_a+\alpha_b\exp\left[-t\left(\alpha_a+\alpha_b\right)\right]\right\}$$

The probability that a particle initially at $\left(b,-\frac{t}{2}\right)$ will be found at $\left(a,\frac{t}{2}\right)$

$$P\left(a,\frac{t}{2};b,-\frac{t}{2}\right)=\left(\frac{U_a''}{2\pi D}\right)^{1/2}\frac{\alpha_b}{\alpha_a+\alpha_b}\left\{1-\exp\left[-t\left(\alpha_a+\alpha_b\right)\right]\right\}$$

The probability that a particle initially at $\left(a,-\frac{t}{2}\right)$ will be found at $\left(b,\frac{t}{2}\right)$

$$P\left(b,\frac{t}{2};a,-\frac{t}{2}\right)=\left(\frac{U_b''}{2\pi D}\right)^{1/2}\frac{\alpha_a}{\alpha_a+\alpha_b}\left\{1-\exp\left[-t\left(\alpha_a+\alpha_b\right)\right]\right\}$$

The probability that a particle initially at $\left(a,-\frac{t}{2}\right)$ will be found at $\left(a,\frac{t}{2}\right)$

$$P\left(a,\frac{t}{2};a,-\frac{t}{2}\right)=\left(\frac{|U_a''|}{2\pi D}\right)^{1/2}\frac{1}{\alpha_a+\alpha_b}\left\{\alpha_b+\alpha_a\exp\left[-t\left(\alpha_a+\alpha_b\right)\right]\right\}$$

5.0.2 Transitions between Arbitrary Points

The long time limit of the probability density

$$P\left(x, \frac{t}{2}; x_0, -\frac{t}{2}\right)$$

The formula has been derived for the case where the two positions x and x_0 belong to the harmonic regions (a) and (b). This is because asymptotically these regions will be populated.

Consider the case where

$$\begin{aligned} x_0 &\in (b) \\ x &\in (a) \end{aligned}$$

The first trajectory contributing to the path integral is the direct connection between b and a. The contribution to K is

$$K^{direct}\left(x, \frac{t}{2}; x_0, -\frac{t}{2}\right) \sim \exp\left(-\frac{S_{b0} + S_{0a}}{D}\right)$$

Since the points of start and/or arrival x_0, x are different of b, a, there is finite (i.e. non exponentially small) slope of the solution, and there is nomore degeneracy with respect to the translation of the center of the instantons. then there will be not a proportionality with time in the one-instanton term.

The terms connecting x_0 with x with only one intermediate step in either b or a have comparable contributions to the action.

Now we consider the term with two intermediate steps

$$x_0 \rightarrow b \rightarrow a \rightarrow x$$

In the very long time regime, $K^{(1)}$ can be factorized

$$\begin{aligned} K^{(1)}\left(x, \frac{t}{2}; x_0, -\frac{t}{2}\right) &= K_{harm}\left(x, \frac{t}{2}; a, t'\right) \\ &\times \left(\frac{2\pi D}{U''_a}\right)^{1/2} K^{(1)}\left(a, t'; b, t''\right) \left(\frac{2\pi D}{U''_b}\right)^{1/2} \\ &\times K_{harm}\left(b, t''; x_0, -\frac{t}{2}\right) \end{aligned}$$

This equation is independent of t' and t'' as long as

$$t_K \gg \frac{t}{2} - t' \text{ and } t'' - \left(-\frac{t}{2}\right) \gg t_S$$

Then

$$K_{harm}\left(x,\frac{t}{2};a,t'\right)\simeq\left(\frac{U''_a}{2\pi D}\right)^{1/2}\exp\left[-\frac{U(x)-U_a}{2D}\right]$$

There is also the approximate equality

$$K^{(1)}\left(a,t';b,t''\right)\approx K^{(1)}\left(a,\frac{t}{2};b,-\frac{t}{2}\right)$$

It results that all the contributions are included if in the expression of $K^{(1)}\left(x,\frac{t}{2};x_0,-\frac{t}{2}\right)$ we replace the middle factor, which has become $K^{(1)}\left(a,\frac{t}{2};b,-\frac{t}{2}\right)$ by the full $K\left(a,\frac{t}{2};b,-\frac{t}{2}\right)$. It is obtained

$$\begin{aligned}P\left(x,\frac{t}{2};x_0,-\frac{t}{2}\right) &= \exp\left[-\frac{U(x)-U_a}{2D}\right]P\left(a,\frac{t}{2};b,-\frac{t}{2}\right)\\ &= \exp\left[-\frac{U(x)-U_a}{2D}\right]\left(\frac{U''_a}{2\pi D}\right)^{1/2}\frac{\alpha_b}{\alpha_a+\alpha_b}\left\{1-\exp\left[-t\left(\alpha_a+\alpha_b\right)\right]\right\}\end{aligned}$$

The case where both x and x_0 belong to the harmonic region (b)
The same calculation shows that

$$P\left(x,\frac{t}{2};x_0,-\frac{t}{2}\right)=\exp\left[-\frac{U(x)-U_b}{2D}\right]\left(\frac{U''_b}{2\pi D}\right)^{1/2}\frac{1}{\alpha_a+\alpha_b}\left\{\alpha_a+\alpha_b\exp\left[-t\left(\alpha_a+\alpha_b\right)\right]\right\}$$

6 Results

Using the expressions obtained above we can investigate numerically the time behavior of populations of particles initially placed in regions around one or another minimum of the potential. For different values of the parameters we can see the faster or slower decay of the density from one region, accompagned by the corresponding increase in the other region. The time of attaining the stationarity is also of physical interest.

In the following we reproduce the graphs of the time-dependent probability distribution of a system governed by the basic Langevin equation. Various initialization are considered, showing rapid redistribution of the density of the probability of presence of the system. The speed of redistribuition is, naturally, connected to the asymmetry of the potential. Each figure consists of a set of graphs:

- the potential functions $V(x)$, $U(x)$ and $W(x)$
- four probabilities of passage from and between the two equilibrium points a and b;
- the average value of the position of the system, as function of time, for the two most characteristic initializations

- the time evolution of the density of particles initialised in the minimum a
- the time evolution of the density of particles initialised in the minimum b

The two parameters a and D take different values, for illustration.

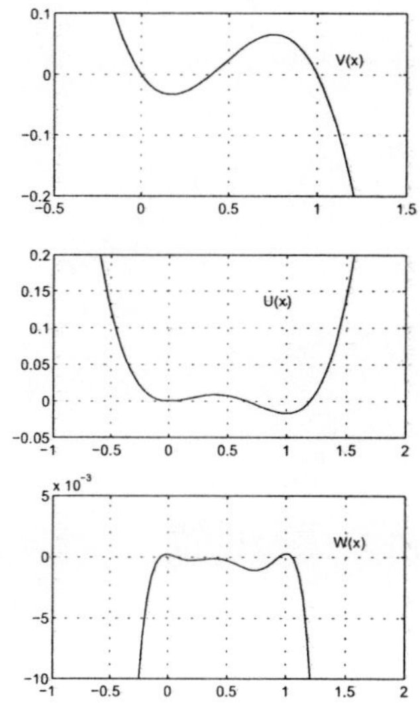

(a) Form of the potential

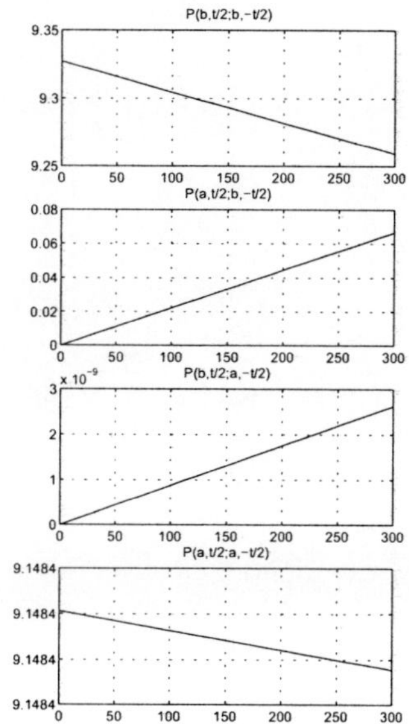

(b) Evolution of probability

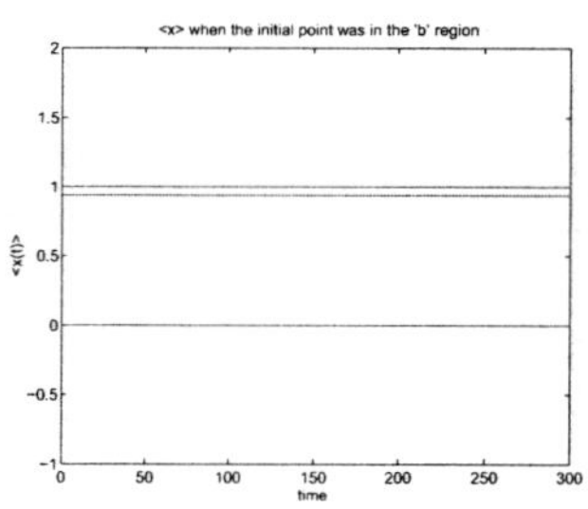

(c)

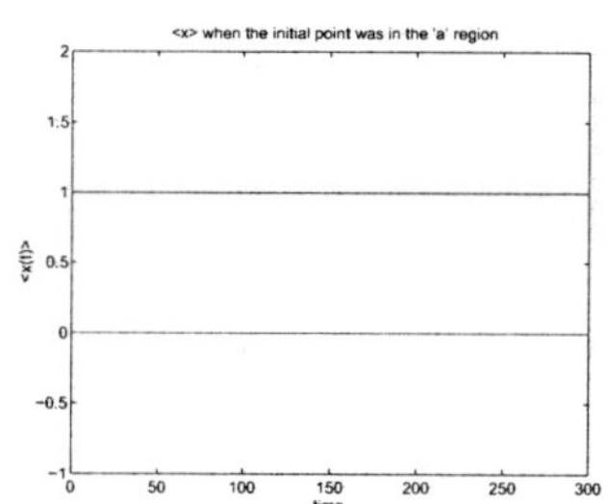

(d)

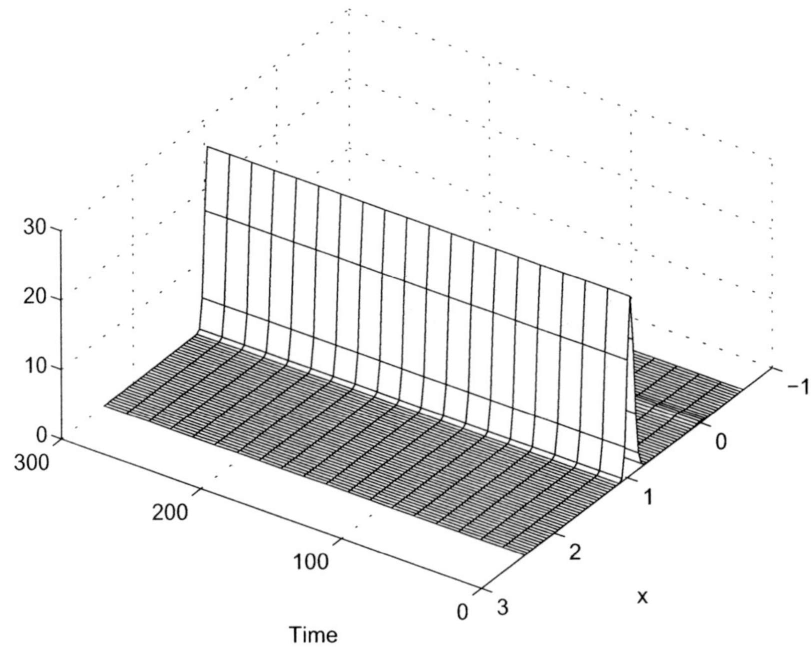

(e)

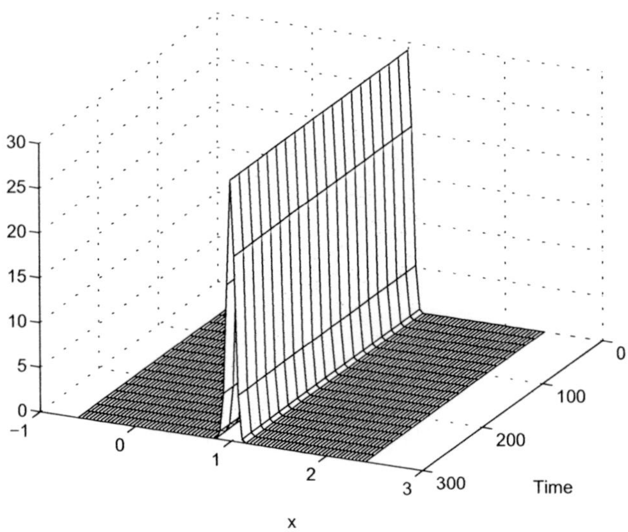

(f)

Figure 5: Parameters: $a = 0.4$ and $D = 0.001$

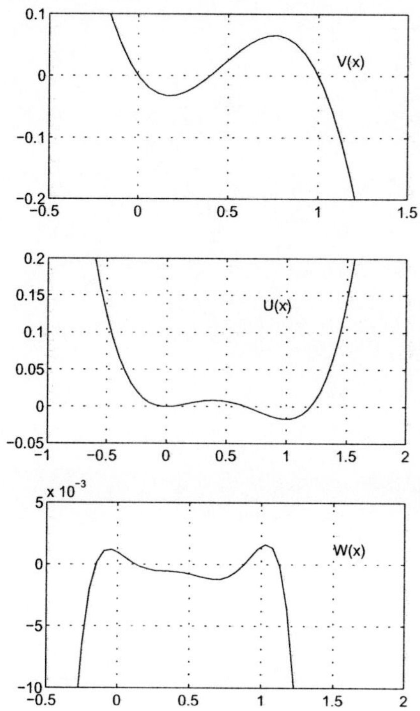

(a) Form of the potential

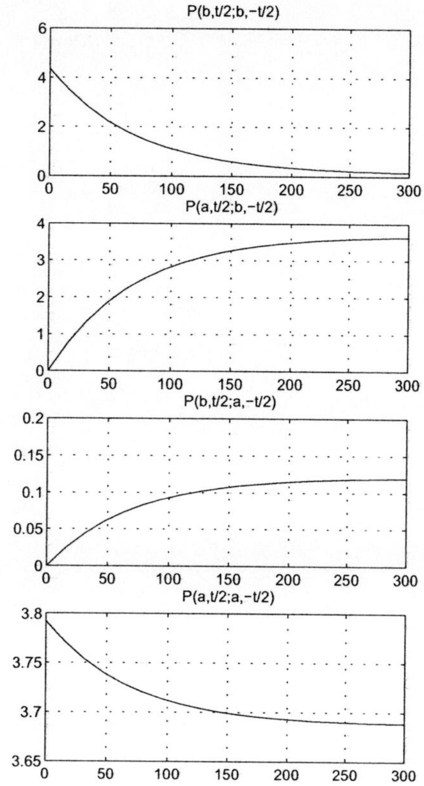

(b) Evolution of probability

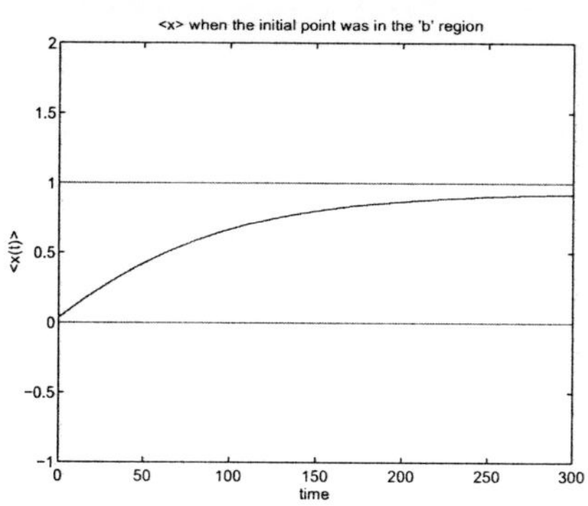

(c)

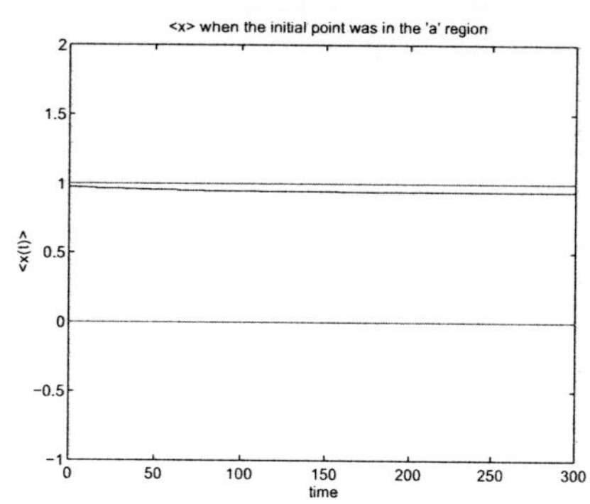

(d)

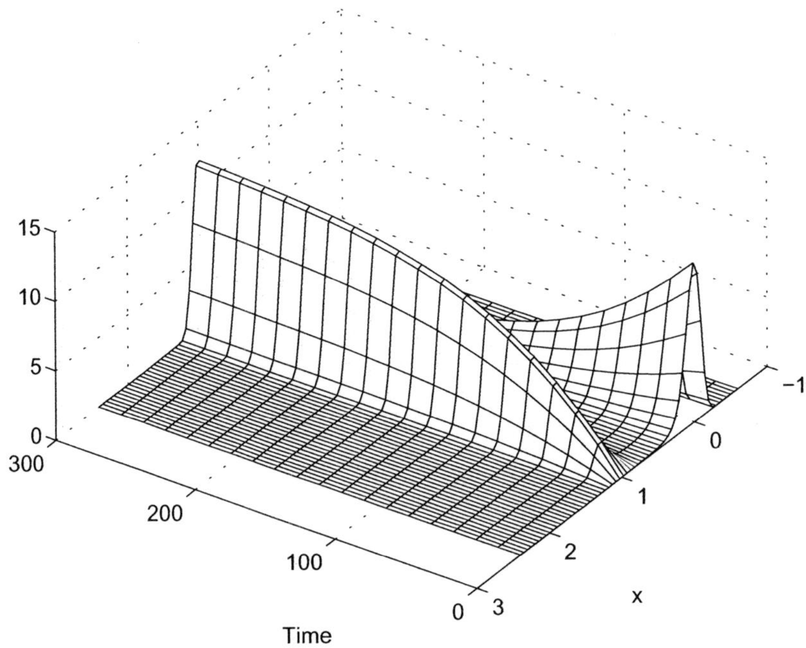

(e)

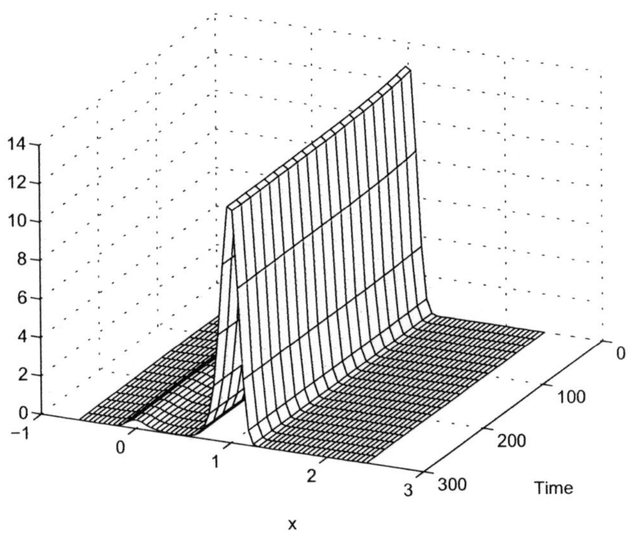

(f)

Figure 6: Parameters: $a = 0.4$ and $D = 0.005$

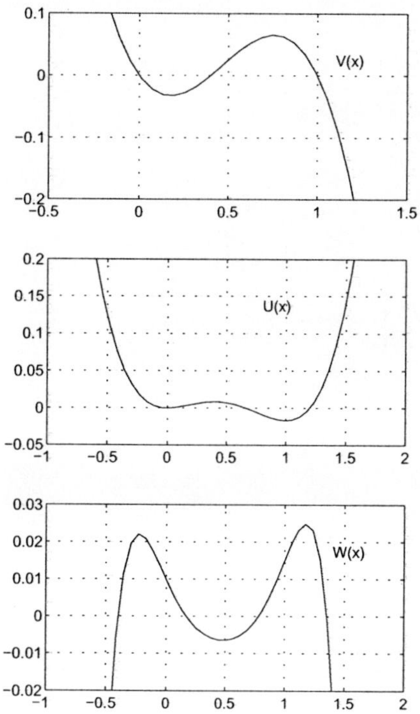

(a) Form of the potential

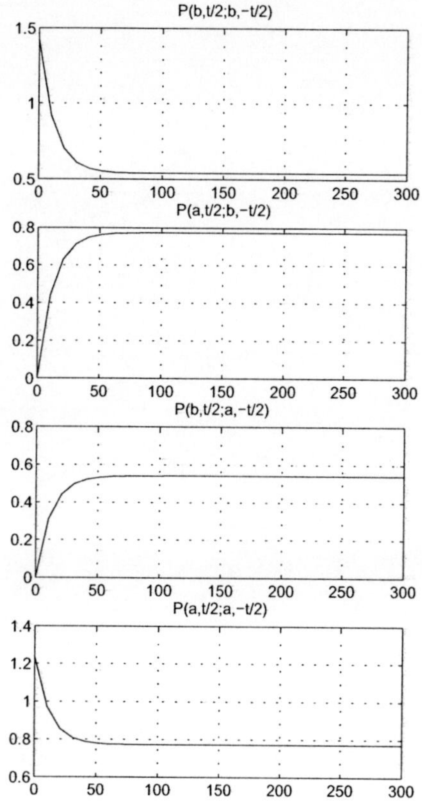

(b) Evolution of probability

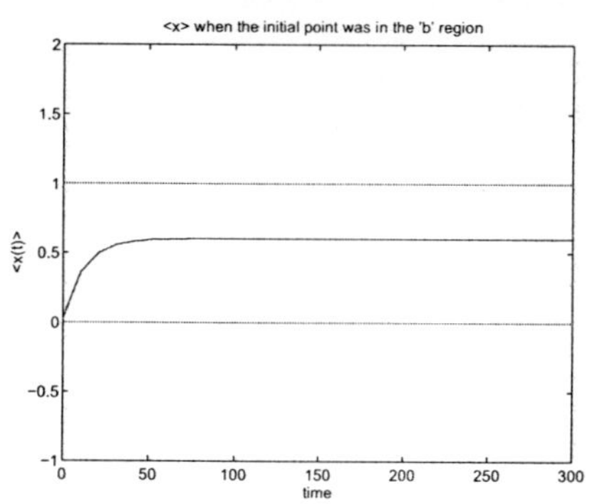

(c)

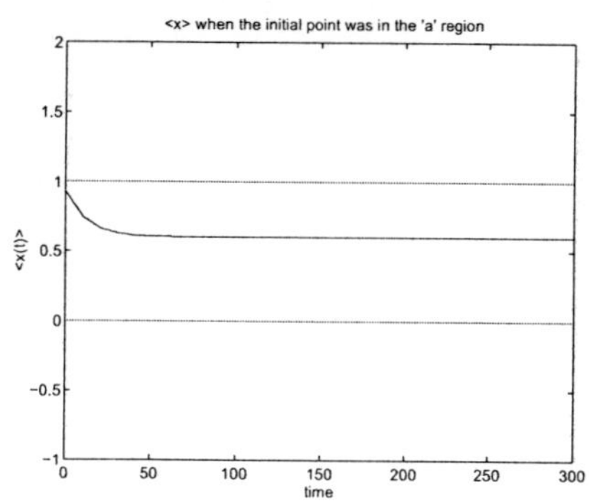

(d)

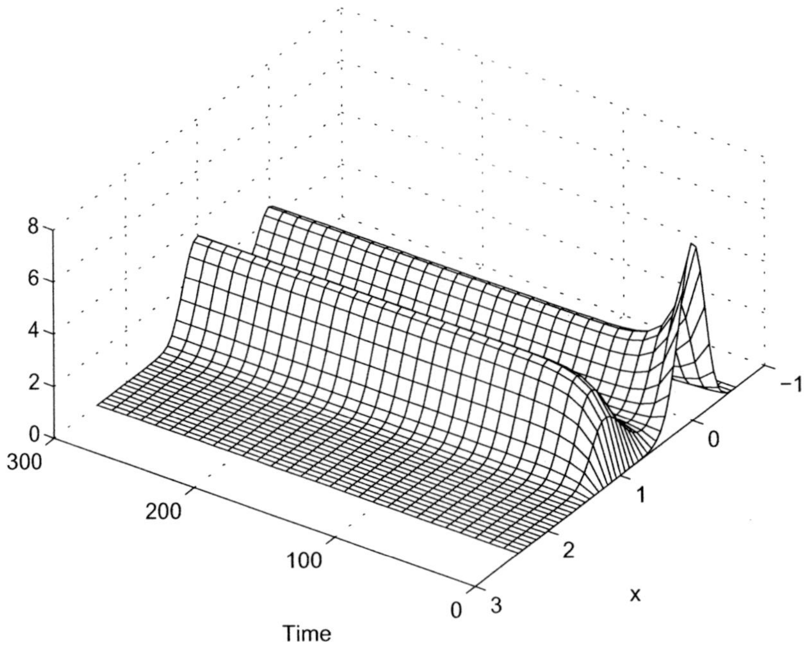

(e)

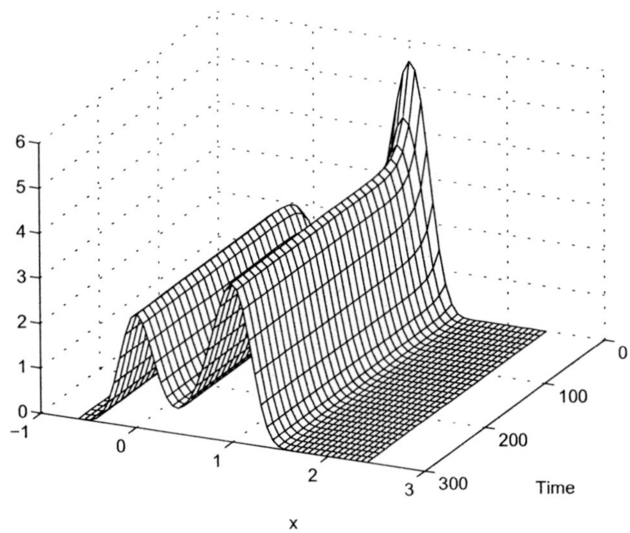

(f)

Figure 7: Parameters: $a = 0.4$ and $D = 0.05$

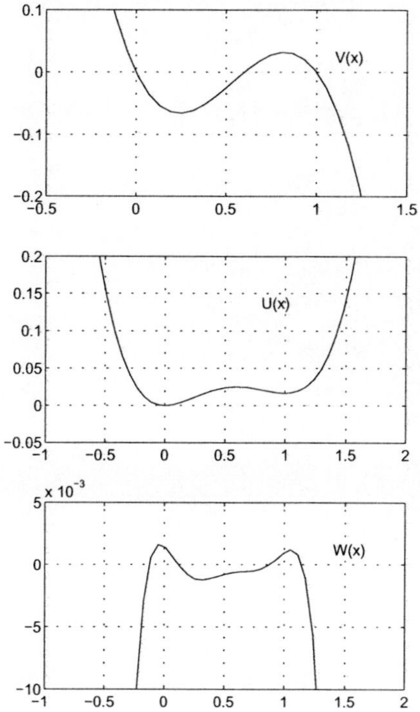

(a) Form of the potential

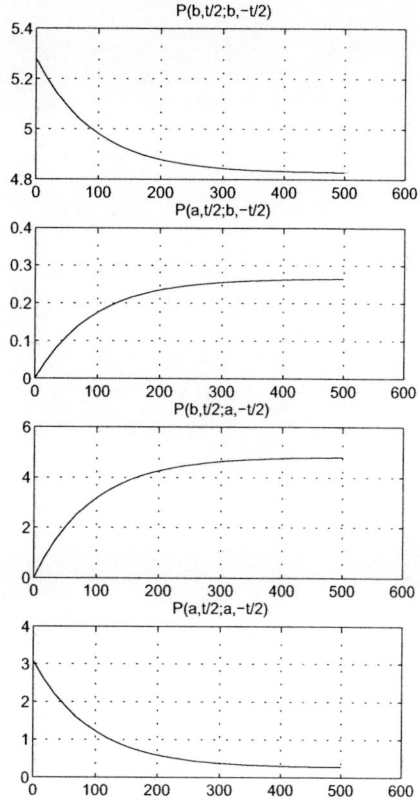

(b) Evolution of probability

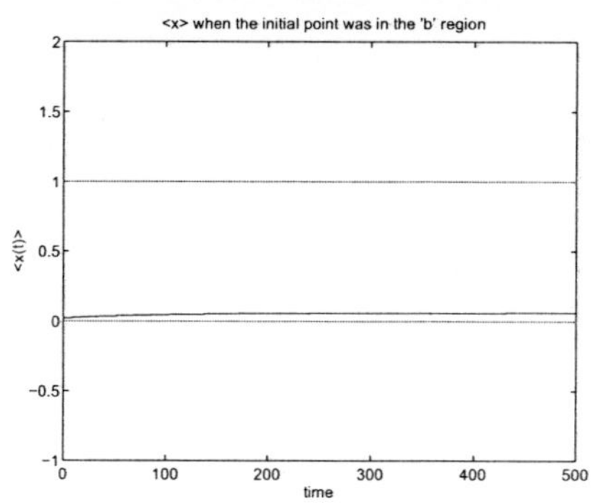

(c)

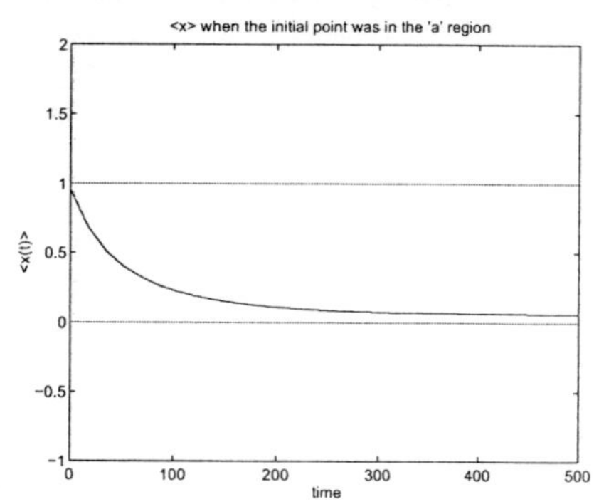

(d)

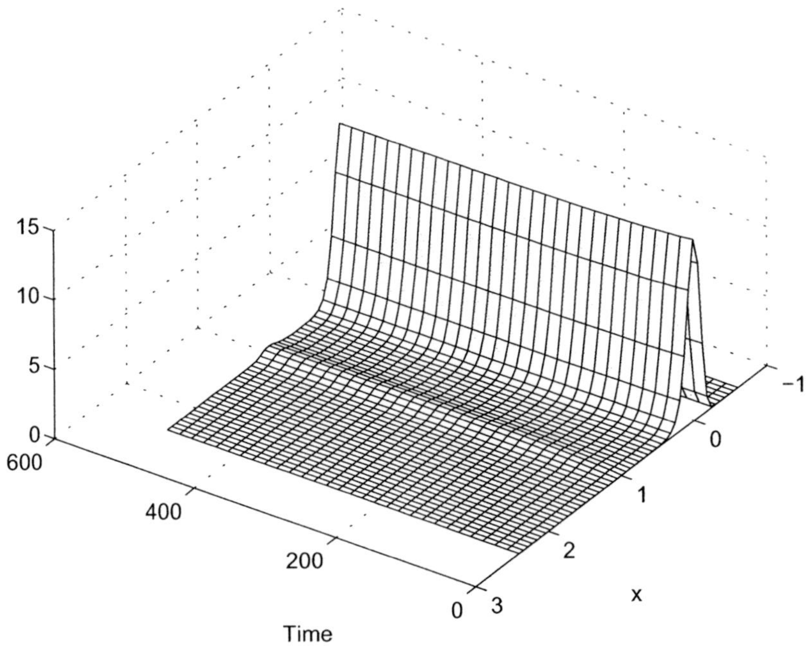

(e)

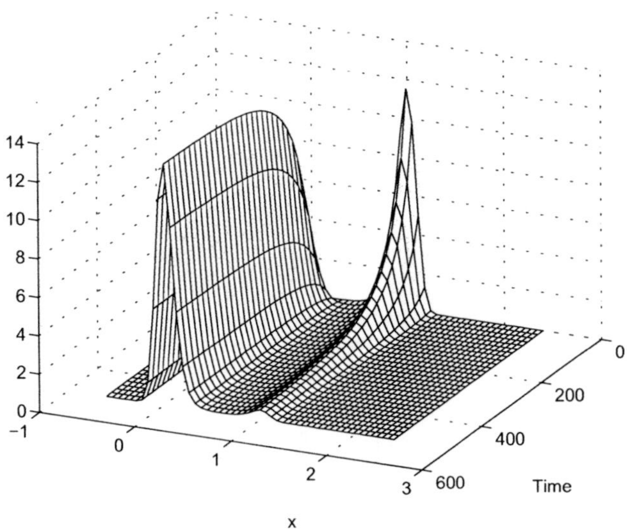

(f)

Figure 8: Parameters: $a = 0.6$ and $D = 0.005$

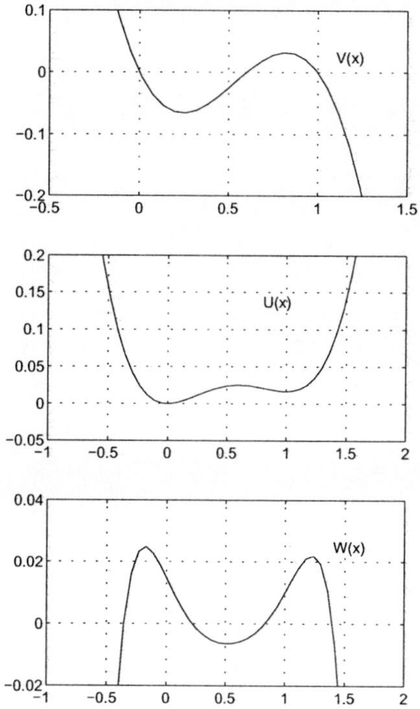

(a) Form of the potential

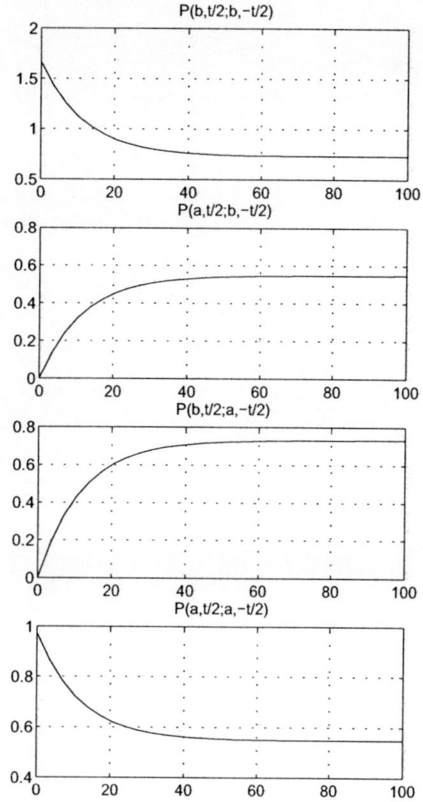

(b) Evolution of probability

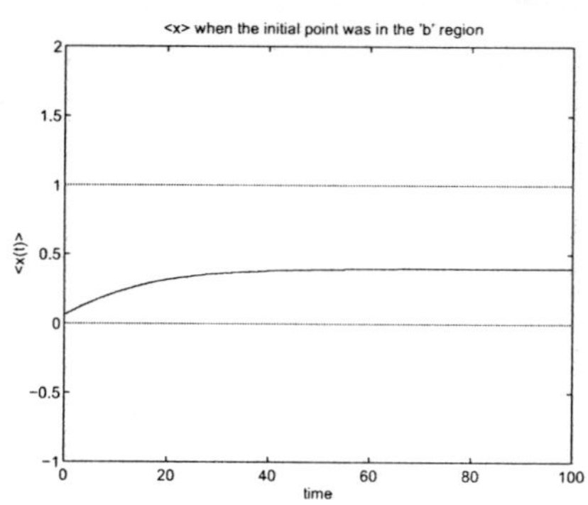

(c)

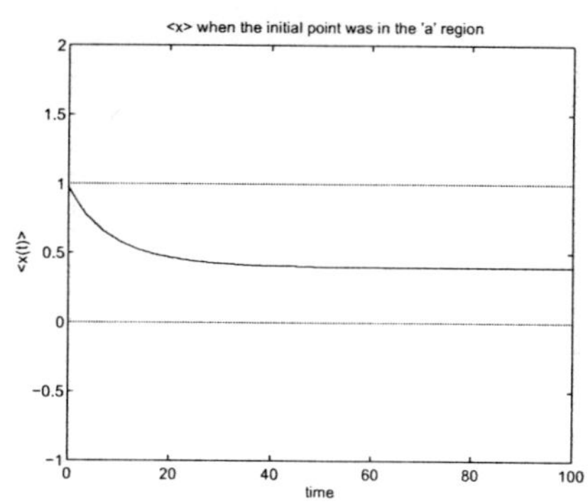

(d)

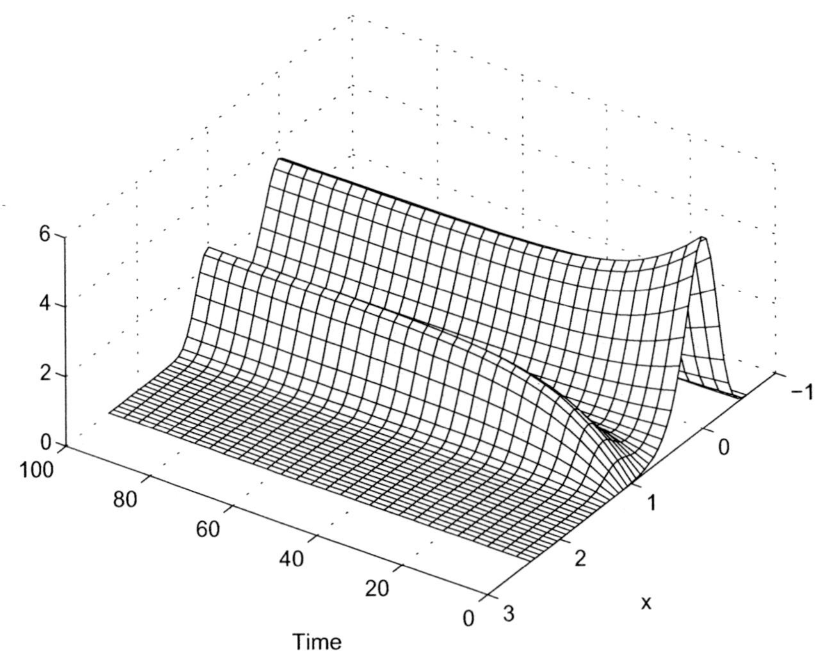

(e)

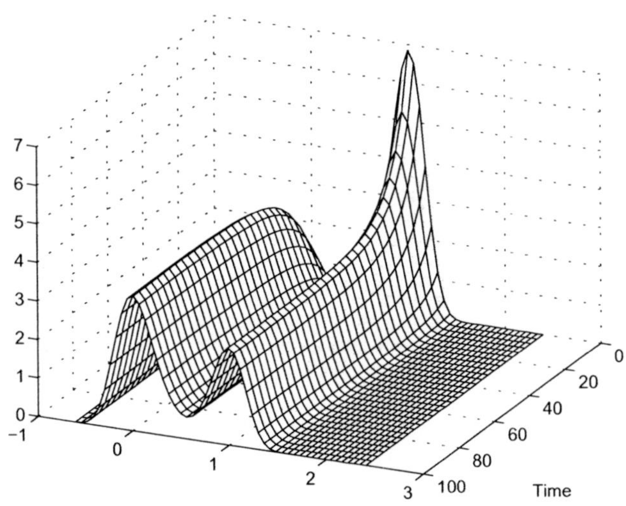

(f)

Figure 9: Parameters: $a = 0.6$ and $D = 0.05$

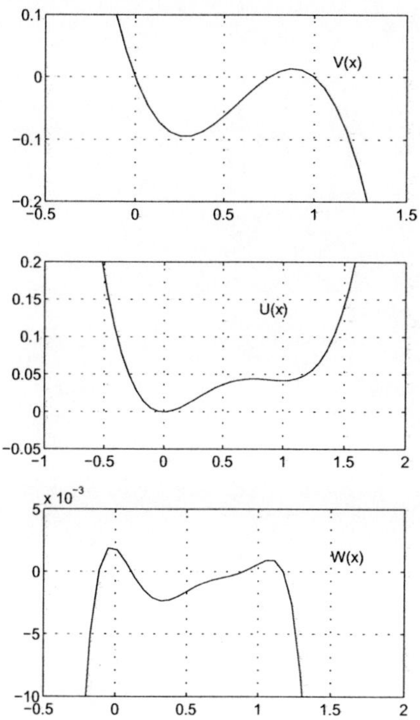

(a) Form of the potential

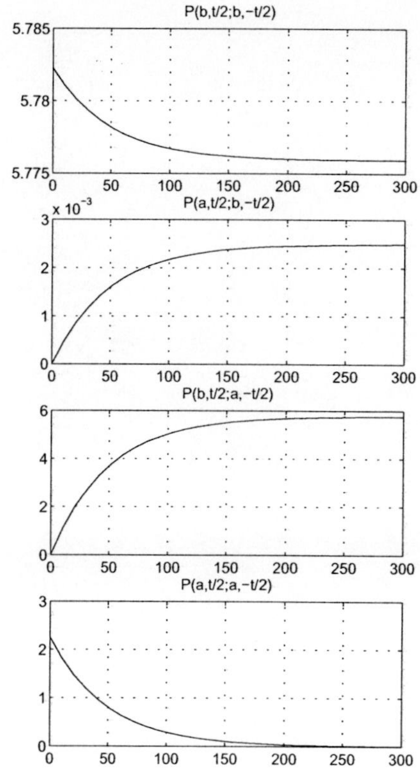

(b) Evolution of probability

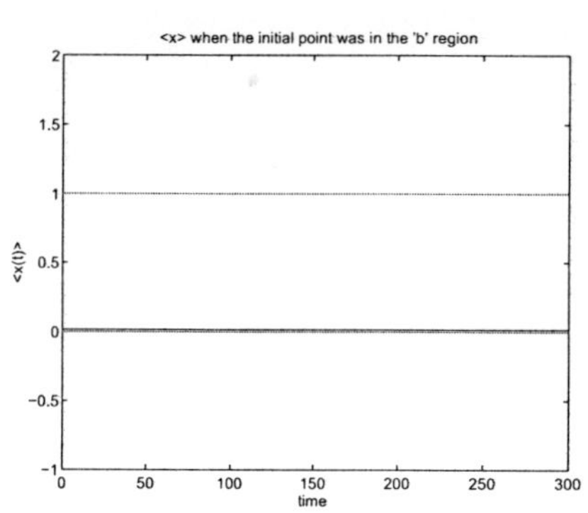

(c)

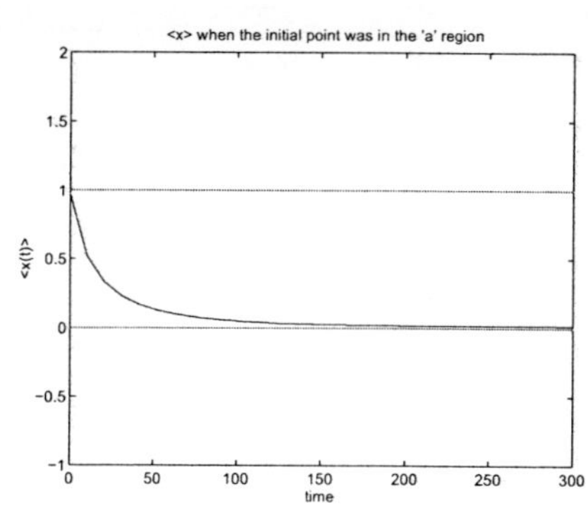

(d)

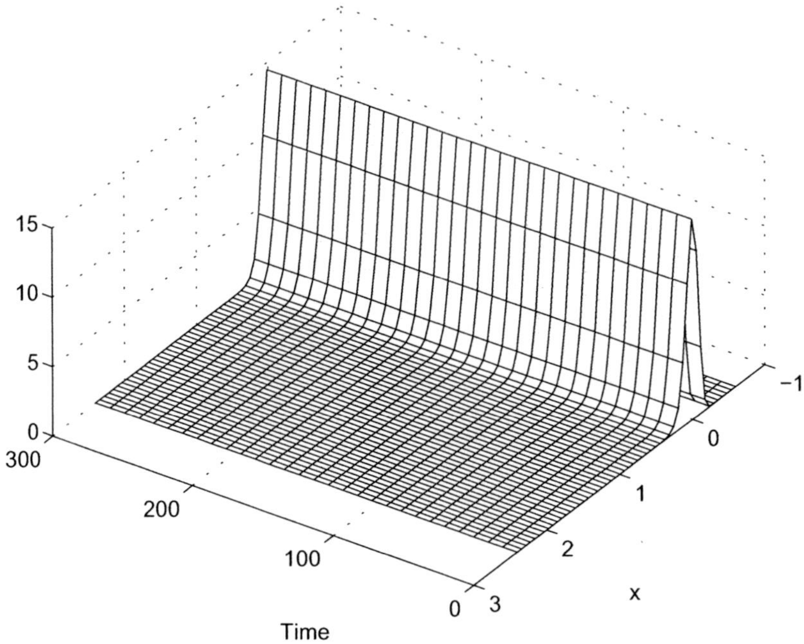

(e)

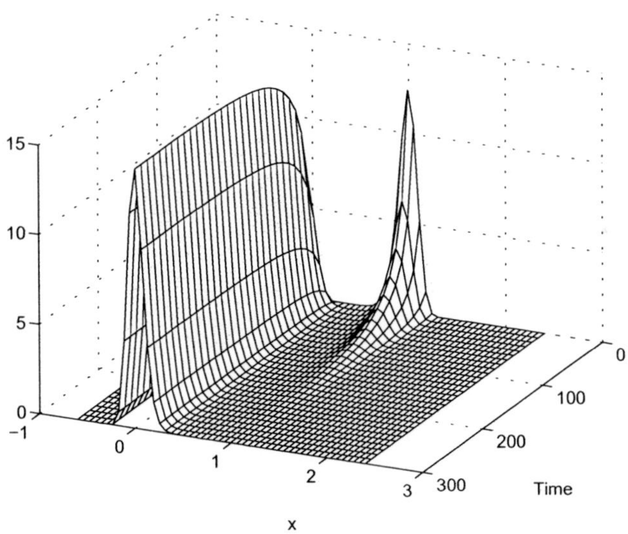

(f)

Figure 10: Parameters: $a = 0.75$ and $D = 0.005$

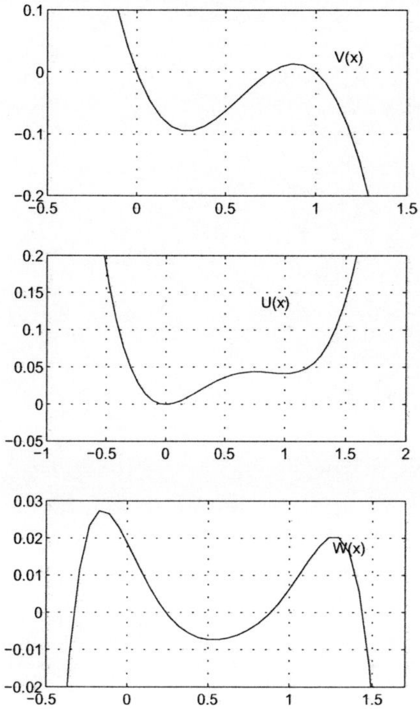

(a) Form of the potential

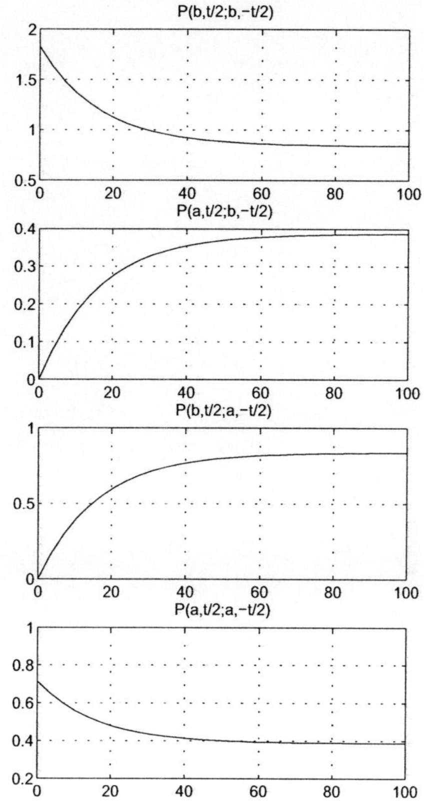

(b) Evolution of probability

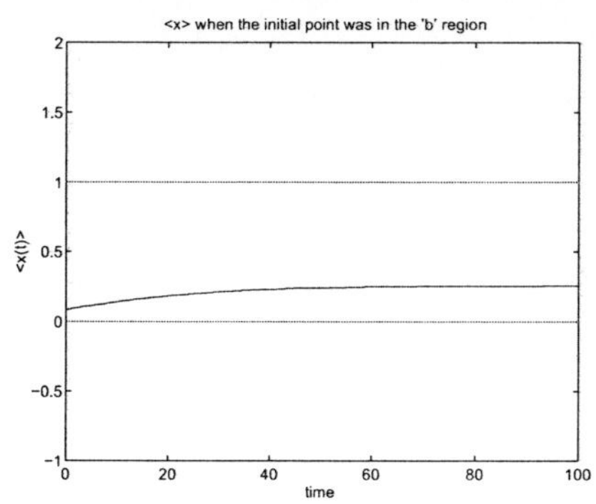

(c)

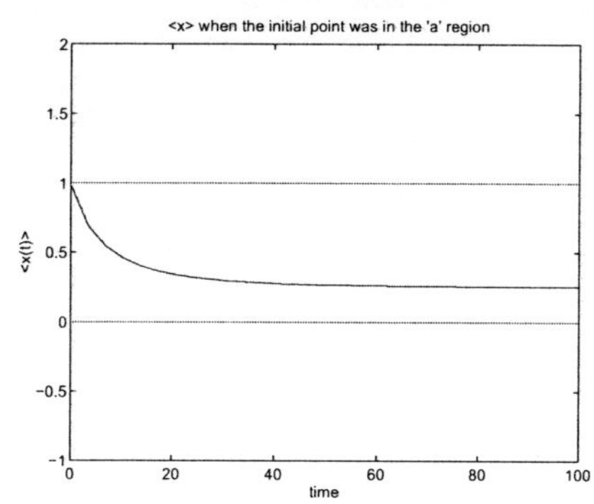

(d)

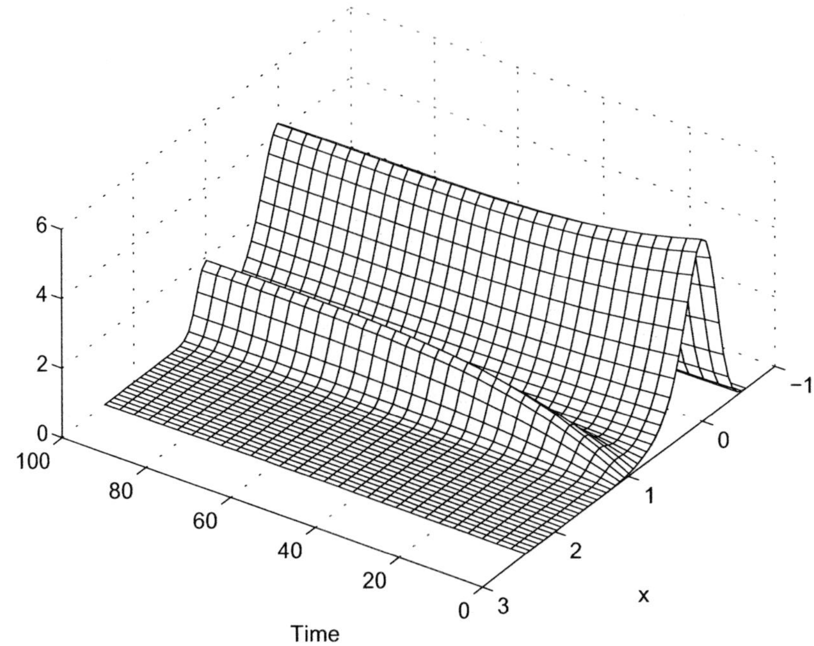

(e)

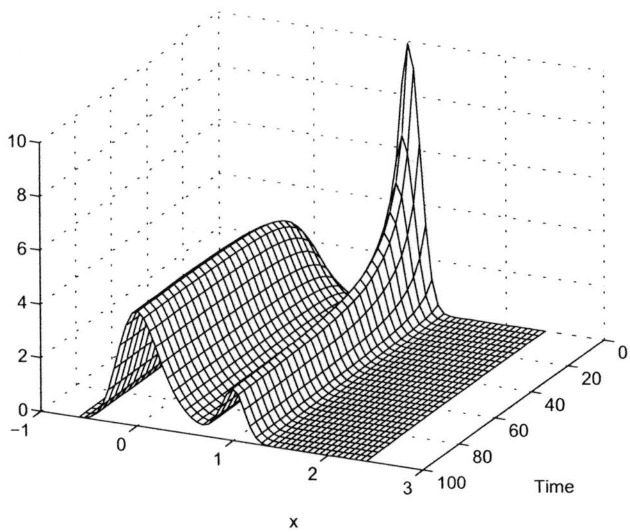

(f)

Figure 11: Parameters: $a = 0.75$ and $D = 0.05$

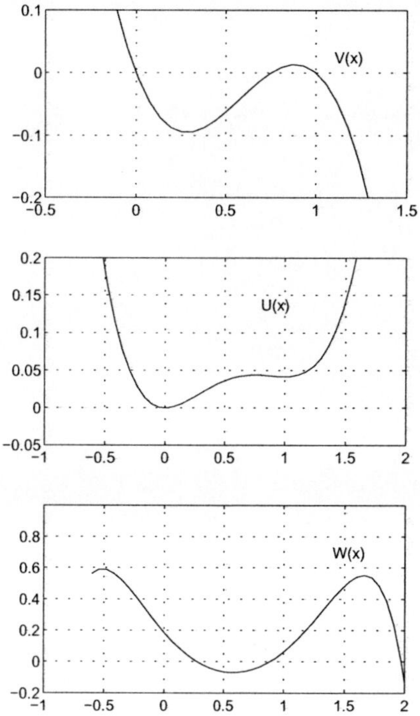

(a) Form of the potential

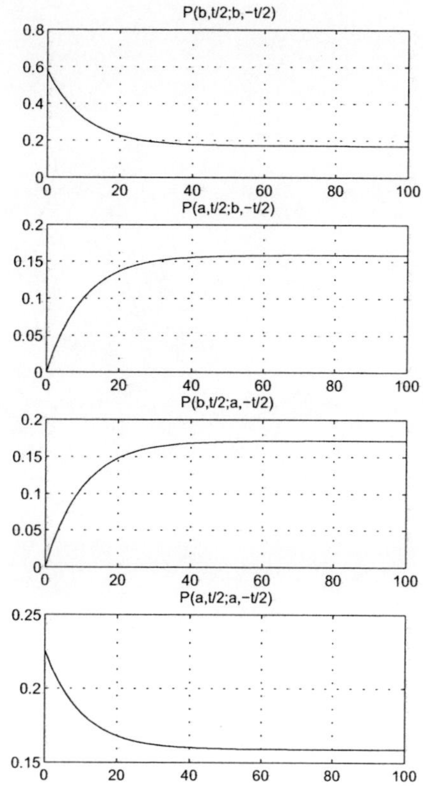

(b) Evolution of probability

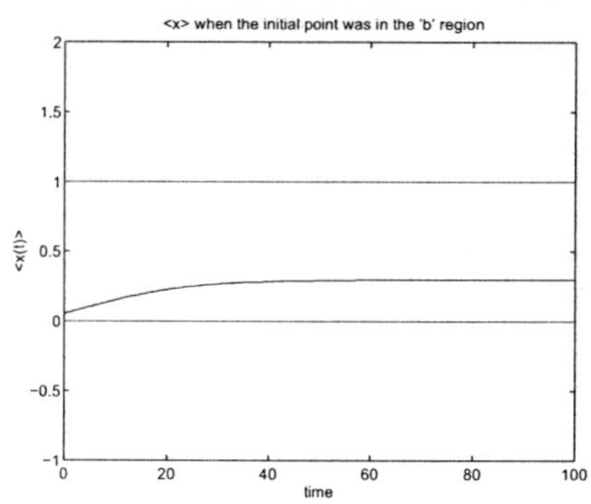

(c)

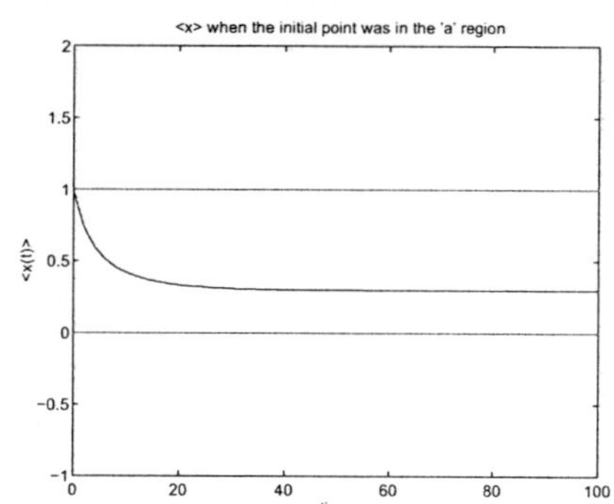

(d)

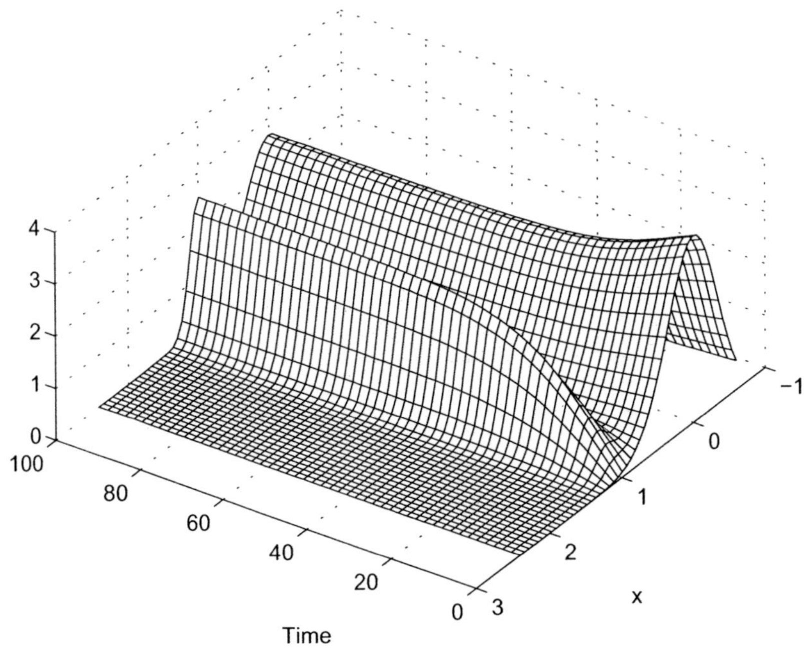

(e)

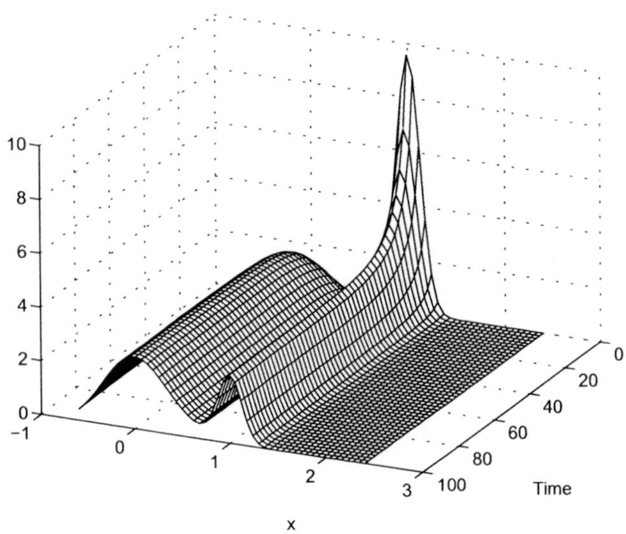

(f)

Figure 12: Parameters: $a = 0.75$ and $D = 0.5$

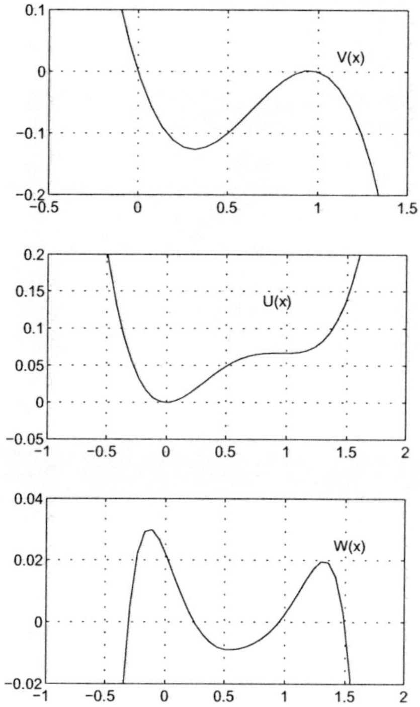

(a) Form of the potential

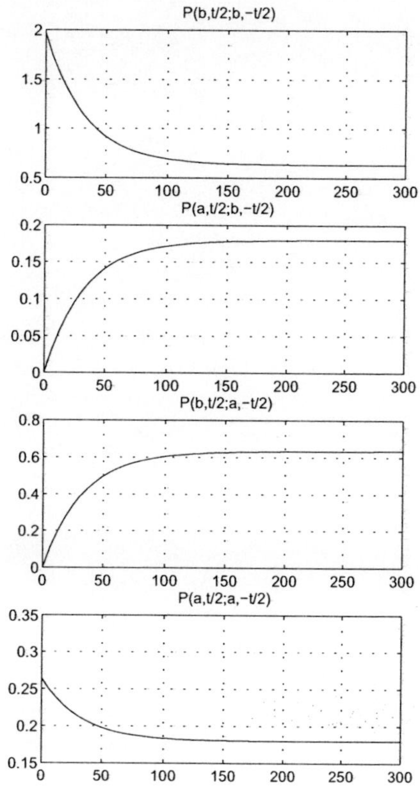

(b) Evolution of probability

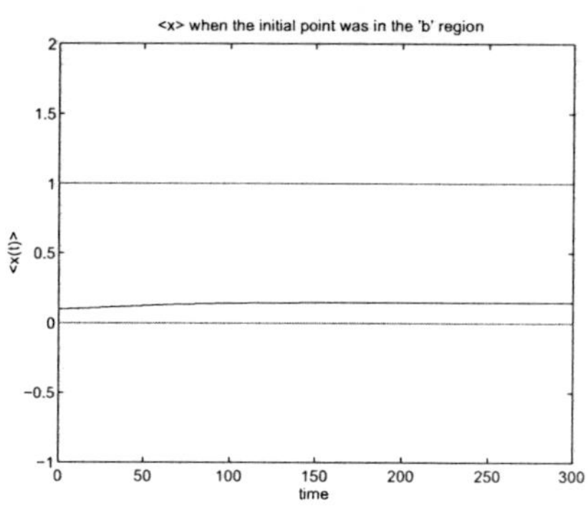

(c)

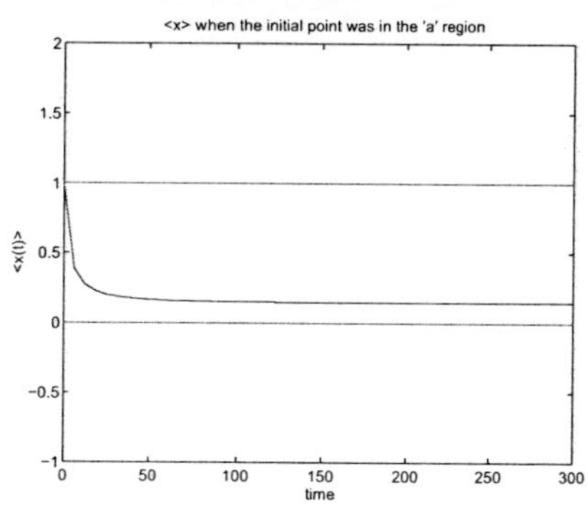

(d)

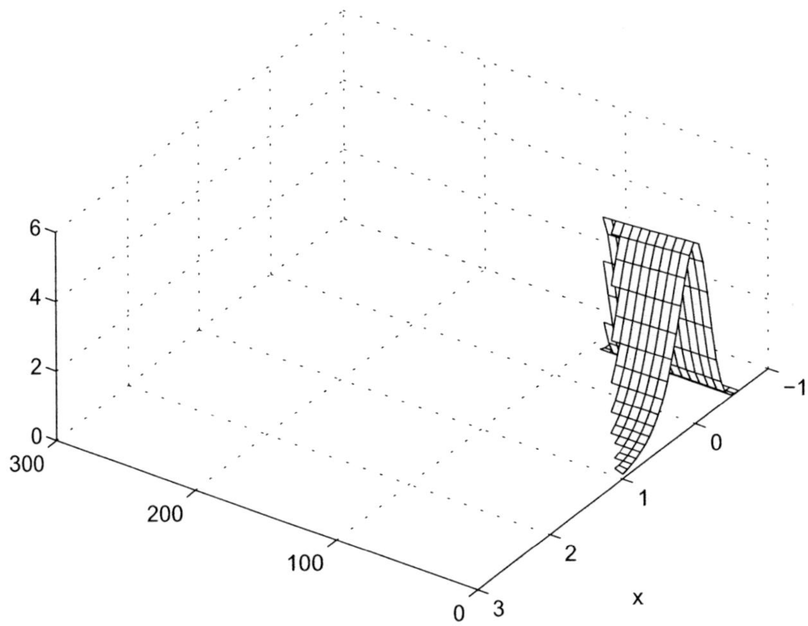

(e)

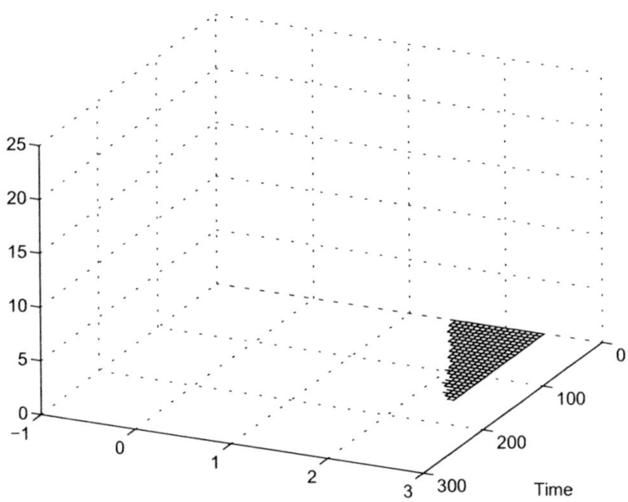

(f)

Figure 13: Parameters: $a = 0.9$ and $D = 0.05$

Appendix A : Time Variation for General Form of the Potential

The stochastic motion is described in terms of the conditional probability $P(x,t;x_i,0)$ that the particle initially $(t=0)$ at x_i to be at the point x at time t. We will use the notation that suppress the 0 as the initial time, $P(x,t;x_i)$. The conditional probability obeys the following Fokker-Planck equation

$$\frac{\partial P}{\partial t} = -\frac{\partial}{\partial x}\left[U'(x)P\right] + D\frac{\partial^2 P}{\partial x^2}$$

where the velocity function is here derived from the potential U

$$F[x(t),t] \equiv -U'(x)$$

since there is no drive. The initial condition for the probability function is

$$P(x,0;x_i) = \delta(x-x_i)$$

The simple form of the potential $U(x)$ allows the introduction of two constants

$$\begin{aligned}\omega_0 &\equiv \left|U''(x=0)\right| \\ \omega_1 &\equiv U''(x=1)\end{aligned}$$

The following orderings are assumed

$$\begin{aligned}D &\ll \omega_0 \\ D &\ll \omega_1\end{aligned}$$

The solution of the Fokker-Planck equation is given in terms of the following path-integral

$$P(x,t;x_i) = \frac{\exp[-U(x)/2D]}{\exp[-U(x_i)/2D]}K(x,t;x_i)$$

where

$$K(x,t;x_i) = \int D[x(\tau)]\exp\left[-\frac{S}{D}\right] \tag{A.1}$$

where the functional integration is done over all trajectories that start at $(x_i,0)$ and end at (x,t). The action functional is given by

$$S(x,t;x_i) = \int_0^t d\tau\left[\frac{1}{4}\dot{x}^2 + W(x)\right]$$

with the notation

$$W(x) \equiv \frac{1}{4}U'^2(x) - \frac{D}{2}U''(x)$$

In order to calculate explicitely the functional integral we look first for the paths $x(t)$ that extremize the action. They are provided by the Euler-Lagrange equations, which reads

$$\frac{\delta S}{\delta x_c} = -\frac{1}{2}\ddot{x}_c + W'(x_c) = 0$$

with the boundary condition

$$\begin{aligned} x_c(0) &= x_i \\ x_c(t) &= x \end{aligned}$$

The first thing to do after finding the extremizing paths is to calculate the action functional along them, S_c. After obtaining these extremum paths we have to consider the contribution to the functional integral of the paths situated in the neighborhood and this is done by expanding the action to second order. The argument of the expansion is the difference between a path from this neigborhood and the extremum path. The functional integration over these differences can be done since it is Gaussian and the result is expressed in terms of the determinant of the operator resulting from the second order expansion of S. Then the expression (A.1) becomes

$$K = N\frac{1}{\left|\det \widehat{L}(x_c)\right|^{1/2}}\exp\left(-\frac{S_c}{D}\right) \tag{A.2}$$

where

$$\widehat{L}(x_c) \equiv \left.\frac{\delta^2 S}{\delta x(\tau)^2}\right|_{x(\tau)=x_c(\tau)} = -\frac{1}{2}\frac{d^2}{d\tau^2} + W''(x_c) \tag{A.3}$$

The constant N is for normalization.

In order to find the determinant, one has to solve the eigenvalue problem

$$\left[\widehat{L}(x_c) - \lambda_n\right]\phi_n(\tau) = 0 \tag{A.4}$$

with boundary conditions for the differences between the trajectories in the neighborhood and the extremum trajectory

$$\phi_n(0) = \phi_n(t) = 0$$

It can be shown that

$$\frac{N^2}{\left|\det \widehat{L}(x_c)\right|} = \frac{1}{2\pi D\dot{x}_c(0)}\frac{\partial E}{\partial x}$$

where E is the energy of the path $x_c(\tau)$. (see Dashen [36]).

There is a fundamental problem concerning the use of the equation (A.2). It depends on the possibility that all the eigenvalues from the Eq.(A.4) are determined. However, a path connecting a point in the neigborhood of the maximum of $U(x)$,

$$x_i \sim 0 + O\left(\sqrt{D}\right)$$

with a point in the neighborhood of one of the minima

$$x \sim \pm 1 + O\left(\sqrt{D}\right)$$

is a kink-like solution. This solution (which will be replaced in the expression of the operator in Eq.(A.4) contains a parameter, the "center" of the kink, t_0. This is an arbitrary parameter since the moment of traversation is arbitrary. The equation for the eigenvalues has therefore a symmetry at time translations of t_0 and has as a consequence the appearence of a zero eigenvalue. Then the expression of the determinant would be infinite and the rate of transfer would vanish. Actually, the time translation invariance is treated by considering the arbitrary moment t_0 as a new variable and performing a change of variable, from the set of functions ϕ_n to the set $(t_0, \phi_{n,n\neq 0})$. The trajectory that extremizes the action is a kink-like solution $\overline{x}(\tau, t_0)$ (is not exactly a kink since the shape of the potential is not that of the $\tanh$ solution) connecting the points :

$$\begin{aligned} \overline{x}(0, t_0) &= x_i \\ \overline{x}(t, t_0) &= x \end{aligned}$$

This treatment then consists of considering t_0 as a collective coordinate (see Gervais & Sakita [37]). The new form of Eq.(A.2) is

$$K = N \int_0^t dt_0 J \frac{\exp\left(-\overline{S}/D\right)}{\left|\det' \widehat{L}\left(\overline{x}\right)\right|^{1/2}} \tag{A.5}$$

where $\overline{S}$ is the action of the path $\overline{x}(\tau, t_0)$. The fact that the zero eigenvalue has been eliminated is indicated by the $'$.

The factor in the integral is approximately

$$J^2 = \frac{1}{2\pi D}\left[U(x_i) - U(x)\right]$$

Two limits of time are important. The first is

$$t_s = \frac{1}{2\omega_0} \ln\left(\frac{\omega_0}{D}\right)$$

For the times

$$0 \leq t \lesssim t_s$$

for a particle initially in the region of the unstable state, $x_i \sim O\left(\sqrt{D}\right)$, the region around the stable state, ± 1, or: $x \sim \pm 1 + O\left(\sqrt{D}\right)$ is insignificant for the calculation of the probability $P(x, t; x_i)$. Then the expression (A.2) can be used.

Much more important is the subsequent time regime

$$t > t_s$$

where the particles leave the unstable point and the equilibrium state with density around the stable positions ± 1 is approached. Then the general formula (A.5) should be used.

The expression of the potential in general does not allow an explicit determination of the extremizing trajectories. Approximations are necessary.

The energy E of the path $x_c(\tau)$ connecting the points $(x_i = 0, 0)$ and (x, t) can be approximated from the expression (see Caroli [34])

$$t = \int_0^x dx' \frac{1}{[2(W_0(x') + E)]^{1/2}} - \Theta_0(x, 0)$$

where

$$\Theta_0(z, y) = \int_y^z dx' \left(\frac{1}{|U_0'(x')|} - \frac{1}{|U'(x')|} \right)$$

The notations are

$$W_0 = \frac{1}{4} U_0'^2 - \frac{D}{2} U_0''$$

$$U_0'(x) = -\omega_0 x$$

The last expressions are the harmonic approximation of V_0 and the potential around the unstable point $x = 0$.

The additional term $\Theta(x, 0)$ is the amount of time accounting for the deviation of V from the parabolic profile, the anharmonic deviation $V - V_0$.

The result for this simpler case is

$$P(x, t; 0, 0) = \frac{\omega_0}{\sqrt{\pi}} \frac{1}{|U'(x)|} \sqrt{G} \exp(-G)$$

where

$$G = \frac{\omega_0}{D} \frac{x^2}{\exp[2\omega_0 T_0(x)] - 1}$$

$$T_0(x) \equiv t + \Theta_0(x, 0)$$

Two time regimes are interesting. First, the very short time, less than the period of the particle in the potential,

$$0 \leq t \lesssim \frac{1}{2\omega_0}$$

the result reduces to the harmonic problem around $x = 0$.

For longer times,

$$\frac{1}{2\omega_0} \ll t \lesssim t_s$$

the result is

$$P(x, t; 0, 0) = \frac{1}{\sqrt{\pi}} F' \exp(-F^2)$$

where

$$F = \left(\frac{\omega_0}{D}\right)^{1/2} x \exp\left[-\omega_0 T_0 (x)\right]$$

This is the scaling solution of Suzuki.

The general expression of the function K in the case where we include the time regimes beyond the limits given above. It has the form of a convolution

$$K = \int_0^t dt_0 \dot{\overline{x}}_0 (t_0)\, K (x, t_1; x_m)\, K (x_m, t_0; x_i)$$

Here the approximations for the ratio $N/\left[\det \widehat{L}(x_c)\right]^{1/2}$ are obtained by the same method as before. The relation between the energy and the time is used for the two main regions: around the stable and the unstable (initial) positions

$$t_0 = \int_{x_i}^{x_m} dx' \frac{1}{\left[2 (W_0 + E_0)\right]^{1/2}} - \Theta_0 (x_m, x_i)$$

$$t_1 = \int_{xm}^{x} dx' \frac{1}{\left[2 (W_1 + E_1)\right]^{1/2}} - \Theta_1 (x, x_m)$$

References

[1] S.-I. Itoh, K. Itoh, M. Yagi, M. Kawasaki and A. Kitazawa: *Physics of Plasmas* **9** (2002) 1947.

[2] S.-I. Itoh, A. Kitazawa, M. Yagi and K. Itoh: *Plasma Phys.Control. Fusion* **44** (2002) 1311.

[3] S.-I. Itoh and K. Itoh: *Plasma Phys.Control. Fusion* **43** (2001) 1055.

[4] S.-I. Itoh and K. Itoh: *J. Phys. Soc. Japan* **68** (1999) 1891.

[5] S.-I. Itoh and K. Itoh: *J. Phys. Soc. Japan* **68** (1999) 2611.

[6] S.-I. Itoh and K. Itoh: *J. Phys. Soc. Japan* **69** (2000) 408.

[7] S.-I. Itoh and K. Itoh: *J. Phys. Soc. Japan* **69** (2000) 427.

[8] S.-I. Itoh and K. Itoh: *J. Phys. Soc. Japan* **69** (2000) 3253.

[9] K. Itoh, S.-I. Itoh, F. Spineanu, M.O. Vlad and M. Kawasaki: *Plasma Phys. Control. Fusion* **45** (2003) 1.

[10] P. H. Rebut and M. Hugon: in *Plasma Physics and Controlled Nuclear Fusion Research* 1984 (IAEA, 1985, Vienna) Vol.2, 197.

[11] B. D. Scott: *Phys. Fluids* **B4** (1992) 2468.

[12] J. F. Drake, A. Zeiler and D. Biskamp: *Phys. Rev. Lett.* **75** (1995) 4222.

[13] D. Biskamp and A. Zeiler: *Phys. Rev. Lett.* **74** (1995) 706.

[14] D. Biskamp and M. Walter: *Phys. Lett.* **109**A (1985) 34.

[15] R. E. Waltz: *Phys. Rev. Lett.* **55** (1982) 1098.

[16] B. A. Carreras *et al.*: *Phys. Fluids B* **4** (1992) 3115.

[17] H. Nordman, V. P. Pavlenko and J. Weiland: *Phys. Plasmas B* **5** (1993) 402.

[18] A. Yoshizawa, S. -I. Itoh and K. Itoh: *Plasma and Fluid Turbulence* (IOP, England, 2002).

[19] R. P. Feynman and A. R. Hibbs: *Quantum Mechanics and Path Integrals* (McGraw-Hill, New York, 1965).

[20] P. Hänggi, P. Talkner and M. Borcovec: *Rev. Mod. Phys.* **62** (1990) 251.

[21] H. Risken: *The Fokker-Planck equation*, 2nd. ed. (Springer, Berlin, 1989).

[22] J. S. Langer: *Ann. Phys. (N.Y.)* **41** (1967) 108.

[23] U. Weiss and W. Haeffner: *Phys. Rev. D* **27** (1983) 2916.

[24] U. Weiss, H. Grabert, P. Hänggi and P. Riseborough: *Phys. Rev B* **35** (1987) 9535.

[25] U. Weiss: *Phys. Rev. A* **25** (1982) 2444.

[26] P. C. Martin, E. D. Siggia and H. A. Rose: *Phys. Rev. A* **8** (1973) 423.

[27] R. J. Jensen: *J. Statistical Mech.* **25** (1981) 183.

[28] G. B. Crew and T. Chang: *Phys. Fluids* **31** (1988) 3425.

[29] F. Spineanu, M. Vlad and J. H. Misguich: *J. Plasma Physics* **51** (1994) 113.

[30] F. Spineanu and M. Vlad: *J. Plasma Physics* **54** (1995) 333.

[31] W. Lehmann, P. Riemann and P. Hänggi: *Phys. Rev.* **E62** (2000) 6282.

[32] S. Coleman: *Phys. Rev.* **D15** (1977) 2929.

[33] C. Callan and S. Coleman: *Phys. Rev.* **D16** (1977) 1762.

[34] B. Caroli, C. Caroli and B. Roulet: *J. Statistical Mech.* **26** (1981) 83.

[35] D. W. McLaughlin, *J. Math. Phys.* **13** (1972) 1099.

[36] R. F. Dashen, B. Hesslacher and A. Neveu, *Phys. Rev.* **D10** (1974) 4114.

[37] J.-L. Gervais and Sakita, *Phys. Rev.* **D11** (1975) 2943.

[38] W. T. Coffey, D. S. F. Crothers and Yu. P. Kalmykov: *Phys. Rev.* **E55** (1997) 4812.

[39] C. De Dominicis and L. Peliti: *Phys. Rev.* **B 18** (1978) 353.

[40] P. F. Byrd and M. D. Friedman: *Handbook of elliptic integrals for engineers and scientists*, 2nd ed. (Springer, New York, 1971)

[41] D. Waxman and A. J. Leggett: *Phys. Rev.* **B 32** (1985) 4450.

In: Trends in Chemical Physics Research
Editor: A.N. Linke, pp. 71-82

ISBN 1-59454-483-2

Chapter 2

OSCILLATORY BEHAVIOR IN THE FE/4 M H_3PO_4 SYSTEM: EXISTENCE OF HIDDEN NEGATIVE FARADAIC IMPEDANCE

*Ali Eftekhari**
Electrochemical Research Center,
P. O. Box 19395-5139, Tehran, Iran

Abstract

Anodic polarization behavior of iron in phosphoric acid displays oscillatory phenomena at different electrolyte concentrations. However, oscillatory behavior at low values of phosphoric acid concentration is different, as the current oscillations do not occur at the negative branch of the polarization curve (where faradaic impedance is negative to fulfill the necessary condition for the occurrence of electrochemical oscillations). For investigating the sign of faradaic impedance, electrochemical impedance spectroscopy was employed. The frequency response or impedance spectra showed the existence of negative faradaic impedance in hidden form. This is the first report on the observation of so-called "hidden negative faradaic impedance" during electrochemical oscillators involving metal dissolution under potentiostatic condition.

Keywords: oscillatory phenomena, negative faradic impedance, Fe dissolution, impedance spectroscopy, stability diagram

1 Introduction

As summarized in recent reviews [1-5], electrochemical oscillations may occur in various electrochemical systems during electrocatalysis [6,7], electroplating [8], electropolishing [9], electromachining [10], and in the active-passive transitions of many metals [1-5,11-14]. Among different electrochemical oscillators involving metal dissolution, electrodissolution of

* E-mail address: eftekhari@elchem.org. Tel.: +98 21 204 2549; fax: +98 21 205 7621

metals in phosphoric acid has been widely studied [1-5]. On the other hand, electrodissolution of Fe in different acidic media was the subject of many researches in this context [11-14]. However, less attention has been paid to the Fe/H_3PO_4 system, and it can be considered as a relatively new electrochemical oscillator, as was first reported by Sazou [15]. In addition to the first report of this system, which was related to moderate concentrations of phosphoric acid, oscillatory behavior has also been reported for high [16-18] and low [19] values of phosphoric acid concentration in this system. For the sake of clarity, $x > 10$ M, $10 \text{ M} \geq x > 4$ M and $4 \text{ M} \geq x$ in the Fe/x M H_3PO_4 system is referred to as high, moderate and low concentrations, respectively.

According to the generalization made by Koper et al. [20-22], the first condition for the occurrence of oscillations is the negativity of faradic impedance, Z_F, which is in parallel with a double layer capacitance, C_d, in an equivalent circuit. The second one is fulfillment of the inequality $R_s > |Z_F|$ (R_s is the external resistance including the ohmic cell resistance). It has been shown that electrochemical systems can also oscillate in the positive slope of the polarization curve in the presence of so-called hidden negative faradaic impedance (which is recognizable in impedance spectra). Depending on the kinetic nature of negative faradaic impedance, two different categories were classified by Koper [20,21]: *(i)* oscillations occurring only under potentiostatic conditions with sufficiently large ohmic drop, and *(ii)* oscillations occurring under both galvanostatic and under potentiostatic conditions with sufficiently large ohmic drop. However, only a few examples in comparison with the numerous electrochemical oscillators, have been reported to illustrate these categories. That is understandable because only a few systems have been studied under both galvanostatic and potentiostatic conditions by using sufficiently large external resistances. It comes from the fact that the most of electrochemical oscillations have been found during anodic electrodissolution of metals. As oscillatory behavior during metals dissolution is due to the transition between the active and the passive states, only potentiostatic current oscillations can occur. Of course, it has been reported that some metal dissolutions such as electrodissolution of nickel [23,24] displays galvanostatic potential oscillations, but this behavior is limited to a few systems. The necessity of the negativity of faradaic impedance as hidden form for the occurrence of electrochemical oscillations has been studied for various systems [25-27]. Li et al. [26,27] have shown the existence of hidden negative faradaic impedance for the oscillations occurred at the branch of polarization curve with both positive slope and zero slope (limiting current plateau) during the reduction of anions. In the present paper, we wish to report for the first time the appearance of negative faradaic impedance in hidden form for the potentiostatic current oscillations found during an electrochemical oscillator involving metal electrodissolution.

2 Experimental

The experiments were performed at room temperature in a conventional three-electrode electrochemical cell. The working electrode was prepared from an iron wire as long as 10 cm with 0.008 inch diameter, which was spiraled, with 0.5 cm in diameter. The Fe was obtained from Hopkins & Williams with purity better than 99.8 %. A platinum rod and an Ag/AgCl | sat. KCl, were used as counter and reference electrode, respectively. A volume of 50 ml electrolyte solution was used each time. Due to the shape of the Fe electrode and short distance of the electrodes in the electrochemical cell, Luggin capillary was not used, but design of the cell was in a manner to minimize the ohmic drop. The electrolyte solutions were prepared from doubly distilled water and phosphoric acid (Merck, 85%). No action was taken to remove oxygen from solutions. In experiments related to bifurcation analysis, an adjustable external resistance, R_S, was connected in series with the working electrode and a fixed potential difference across the cell and R_S was applied.

The voltammetric measurements were carried out using an EG&G (Princeton Applied Research, Princeton, NJ) Model 273A potentiostat/galvanostat. The impedance spectroscopic measurements were performed with a HF frequency response analyser (Solartron 1265) connected to a potentiostat/galvanostat (EG&G 273A). The impedance measurements were made in the frequency range 0.005 Hz to 100 kHz and only reproducible data are presented.

3 Results

3.1 Voltammetric and Impedance Studies of the Fe/10 M H_3PO_4 System

Figure 1 shows a typical polarization behavior of the Fe/10 M H_3PO_4 system. In an electrochemical system and particularly during the anodic electrodissolution of metals, the strongest oscillations appear when mass transport proceeds in a diffusion layer the thickness of which is constant, *i.e.* under steady state conditions. Nevertheless, as steady state voltammetry is based on various points obtained by measuring current at a fixed potential, no transition is observable. In fact, quasi steady-state condition is used for the investigation of electrochemical oscillations instead of totally steady-state condition (sweeping the potential with a slow scan instead of interrupted one with zero scan rate). The polarization curve of the system under investigation was obtained under quasi-steady state condition by applying a low scan rate of 1 mV s^{-1} for the potential region located between –0.5 and 1.5 V. As it has been reported [15], five main regions can be observed at the polarization curve.

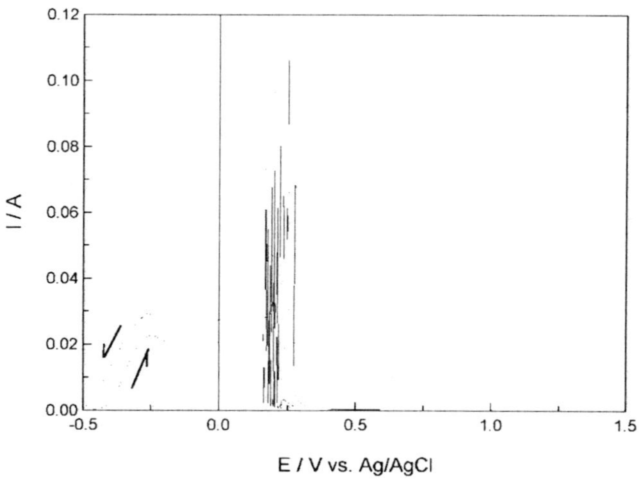

Fig. 1. Typical cyclic voltammogram of the Fe/10 M H_3PO_4 system with $\nu = dE/dt = 1$ mV s^{-1}.

There is an oscillatory region, which occurs two different times during the forward and reverse potential scans. During the forward scan, the oscillation amplitude is smaller than the oscillation of the reverse scan. In addition, the width of the oscillatory region of the reverse oscillation is higher. Lower boundary of the oscillatory region is indicated by the passivation potential during the forward scan and upper boundary by the activation potential during the reverse scan. A typical impedance spectrum of the Fe/10 M H_3PO_4 at a fixed potential selected from the oscillatory region is shown in Fig. 2. As expected the curve enters to the second and third quadrants respectively, which indicates negativity of the faradaic impedance.

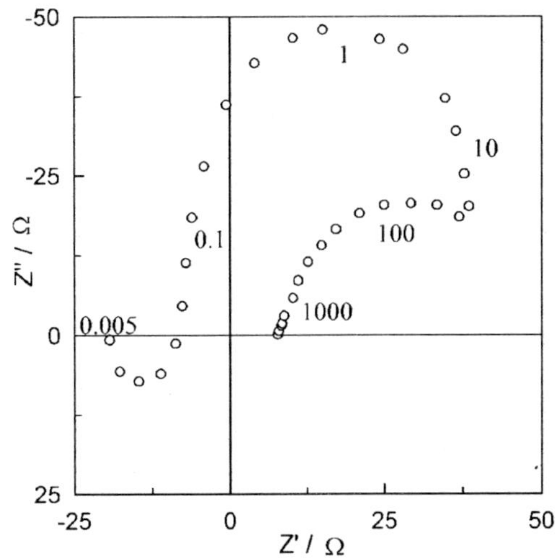

Fig. 2. Impedance diagram of the Fe/10 M H_3PO_4 system at $E = 200$ mV. Frequencies are noted in Hz

3.2 Voltammetric and Impedance Studies of the Fe/4 M H_3PO_4 System

The cyclic voltammograms of the Fe/H_3PO_4 system at low values of phosphoric acid concentration display an oscillatory behavior (Fig. 3). The appearance of these new current oscillations has been reported previously [19]. It should be taken into account that novelty of this oscillatory behavior is related to its occurrence (as it has been reported [15] that the Fe/*x* M H_3PO_4 system does not show oscillatory behavior for 4 M $\geq x$), and the type of the current oscillations from nonlinear dynamic point of view is not addressed here. As these oscillations are stronger in comparison with the former case, the curves were obtained by using a higher scan rate but still under quasi steady-state condition (5 mV s^{-1}). In the low values of phosphoric acid concentration (lower than 6 M), the shape of cyclic voltammograms is changed, and the reactivation occurs at lesser positive potential than passivation potential ($E_{act} < E_{pass}$). The peak of the pre-limiting current region disappears and the oscillatory region and the limiting current plateau coincide on each other. There is a jump transition from higher to lower values of current in the forward scan and there is also a jump transition from lower to higher values of current in the reverse scan. Between these two jump transitions, a hysteresis loop is formed. In this hysteresis loop, the current is not stable and oscillates. The width of the hysteresis loop is strongly dependent on the phosphoric acid concentration and decreases by decreasing the phosphoric acid concentration. There are only three regions for the Fe/*x* M H_3PO_4 system when ($x < 6$ M) instead of the five regions observed for $x > 6$ M: *i)* the active region that the current changes monotonically by potential scanning, *ii)* the oscillatory region or hysteresis loop which is indicated by jump transitions of the current, *iii)* the passive region, where the electrode surface is covered by passive films of iron phosphate and oxides and the current takes very low value.

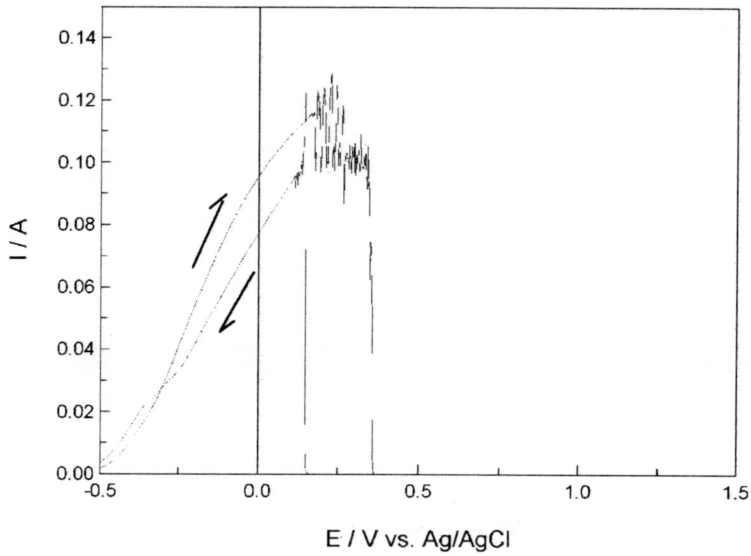

Fig. 3. Polarization behavior of the Fe/4 M H_3PO_4 system obtained with $\nu = dE/dt = 5$ mV s^{-1}.

In contrast with the Fe/10 M H_3PO_4 system (Fig. 1), the amplitude of the current oscillations of the forward scan is larger than that of the reverse scan. The oscillations of the forward scan do not occur in a part of negative branch, but the negative slope of the polarization curve is indicated by difference of the current in the beginning and end points of the oscillatory region. Indeed, the width of the oscillatory region is equal to the width of the hysteresis loop. On the contrary of the oscillations occurred during the forward scan, the oscillations of the reverse scan occur in the positive slope of polarization curve. As it has been described [21], hidden negative faradaic impedance can be observed in frequency response of such systems. Figure 4 presents impedance spectrum of the Fe/4 M H_3PO_4 oscillator at a fixed potential selected from the oscillatory region. By comparing this plot with the impedance spectrum obtained for high H_3PO_4 concentration as a typical behavior of the system, it can be seen that the spectrum has changed its shape dramatically and visits all four quadrants. It is seen that between 0.5 and 0.04 Hz, the real impedance is negative, which is the hidden negative faradaic impedance that destabilizes the system in the presence of a large series resistance or under galvanostatic condition. For the lowest frequencies, the real impedance is positive, which is in consistent with the positive slope of the polarization curve.

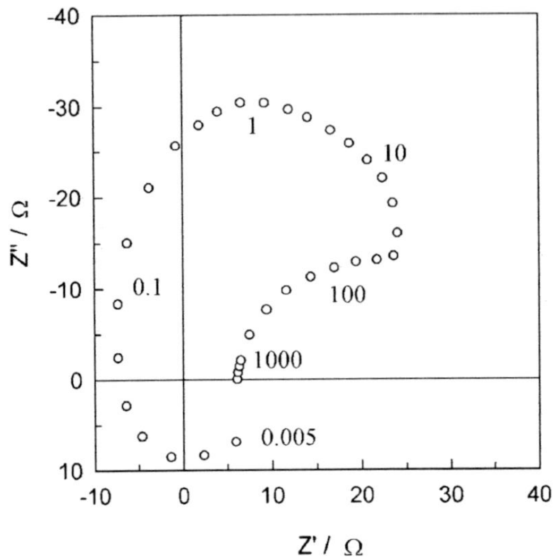

Fig. 4. Impedance diagram for the Fe/4 M H_3PO_4 system at $E = 200$ mV. Frequencies are noted in Hz.

3.3 Oscillatory Boundaries of Both Systems

For further investigation of the dynamical behavior of the Fe/H_3PO_4 system, oscillatory boundaries of both systems were estimated (Fig. 5 and Fig. 6). The plots were obtained for different values of two important controllable parameters namely applied potential and external resistance. They indicate oscillation boundaries and possible characteristics of the systems for the oscillations to occur. Similar to the difference observed between the cyclic voltammograms of two systems, there is a difference between their two-parameter phase diagrams.

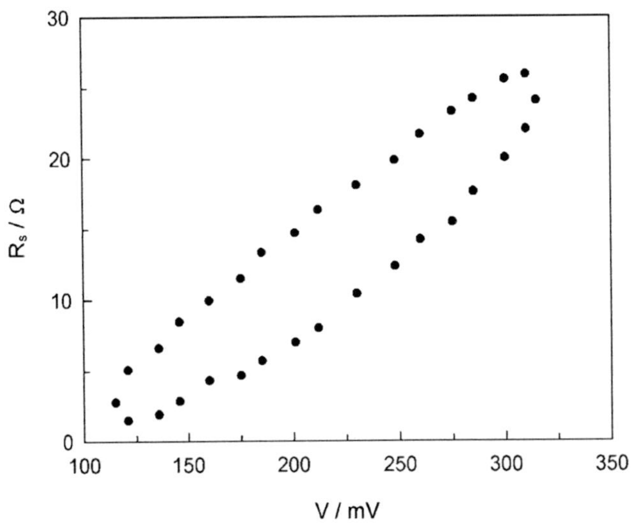

Fig. 5. Two-parameter phase diagram of the Fe/10 M H_3PO_4 system.

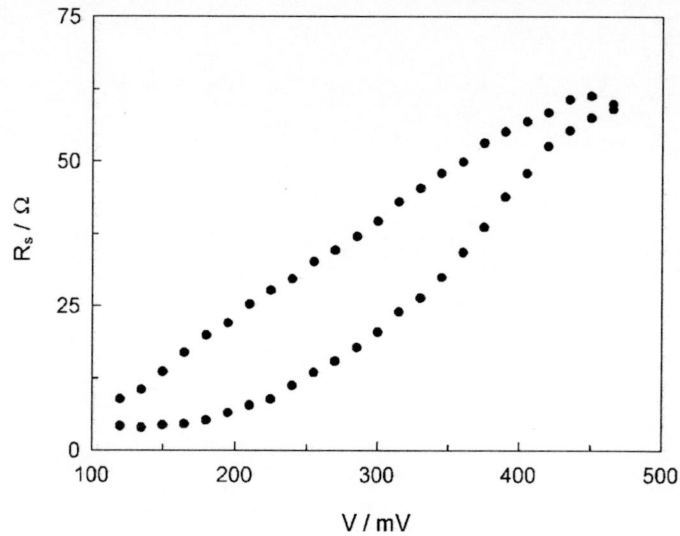

Fig. 6. Two-parameter phase diagram of the Fe/4 M H_3PO_4 system.

As the system can oscillate at some values of potential in the present of a constant fixed external resistance and *vice versa*, it is of interest to study the nature of the oscillations at these different values. The variation of the oscillation frequency for different values of external resistance at a typical potential is presented in Fig. 7. Similar curve is illustrated in Fig. 8 for the distance from the bifurcation potential of the system at a constant external resistance. The bifurcation potential is referred to the potential which the passive steady-state changes to the oscillatory region, and indeed it is the beginning point of the oscillatory region.

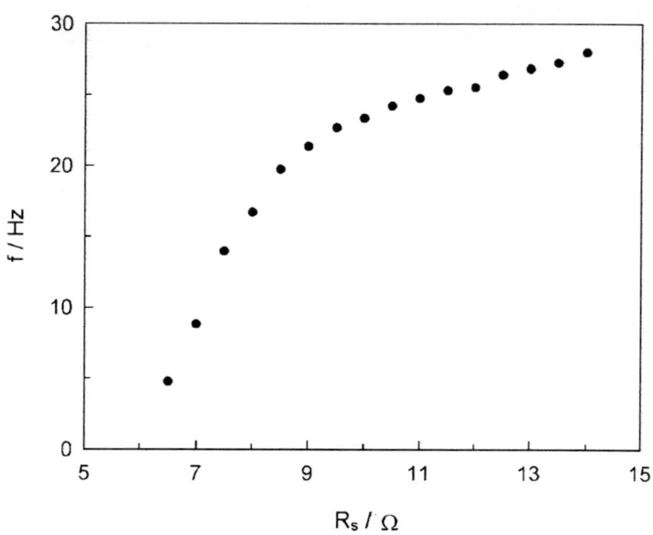

Fig. 7. Effect of external resistance on the oscillation frequency of the Fe/ 4 M H_3PO_4 system, at a typical potential of 200 mV.

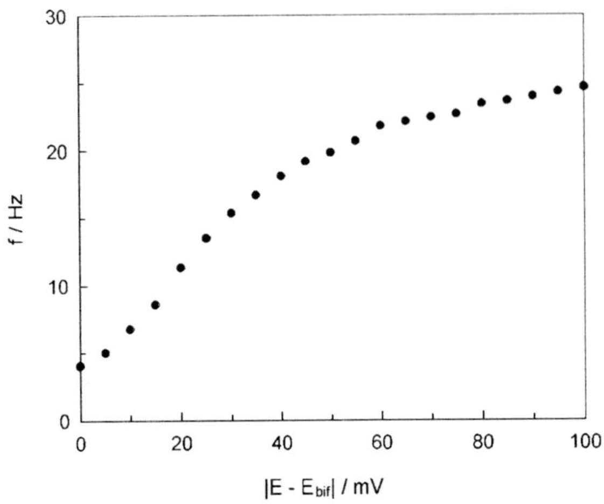

Fig. 8. Dependence of the oscillation frequency on the distance from the bifurcation point. No external resistance was used, as the ohmic drop is a sufficient system resistance for the occurrence of the oscillatory behavior.

4 Discussion

The anodic polarization curves of the Fe/H_3PO_4 system at low values of phosphoric acid concentrations show an oscillatory region. The current oscillations are observed within a hysteresis loop formed by the jump transitions of current. According to the results reported by Sazou [15], the current oscillations disappear by decreasing the phosphoric acid concentration. The results presented here are in agreement with the former report, but the current oscillations are found in the Fe/H_3PO_4 system at low phosphoric acid concentrations. The occurrence of these new oscillations is related to instability of the formed hysteresis loop. The typical cyclic voltammogram for moderate phosphoric acid concentration (Fig. 1) is similar to that presented previously [15].

Although, typical cyclic voltammograms of the Fe/10 M H_3PO_4 system in the former and the latter cases are similar but at low phosphoric acid concentrations, polarization behaviors of the system are different. It is due to different experimental conditions and particularly the electrode shape. The geometrical effects of the electrode on oscillatory behavior of electrochemical systems have been investigated in the literature [28,29]. The conventional polarization curves of metals electrodissolution-based oscillators contain three general regions, the active region, the oscillatory region and the passive region. In middle of the polarization curves, there is a transition region which its boundaries are indicated by the passivation potential and the activation potential. In the forward scan, after the active region, in a certain potential namely passivation potential, the electrode surface becomes passive, and in the backward scan, after the passive region in a certain potential namely activation potential, reactivation of the electrode surface occurs. At the transition region located between the activation and the passivation potentials, the system does not reach a complete active or passive state. The current oscillations during the anodic electrodissolution of metals usually occur in this region. In the Fe/H_3PO_4 system, both the passivation and activation potentials shift towards more negative potentials by decreasing the phosphoric acid concentration. This is due to the surface reaction of the Fe electrode, which is dependent on the electrolyte properties such as concentration and viscosity. Franck [30] reported the dependence of the passivation potential (Flade potential) to pH for the Fe/H_2SO_4 system:

$$V_F = 0.58 - 0.058 \text{ pH (V } vs. \text{ NHE)}$$

Sazou [15] has reported that by further decreasing the phosphoric acid concentration (lower than 4 M) the passivation potential suddenly shifts towards more positive potential even higher than the activation potential and consequently the place between the passivation and activation potentials disappears and no current oscillation is occurred. As it can be seen in Fig. 3, the activation potential occurs at more negative potentials than the passivation potential. In this case, there is another place between the activation and passivation potentials but with reformed boundaries, so-called hysteresis loop.

In the Fe/H_3PO_4 system, when H_3PO_4 has low values of concentration, the passivation and activation potentials are generated by jump transition of current. During the forward scan, at the end of the active region, a passive film is suddenly formed on the electrode surface, which causes the jump transition of current from its higher values to lower ones. Similarly, during the backward scan, after the passive region where the electrode surface is covered by a passive film and current is constant, the passive film breaks down suddenly and reactivation of the Fe electrode causes the jump transition of current from its lower values to higher ones. In this case, the passivation and activation potentials can be clearly recognized by these jump transitions of current. Although, the shape of cyclic voltammogram of the Fe/4 M H_3PO_4 system is similar to that reported in the literature [15], the main difference is the appearance of the current oscillation at the hysteresis loop of the polarization behavior.

Ohmic drop cannot be responsible for the behavior observed, as the resistance of the Fe/4 M H_3PO_4 systems is lesser than that of the Fe/10 M H_3PO_4 system. The measurements were carried out in similar cells before electrochemical experiments. It can also be concluded from the starting point of the impedance spectra (Fig. 2 and Fig. 4) at high frequencies, as suggest solution resistances of 5.5 and 7.5 Ω for two systems, respectively (just as a comparative measurement).

The behavior observed can be attributed to stability of the passive film on the Fe electrode surface. Due to lesser viscosity of the electrolyte solution in the 4 M H_3PO_4 system, the growth and breakdown of the passive film is easier. Consequently, formation and breakdown of this film is faster in comparison of those for the Fe/10 M H_3PO_4 system, which this is in consistent with higher oscillation frequency for the Fe/4 M H_3PO_4 system in comparison with the Fe/10 M H_3PO_4 system.

It is noticeable that this behavior is stronger for a spiraled Fe electrode in comparison with a Fe disk. This is the geometrical effect, which is addressed throughout the paper. As breakdown of the passive film and diffusion towards the bulk solution is related to the concentration gradient, local concentration of Fe ions around the electrode surface is strongly effective on this process. It is clear that breakdown of the passive film from spiraled Fe electrode, which is spread in a vast area of the electrolyte solution, is easier in comparison with Fe disk. Therefore, the passive film growth on a spiraled electrode is less stable in comparison with a disk electrode. This is the essential properties of the system under investigation (and generally similar systems) and the mentioned processes occur before the electrochemical experiments. The effect of initial condition and passivity of the electrode surface before and during the electrochemical measurements on the oscillatory behavior of electrochemical systems has been discussed elsewhere [31,32].

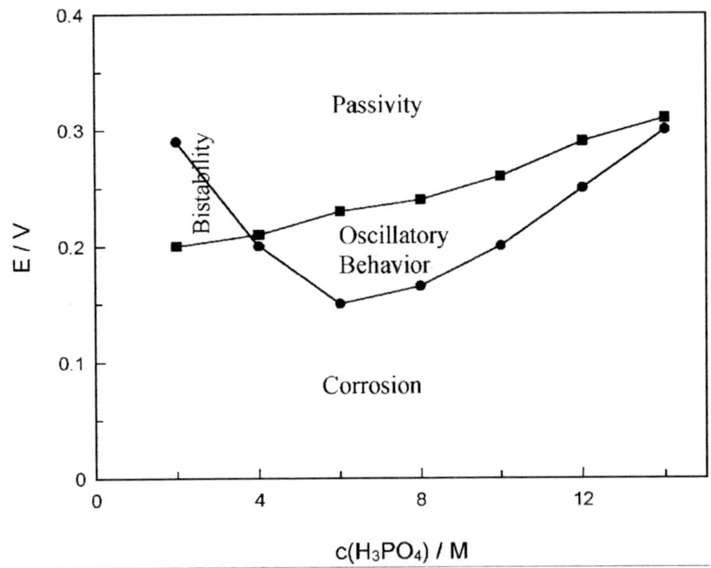

Figure 9(a).

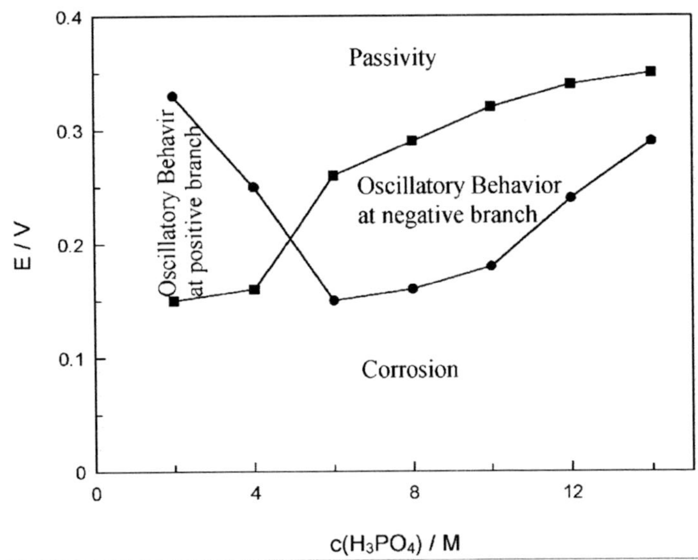

Figure 9(b).

Fig. 9. E-boundaries of the Fe/*x* M H_3PO_4 system based on (a) the Fe disk electrode and (b) the spiraled Fe electrode as a function of the H_3PO_4 concentration indicated by the passivation potential (•) and the activation potential (■).

To examine this hypothesis, the E-boundaries of the Fe/4 M H_3PO_4 system based on a spiraled Fe electrode and a Fe disk electrode were recorded and compared. Fig. 9 compares the possible oscillatory regions for the Fe/*x* M H_3PO_4 system using two different Fe electrodes. It is obvious that the oscillatory region is wider for the system using the spiraled Fe electrode. Similarly, the current oscillations recorded for the system under investigation based on the spiraled Fe electrode are accompanied by higher frequencies and larger amplitudes (not shown). In the other words, the oscillatory behavior for the Fe/*x* M H_3PO_4 is stronger when a spiraled Fe electrode is employed. As seen, the shape of the working electrode has no significant effect on the passivation potential, but its effect on the activation potential is significant. This indicates that the special shaped of the working electrode (spiraled Fe) strongly assist the passive film to dissolve in the acidic medium. Interestingly, in agreement with Sazou [15], the current of the Fe disk electrode does not oscillate for the Fe/4 M H_3PO_4 system.

The main consequence of the results obtained in this research is the existence of hidden negative faradic impedance in the generation of the potentiostatic current oscillations. Interestingly, it is the first report on the appearance of "hidden negative faradaic impedance" in electrochemical oscillators involving metals dissolution. Thus, it is a typical example for possibility and usefulness of this classification even for metals dissolution-based electrochemical oscillations. It is worth noting that "hidden negative faradaic impedance" has also been reported for the oscillatory phenomena during electrodissolution of nickel [3,33]. However, that report is not belonged to the present category as the Ni/H_2SO_4 system oscillates under galvanostatic condition and its "hidden negative faradaic impedance" is related to the oscillations occurred during the galvanodynamic polarization [23,24].

The most reasonable interpretation of this negative faradaic impedance is the formation of a superficial (hydr)oxide film, which leads to the passivation of the electrode surface with increasing anodic polarization and thereby reduces the overall oxidation current. The results obtained are mainly due to the special shape of the Fe electrode (Spiraled). It is clear that passivation, which is the main process for the occurrence of the oscillatory behavior in the system, is different for spiraled Fe electrode and Fe disk. Therefore, the difference in the formation of passive film, which is the subject of our future work, causes the appearance of this behavior.

5 Conclusion

The analysis of the oscillatory behavior observed during the anodic electrodissolution of Fe in phosphoric acid provided some interesting results. The system under investigation can exhibit oscillatory phenomena at low values of phosphoric acid concentration. On the contrary of electrochemical oscillators involving metals dissolution, the current oscillations were found in the positive branch of the polarization curve where real faradaic impedance is not negative. The negativity of the faradaic impedance as a necessary condition for the occurrence of electrochemical oscillations was fulfilled in the system under investigation, by the observation of "hidden negative faradaic impedance". It is the first report on the possibility of the appearance of hidden negative faradaic impedance in electrochemical systems involving metals electrodissolution.

Acknowledgments

The author would like to thank Dr. M. G. Mahjani (K N Toosi University of Technology) for his supports for this research, and Dr. M. T. M. Koper (The Eindhoven University of Technology, The Netherlands) for his useful discussion.

References

[1] J. Wojtowicz, in :*Modern Aspects of Electrochemistry*, J.O'M. Bockris, B.E. Conway (Eds.), vol. 8, p. 47, Plenum Press, New York, 1972.
[2] J.H. Hudson, T.T. Tsotsis, *Chem Eng. Sci.* **49** (1994) 1493.
[3] T.Z. Fahidy, Z.H. Gu, in: *Modern Aspects of Electrochemistry*, R.E. White, J.O'M. Bockris, B.E. Conway (Eds.), vol. 27, p. 383, Plenum Press, New York, 1995.
[4] M.T.M. Koper, *Adv. Chem. Phys.* **92** (1996) 161.
[5] J.L. Hudson, M.R. Bassett, *Rev. Chem. Eng.* **7** (1991) 109.
[6] J. Wojtowicz, N. Marincic, B.E. Conway, *J. Chem. Phys.* **48** (1968) 4333.
[7] H. Tributschh, Ber. Bunsen-Ges. *Phys. Chem.* **79** (1975) 570.
[8] M. Clarke, J.A. Bernie, *Electrochim. Acta* **12** (1967) 205.
[9] L.T. Tsitsopoulos, T.T. Tsotsis, I.A. Webster, *Surf. Sci.* **191** (1987) 225.
[10] J.F. Cooper, R.H. Muller, C.W. Tobias, *J. Electrochem. Soc.* **127** (19880) 1733.
[11] U.F. Franck, R. FitzHugh, *Z. Elektrochem.* **65** (1961) 156.
[12] J.J. Podesta, R.C.V. Piatti, A. Arvia, *J. Electrochem. Soc.* **126** (1979) 1363.
[13] M.R. Bassett, J.L. Hudson, *J. Electrochem. Soc.* **137** (1990) 922, 1815.
[14] B. Rush, J. Newman, *J. Electrochem. Soc.* **142** (1995) 3770.
[15] D. Sazou, *Electrochim. Acta* **42** (1997) 627.
[16] D. Sazou, M. Pagitsas, *J. Electroanal. Chem.* **451** (1998) 77.
[17] M. Bojinov, I. Betova, G. Fabricius, T. Laitinen, R. Raicheff, *J. Electroanal. Chem.* **475** (1999) 58.
[18] A. Eftekhari, *197th ECS Meeting,* Toronto, Canada, May 14-18, 2000, Abst. 0034.
[19] A. Eftekhari, *198th ECS Meeting,* Phoenix, Arizona, October 22-27, 2000, Abst. 0662.
[20] M.T.M. Koper, J. Chem. Soc., *Faraday Trans* **94** (1998) 1369.
[21] M.T.M. Koper, *J. Electroanal. Chem.* **409** (1996) 175.
[22] M.T.M. Koper, J.H. Sluyters, *J. Electroanal. Chem.* **371** (1994) 149.
[23] O. Lev, A. Wolffberg, M. Sheintuch, L.M. Pismen, *Chem. Eng. Sci.* **43** (1988) 1339.
[24] O. Lev, A. Wolffberg, L.M. Pismen, M. Sheintuch, *J. Phys. Chem.* **93** (1989) 1661.
[25] W. Wolf, M. Lubke, M.T.M. Koper, K. Krischer, M. Eiswirth, G. Ertl, *J. Electroanal. Chem.* **399** (1995) 185.
[26] Z. Li, J. Cai, S. Zhou, *J. Electroanal. Chem.* **436** (1997) 195.
[27] Z. Li, J. Cai, S. Zhou, *J. Electroanal. Chem.* **432** (1997) 111.
[28] Z. Li, J. Cai, S. Zhou, *J. Phys. Chem. B* **102** (1998) 1539.
[29] Y. Wang, J.L. Hudson, *AIChE J.* **37** (1991) 1833.
[30] U.F. Franck, Z. *Naturforsch.* 4a (1949) 378.
[31] A. Eftekhari, *Fluctuation Noise Lett.* **1** (2001) L171.
[32] A. Eftekhari, *Physics Lett. A*, **332** (2004) 382.
[33] M. Keddam, H. Takenouti, N. Yu, *J. Electrochem. Soc.* **132** (1985) 2561.

In: Trends in Chemical Physics Research
Editor: A.N. Linke, pp. 83-93
ISBN 1-59454-483-2

Chapter 3

METRIC CONTRIBUTIONS IN REACTION RATES AND THEIR INFLUENCE ON THE RELAXATION EFFICIENCY ON COMPLEX POTENTIAL ENERGY SURFACES

F. Despa and R.S. Berry
Department of Chemistry, The University of Chicago,
Chicago, Illinois 60637

Abstract

Means are presented for incorporating metric contributions into transition rates in the attempt to relate more specific properties of the potential energy surface at stationary points with the relaxation towards the global minimum. Metric contributions are corrections due to reducing the reaction coordinate from ordinary coordinates to internal degree of freedom. These mentain the signature of corresponding geometric arrangements at stationary points on the canonical behavior of the system. As an application, we work out the *four*-level system of the H_2CO surface and finger out the level at which the metric contributions may affect the accuracy of characterizing the ability of the system to find the global minimum.

The output of the present work can lead to identify a way to enrich our knowledge about the ability of the complex systems to relax only to preferential structures.

1 Introduction

The different aspects of the topographies and topologies of multidimensional potential surfaces of polyatomic molecules, clusters, and nanoscale particles and the way they govern the dynamics and phaselike behavior of these systems are currently receiving a great interest; for reviews, see [1, 2, 3, 4] and references therein. As a result, a new picture of their energy landscape is emerging which greatly enhances our understanding in this pressing issue. However, despite the enormous power to do computations and great disposal of methods for extracting useful informations, the global optimization problem of complex systems is far from a complete solution.

For complex systems, development of better potential functions still remains clearly a priority. We can say that, at least, starting from available expressions for the interaction between and among the system constituents, accurate potential energy surfaces (PES) may accurately be built up. Finding minima and saddles on potential surfaces has become a straightforward matter by employing some of the most widely used algorithms or combinations thereof.[5, 6, 7, 8, 9] The spinous problem shifted from the how to find minima efficiently to what method should be employed to provide a better connection between the topography and the dynamics on the surface. Statistical methods have been used before to study complex kinetics[10, 11, 12] and frequently invoked to illuminate the relation between topography and kinetics.[13, 14, 15, 16, 17] Especially, the statistical approach based on the topography diagnosis procedure[13, 15, 18, 19, 20, 21] proved to encounter a lot of success in carrying the study of complex systems to the point where one can begin to infer dynamics of ow on the surface from the general characteristics of the topography. A serious shortcoming of the existing dynamical studies, however, has to be pointed out: virtually, all of them describe the thermally activated barrier crossing by the traditional transition state theory (TST) rate which deals only with the height of the barrier and the densities of initial and saddle states.[22] No information about the geometrical structures of the initial and saddle sites are contained in this approach. From this point of view, there is a limit in the precision with which we can characterize the ability of complex systems to relax preferentially to only a limited number of geometrical structures from the vastly larger variety that the system might exhibit. Indeed, there are, in turn, other ways to compensate for this lack of information. It was shown, for example, that the topographical properties of the potential energy surface (PES) determine the extent to which the system is either glass-forming or focusing on regular structures.[2, 13, 23] In addition, qualitative interpretations of the interplay between structure and dynamics have been fingered out from the variety the complex systems show for the path length distribution between connected minima along the reaction coordinate.[23, 24]

For a better understanding of dynamics, it is now useful to investigate the relation between the variety of path lengths between neighboring minima along the reaction coordinate of a specific PES and the ow of probability between those minima. In the following, we attempt to get more insight into this particular aspect by improving the procedure to describe the relaxation of the system on the PES at two instants. First, we identify the most suitable pictorial representation of the energy landscape, that is a restricted two dimensional PES which connects the discrete set of energy levels with the corresponding distribution of reaction path lengths (or with the distribution of atomic displacements in the configuration space, the distribution of atomic configurations, etc.).[23, 24] Second, we incorporate into reaction rates more information about the specific atomic rearrangements of the states involved in transitions. The latter can be achieved by an appropriate reconsideration of the (canonical) partition function used to derive the TST escape rate. To this purpose, we include in the usual partition function[25] corrections due to the subsequent reduction of the full phase space arised by transforming the reaction coordinate from ordinary degrees of freedom (coordinates) to the internal degree of freedom (bond lengths, angles, etc.).[11, 26] These corrections are metric contributions which preserve information about the structure of the system at the energetic stationary points involved in transition. We work out a *four-level* system (H_2CO) which displays very different atomic rearrangements at transition points and, therefore, leads to distinct metric contributions into reaction rates.[27, 28] We show that metric contributions change the relaxation dynamics and improve the accuracy with which we can characterize the ability of the system to find the global minimum.

2 Canonical Multidimensional TST Rate

Let x denote the reaction coordinate of a complex system with N degrees of freedom, which is a function of all these degrees of freedom, $x = (\mathbf{q}_1, ...\mathbf{q}_N; \mathbf{p}_1, ...\mathbf{p}_N)$, where $\mathbf{q}_i$'s stand for the configuration coordinates and $\mathbf{p}_i$ for momenta. The interactions between and among all N constituent particles give rise to the potential surface of the system (PES). This is a high-dimensional surface having a complex topology with hills and walleys, for which we know the locations (coordinates) and attitudes (energies) of the bottoms of those walleys and of the saddles that allow passage between all of these walleys. We know precisely for the reaction coordinate x only its values at stationary points of the PES (at which the potential gradient vanishes). These are configurations of the system at the local minima and saddles and we denote them in the following by $x_1,\ x_\alpha,\ x_2,\ x_\varepsilon,\ ...,\ x_k,\ x_\gamma,\ ...$, where indices $1,\ 2,\ ...,\ k,\ ...$ stand for minima and $\alpha,\ \varepsilon,\ ...,\ \gamma,\ ...$ for saddles, respectively.

Dynamic and thermodynamic properties of the PES can generally be described by coupling the reaction coordinate to an environment which can either donate or remove energy from it. Due to such a coupling, the reaction coordinate is not a deterministic dynamic variable, but becomes a stochastic process, i.e., $x\,(t)$ is a random trajectory. The way two local minima on the PES are connected directly to the transition state by paths of steepest descent forms the reaction path. The reaction path still has no dynamical significance. This is a mathematical device in $3N$-dimensional configuration space that follows the gradient of the potential to get from high ground, around the transition state, to low ground where the stable configurations are. Stationary points on the PES are points of decision for the reaction path in what direction to start. A measure of the reaction path is the path length S defined usually as the integrated arc length in $3N$-dimensional nuclear configuration space[24]

$$S = \int ds \equiv \sum_{steps} \sqrt{\sum_{i=1}^{N} \Delta \mathbf{q}_i^2}\ , \tag{1}$$

where $\Delta\mathbf{q}_i$ is the step for coordinate $\mathbf{q}_i$. The outer sum depends on the method involved to search for stationary points; for example, this runs over all steps along the eigenvector-following pathway.[6, 7, 8, 9] By using (1) we can find the arc length between two consecutive stationary points (min-saddle) on the PES, say k and l (S_{kl}). One can define also a distance in nuclear configuration space between these stationary points as

$$D_{k\gamma} = \sqrt{\sum_i \left[\mathbf{q}_i(k) - \mathbf{q}_i(\gamma)\right]^2} \quad , \tag{2}$$

which is shorter than the coresponding integrated path length along the reaction path ($D_{k\gamma} < S_{k\gamma}$). $\mathbf{q}_i(k)$ and $\mathbf{q}_i(\gamma)$ are coordinate values at corresponding stationary points. $S_{k\gamma}$ and $D_{k\gamma}$ may connect, in different ways, two consecutive values of the reaction coordinate, x_k and x_γ. The probability for the system to pass from the local minimum k to the next direct connected local minimum $k+1$ via the saddle γ is subject to statistical weighting of all initial coordinates $\mathbf{q}_i$ and momenta $\dot{\mathbf{q}}_i$. (The overdot refers to a differentiation with respect to time t.)

We can display now sequences of min-sad-min against corresponding path lengths $[S_n]$ which separate them in the configuration space. These sequences lead monotonically to the global minimum. Some examples of these topological maps of the original PES have been worked out in Ref. [23]. The same kind of monotonic sequences can also be build up with respect to the appropriate distribution of distances $[D_n]$ in the nuclear configuration or, simply, by displaying on the *abscissa* the set of reaction coordinate values $x_1,\ x_\alpha,\ x_2,\ x_\varepsilon,\ \ldots,\ x_k,\ x_\gamma,\ \ldots$. By this procedure, the multidimensional PES is reduced to a two-dimensional pictorial representation, and brings the study of complex systems to the point where one can begin easier to infer dynamics of probability ow on that surface. We may observe that the above reduced representation of the PES (RRPES) preserves the main topological and topographical characteristics of the original PES in a manner which should allow us to obtain an accurate dynamic description of the system. We must notice that the present approach discards any quantitative information about the volume of the phase space associated with each minimum. In turn, it contains (on the *abscissa)* elements of the system's evolution through the configuration space.

The general procedure describing the dynamics on any RRPES, as described above, relies in practice on the explicit knowledge of a master equation governing the time dependence for the single-event probability $P_k(x_k, t)$ of the reaction coordinate

$$\frac{dP_k(x_k)}{dt} = \sum_j W_{k-j} P_j(x_j) \quad , \tag{3}$$

where

$$W_{k-j} = w_{k-j} - \delta_{kj} \sum_n w_{n-k} \quad , \tag{4}$$

and w_{k-j} is the rate of transition from state k to state j. We observe that, generally, (3) provides a quantitative description of the density of local minima on the PES (by the number of equations the system contains). The reaction rates w_{k-j} include information about the local sites the system is visiting in its relaxation towards the global minimum. Usually, the rates w_{k-j} are cast in the RRKM (Rice-Ramsperger-Kasse-Marcus) theory[22] within the harmonic approximation for the potential of mean force

$$w_{k-j} = \sigma \frac{\overline{\nu_k}^m}{\overline{\nu_\gamma}^{m-1}} \exp\left[-\beta\left(E_\gamma - E_k\right)\right] \quad , \tag{5}$$

where $\overline{\nu_k}$ and $\overline{\nu_\gamma}$ are geometric mean vibrational frequencies at initial minimum k and saddle γ, m is the number of internal coordinates $(m = 3N - 6)$, σ stands for the eventual density of saddle states (if any) and $E_\gamma - E_k$ is the height of the potential barrier. We may observe that the specific geometrical configurations of the system at the local minimum and saddle point do not intervene in (5) and, the minimum structural information preserved usually by Hessian eigenvalues is actually lost by averaging ν_k and ν_γ. So that, the specific atomic rearrangements of the system at stationary points given by, for example, the corresponding values of the reaction coordinate $(x_1,\ x_\alpha,\ x_2,\ x_\varepsilon,\ ...,\ x_k,\ x_\gamma,\ ...)$ on the *abscissa* of the RRPES are completely neglected. From this point of view, there is a limit in the precision with which we can characterize the ability of complex systems to relax preferentially to only a limited number of geometrical structures from the vastly larger variety that the system might exhibit.

The classical transition state theory (TST) for the reaction rate (see eq. 5) is a direct result of the fact that the potential energy and evolution of the reaction coordinate depend only on the internal coordinates of the system (bond lengths, angles, etc) and, therefore, the statistical weighting is straightforward with respect to these degree of freedom. Corrections due to the initial reduction[11] of the full phase space (coordinates and momenta plus environment) lie beyond the scope of the traditional TST. We show in the following that, actually, these corrections are important for the accurate description of the evolution of the complex system through the configuration space. These are metric contributions which preserve the fingerprint of the atomic arrangements of the states involved in transition.

We start from the observation that if at any point on the reaction path we define a local coordinate system,[29] other than one used to describe the PES, the volume in the phase space underlying the local coordinate system is in direct proportion to the volume in the full phase space. The proportionality constant includes usually the volume of the system obtained by integrating over the center-of-mass coordinates, the result of integration over all the orientations and other metric contributions due to coordinate transforms. The ratio between the constant values at the positions of the saddle and initial potential well is the correction to the reaction rate (5).

The derivation of this correction is straightforward and we need only to write the canonical partition function of the system in one particular walley of the PES, say the minimum k, downhill the saddle γ. This is

$$Z = \int \mathbf{d}^N\mathbf{Q}\mathbf{d}^N\mathbf{P}\theta\left(x_\gamma - x\right) e^{-\beta\left[U(\mathbf{Q}_1,\ldots\mathbf{Q}_N)+\frac{1}{2}\sum_i \mathbf{P}_i^2\right]} \quad , \tag{6}$$

where $\mathbf{Q}_i = \mathbf{q}_i m_i^{1/2}$ and $\mathbf{P}_i = \mathbf{p}_i m_i^{-1/2}$ are the mass-weighted coordinates (m_i stands for the masses of the corresponding configurational degrees of freedom), β has the usual meaning, θ is the *step*-function and $U(\mathbf{Q}_1, ...\mathbf{Q}_N)$ is the configurational (potential) energy. We perfom the integration over the momenta which leads to

$$Z = \left(\frac{\pi}{2\beta}\right)^{\frac{3N}{2}} \int d^N \mathbf{Q} e^{-\beta U(\mathbf{Q}_1, ...\mathbf{Q}_N)} , \tag{7}$$

By performing the above integration, the reaction coordinate is just the coordinate configuration. As the potential energy of the system depends only on the internal coordinates, say $(g_1, ..., g_m)$, it is used to make the coordinate transform $(\mathbf{Q}_1, ...\mathbf{Q}_N) \rightarrow (\mathbf{R}, \mathbf{\Theta}, g_1, ..., g_m)$, where $\mathbf{R}$ is the center-of-mass position and $\mathbf{\Theta}$ stands for all the orientations. In this case, (7) can be further simplified by integrating over $\mathbf{R}$ and $\mathbf{\Theta}$

$$\begin{aligned} Z &= \left(\frac{\pi}{2\beta}\right)^{\frac{3N}{2}} \int \mathbf{dRd\Theta} dg_1...dg_m \frac{\partial(\mathbf{Q}_1, ..., \mathbf{Q}_N)}{\partial(\mathbf{R}, \mathbf{\Theta}, g_1, ..., g_m)} e^{-\beta U(g_1, ..., g_m)} \\ &= \left(\frac{\pi}{2\beta}\right)^{\frac{3N}{2}} \int dg_1...dg_m \mathbf{J}[g] e^{-\beta U(g_1, ..., g_m)} , \end{aligned} \tag{8}$$

where $\mathbf{J}[g]$ is the Jacobian of this transformation integrated over $\mathbf{R}$ and $\mathbf{\Theta}$. Within the steepest-descent approximation for the reaction path, the result of the above integration is

$$Z = \left(\frac{\pi}{2\beta}\right)^{3N-3} J_k \left[\prod_{i=1}^{m} \lambda_i^{(k)}\right]^{-1} e^{-\beta E_k} , \tag{9}$$

where E_k is the potential energy of the system, J_k is the metric contribution at this particular value of the reaction coordinate $x_k = (Q_1...Q_N)_k$ and $\lambda_i^{2(k)}$ are the eigenvalues of the *covariant* second derivative matrix (Hessian)[30, 31] of the effective potential $U_{eff}(\mathbf{g}) = U(\mathbf{g}) - k_B T \ln J[\mathbf{g}]$ at the position of the minimum point k.[11, 26] J_k depends on the values of all the molecular coordinates $\mathbf{g}$.

We can derive in the same manner the steady-state reactive ux at the position of the saddle point x_γ by averaging $\dot{x}_\gamma$ over the canonical probability density

$$\kappa(t) = \frac{1}{Z} \left\langle \delta(x - x_\gamma) \dot{x}_\gamma \theta[x(t)] \right\rangle , \tag{10}$$

where δ is the *delta*-function. This shows how often a system which is continously fed into the initial domain of attraction can be removed by an observer in the neighboring domain of attraction. If the saddle-point placed at x_γ is a point of no return (the Wigner's fundamental assumption[?]) then, eq. (10) represents the equilibrium average of a one-way ux at the transition state

$$\kappa_{TST} = \frac{1}{Z} \left\langle \delta(x - x_\gamma) \dot{x}(x_\gamma) \theta(\dot{x}) \right\rangle \tag{11}$$

which is the main result of the transition-state theory (TST).[32, 33, 34] By taking (9) into account and performing the canonical average in above, one finds the correction to the reaction rate, that is $\frac{J_\gamma}{J_k}$. This is the ratio between the metric contributions due to phase space reductions in the states involved in the transition of the system from $(Q_1...Q_N)_k$ to $(Q_1...Q_N)_\gamma$. With this correction eq. (5) becomes

$$w_{k-j} = \frac{J_\gamma}{J_k}\sigma\frac{\overline{\nu_k}^m}{\overline{\nu_\gamma}^{m-1}}\exp\left[-\beta\left(E_\gamma - E_k\right)\right] \ . \quad (12)$$

3 Application to the H_2CO Surface

In the following we shall apply the above TST rate formula (12) to a relevant archetypal situation. This is the landscape model of low-energy reactions of formaldehyde (H_2CO). The original PES for this molecule has been constructed by Handy and Carter[27] and tested further for relevant physical characteristics by Berry group.[28] The calculus for the *bound* molecule has revealed four minima on the PES: the formaldehyde structure (A), two linear structures (B) and (C) and the cis-hydroxycarbene structure (D) (see Fig.5 in Ref. [28]). The most striking aspect is the appearance of the linear structure (B) in the position of the global minimum. This structure is not easily accessible from areas of the surface around the usual reaction.[28] There are three saddle states (see Fig.6 in Ref. [28]) associated with transitions between these minima: one is nonplanar (AB) and the other two are planar, (BC) and (AD). The highest energy level of the schematic diagram of this surface corresponds to the cis-hydroxycarbene structure (D) and is separated from the formaldehyde structure (A) by the saddle structure AD. The other adjacent minimum to A is the linear structure B, which is the global minimum, and are separated from each other by the saddle AB. The landscape ends by one other linear structure C which can be reached from the global minimum over the saddle BC. For geometrical structures of stationary points and the schematic diagram the reader is refered to Figs. 5, 6 and 7 of the work by Berry group. [28]

The stationary molecular structures on the H_2CO surface are given, as usual, in terms of bond lengths, or angles and bond lengths, and correspond to well defined energy values. There are six bond lengths ($l_{CO}, l_{CH}, l_{CH'}, l_{HH'}, l_{OH}, l_{OH'}$) in order to represent $m = 6$ internal degrees of freedom. The bond lengths l and the coordinates q_i satisfy the following set of equations

$$l_{CO}^2 = \sum_{i=1}^{3}(q_{iC} - q_{iO})^2; \ ..., \ l_{OH'}^2 = \sum_{i=1}^{3}(q_{iO} - q_{iH'})^2 \ . \quad (13)$$

For simplicity, we refer the coordinate system to a molecular orientation with the origin placed on the C atom and with the q_2 axis along the $C - O$ bond, therefore $q_{iC} = 0$, $(i = \overline{1,3})$ and $q_{1O} \equiv q_{3O} = 0$. Three atoms of this molecule must be planar, C, O and one hydrogen atom (H) may be disposed in the same plan, which allow us to impose, for example, $q_{1H} = 0$. All the other coordinates (q_{1H}, q_{2H}, $q_{2H'}$, q_{3H}, $q_{3H'}$, and q_{2O}) can be determined with respect to the bond lengths by using the system of equations (13). The practical problem is to determine the metric contribution J to the *bound* molecule phase

Table 1: Bond lengths and energies for stationary structures of the H_2CO molecule (from Ref. [28]) used in the calculus of the metric contributions J. All lengths are in $\AA$ and the energies in eV.

	A	B	C	D	$A-B$	$A-D$	$B-C$
E	-16.28	-20.50	-16.14	-12.96	-12.93	-10.26	-13.25
$l_{HH'}$	1.899	0.558	0.748	2.958	1.255	2.382	1.032
l_{OH}	2.029	1.941	3.410	2.632	1.939	2.523	1.962
$l_{CH'}$	1.116	1.423	3.014	1.980	1.688	1.468	2.115
l_{CH}	1.116	0.865	2.266	1.267	1.055	1.217	1.094
$l_{OH'}$	2.029	2.499	4.158	0.958	2.391	1.197	2.819
l_{CO}	1.208	1.076	1.143	1.423	1.167	1.308	1.150
J	0.952	1	1	6.810	4.332	1.183	5.493

space due to the coordinate transform (13). The metric contributions are the values of the Jacobian

$$J = \frac{\partial\left(q_{1H}, q_{2H}, q_{2H'}, q_{3H}, q_{3H'}, q_{2O}\right)}{\partial\left(l_{HH'}, l_{OH}, l_{CH'}, l_{CH}, l_{OH'}, l_{CO}\right)} \quad . \tag{14}$$

for each of the above structures and are computed according to the corresponding bond lengths l_{CO}, ..., $l_{OH'}$. (The bond lengths [28] and displayed in Table I.) The atomic rearrangements involved in this landscape have different geometries and therefore produce distinct metric contributions. Obviously, the metric tensor is identity for the linear structures B and C which leads to $J_B \equiv J_C = 1$. The calculated values for the other Jacobians, J_A, J_D, J_{A-B}, J_{B-C} and J_{A-D}, are collected in the last row of Table 1.

We determine now the probability distributions P_A, ..., P_D and investigate the evolution of the molecule in the configuration space. The characteristic kinetic equations for the H_2CO landscape model described above are derived form (3) with the reaction rates given by (12) with $\sigma = 1$. (For parameters entering (12) see Table I and Ref. [28].) The precise canonical behavior of this molecule can be determined uniquely by imposing initial conditions. The dynamical behavior of the system is examined by initially populating the highest energy state, i.e., setting $P_D(0) = 1$ and observing the relaxation towards equilibrium at constant energy $(T = 1000\ K)$. The time evolution of the probability P_B for which the system relaxes into the global minimum is displayed in Fig.1(a). We consider in particular the time required for the global minimum to develop an $\sim 80\%$ probability of ocupation. By analogy to relaxation in proteins we call this the "folding" time, t_f; i.e. $P_B(t_f) = 0.8$. At this time, the potential well D is completely depopulated. The probability of occupation for the potential well A is just the remaining proportion of 20%. The population in the potential well A reaches the maximum almost instantaneously due to rapidly spilling of the population from the potential well D and starts decreasing exponentially towards equilibrium, as can be seen in Fig. 2 (a). The potential well C is practically inaccessible, the system being trapped by the adjacent well pronounced global minimum B. (Actually, the system shows the usual *three*-level system behavior.[35]) The presence of metric contributions J in reaction rates is an important factor which can be proved, simply, by imposing $J = 1$ over all the reaction rates entering (15), and looking how this change is re ected in the relaxation process. The behavior of the system can be seen in Figs.1(b) and 2(b); the neglect of metric contributions leads to a time scale about four times slower. The "folding" time increases from

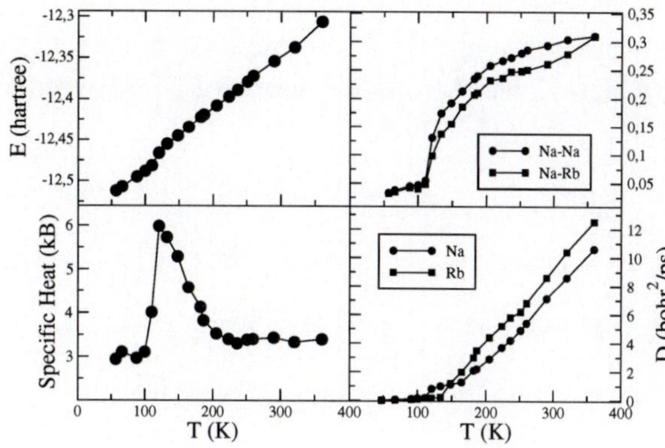

Figure 1: Time evolution of the probability P_B subject of metric corrections (a) and without metric corrections (b). For explanation, see the text.

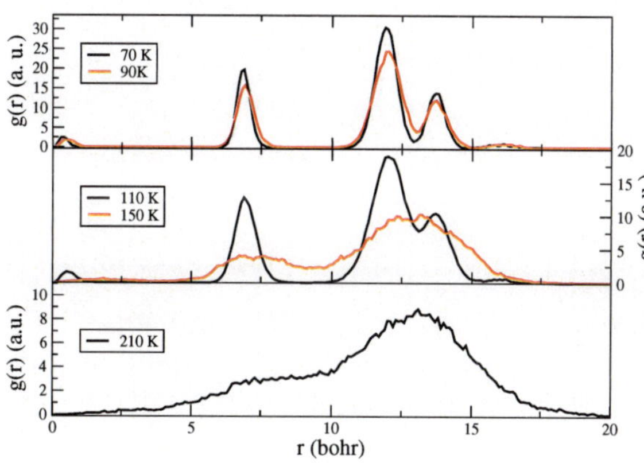

Figure 2: Same as above for the probability P_A.

In conclusion, we can say that the description of the relaxation dynamics on the PES towards the global minimum depends on the metric contributions to the transition rates. The metric contributions are directly related to the atomic rearrangements of the system at transitions through configuration space. Therefore, these play, in concert with the barrier heights, an implicit role in classifying the archetypal energy landscapes.[2, 3, 13, 23] In addition, we can say that, by incorporating the metric contribution to the TST rate, one obtains the appropriate way to combine the height of the barrier and the densities of initial and saddle states with the path length corresponding to the transition of the system between these specific stationary points. The present approach allows the TST rate to be implicitly related to the geometrical structures involved in the min-saddle-min transition of the system. We intend to use the information so obtained to identify a way to enrich the precision with which we can characterize the ability of complex systems to relax, preferentially, to only a limited number of geometrical structures from the vastly larger variety that the system might exhibit.

Moreover, we must observe that, by having the local height and length, the slope of the PES is automatic, which is, apart from the implicit Langevin random force of thermal motion, the effective force driving the system.

References

[1] R.S. Berry, *Int. J. Quantum Chem.* **58** 657 (1996)

[2] K.D. Ball, R.S. Berry, R.E. Kunz, F.-Y. Li, A. Proykova, and D.J. Walles, *Science* **271** 963 (1996)

[3] D.J. Wales, M.A. Miller and T. Walsh, *Nature* **394** 758 (1998)

[4] D.J. Wales and H.A. Scheraga, *Science* **285** 1368 (1999)

[5] W.H. Press, B.P. Flannery, S.A. Teukolsky, and W.T. Vetterling, *Numerical Recipes* (Cambridge, Cambridge, 1986)

[6] D.J. Wales, *J. Chem. Soc. Faraday Trans.* **86** 3505 (1990)

[7] R.S. Berry, H.L. Davis, and T.L. Beck, *Chem. Phys. Lett.* **147** 13 (1988)

[8] H.L. Davis, D.J. Wales, and R.S. Berry, *J. Chem. Phys.* **92** 4473 (1990)

[9] R.J. Hinde and R.S. Berry, *J. Chem. Phys.* **99** 2942 (1993)

[10] J.C. Light, *J. Chem. Phys.* **40** 3221 (1995)

[11] P.M. Rodger and M.G. Sceats, *J. Chem. Phys.* **83** 3358 (1985); M.G. Sceats, *Adv. Chem. Phys.* **70** Vol. 2, 357 (1988)

[12] A. Nitzan, *Adv. Chem. Phys.* **70** Vol. 2, 489 (1988)

[13] R.E. Kunz and R.S. Berry, *J. Chem. Phys.* **103** 1904 (1995)

[14] J.P.K. Doye and D.J. Wales, *J. Chem. Phys.* **105** 8428 (1996)

[15] R.E. Kunz and R.S. Berry, *J. Chem. Phys.* **108** 2576 (1998)

[16] J.P.K. Doye and D.J. Wales, *Phys. Rev Lett.* **80** 1357 (1998)

[17] R.P. White and H.R. Mayne, *J. Chem. Phys.* **112** 7964 (2000)

[18] R.S. Berry and R. Breitengraser-Kunz, *Phys. Rev. Lett.* **74** 3951 (1995)

[19] F. Despa and R.S. Berry, *J. Chem. Phys.* **115** 8274 (2001).

[20] F. Despa, A. Fernandez, R.S. Berry, Y. Levy and J. Jortner, *J. Chem. Phys.* **118**, 5673 (2003).

[21] F. Despa and R.S. Berry, *J. Chem. Phys.* **120**, 5164 (2004).

[22] W. Forst, *Theory of Unimolecular Reactions* (Academic, New York, 1973)

[23] M.A. Miller, J.P.K. Doye and D.J. Wales, *J. Chem. Phys.* **110** 328 (1999)

[24] D.J. Wales, *J. Chem. Phys.* **101** 3750 (1994)

[25] R.M. Lynden-Bell and D.J. Wales, *J. Chem. Phys.* **101** 1460 (1994)

[26] P. Hänggi, P. Talkner, M. Borkovec, *Rev. Mod. Phys.* **62** 251 (1990)

[27] N.C. Handy and S. Carter, *Chem. Phys. Lett.* **79** 118 (1981); N.C. Handy and S. Carter, *Chem. Phys. Lett.* **83** 216 (1981)

[28] H.L. Davis, D.J. Wales and R.S. Berry, *J. Chem. Phys.* **92** 4308 (1990)

[29] The set of normal coordinates orthogonal to the reaction coordinate, the system of molecular center of mass or the molecular orientation system can equally be used. For instance, Marcus (R.A. Marcus *J. Chem. Phys.* **110** 328 (1999)) used a coordinate system in which a point in the configuration space is specified by the distance to the reaction path and the distance along that path to a given point on it.

[30] A. Banerjee and N.P. Adams, *Int. J. Quantum Chem.* **43** 855 (1992)

[31] D.J. Wales, *J. Chem. Phys.* **113** 3926 (2000)

[32] P. Pechukas in *Modern Theoretical Chemistry* edited by W.H. Miller (Plenum, N.Y. 1976), vol. 2B; *Ann. Rev. Phys. Chem.* **29** 59 (1978)

[33] D.J. Chandler, *J. Chem. Phys.* **68** 2959 (1978)

[34] D.G. Truhlar and B.C. Garrett, *Ann. Rev. Phys. Chem.* **35** 159 (1984)

[35] F. Despa and R.S. Berry, *Eur. Phys. J. D* **16** 55 (2001).

In: Trends in Chemical Physics Research
Editor: A.N. Linke, pp. 95-115
ISBN 1-59454-483-2

Chapter 4

ON SUM RULES AND RECURRENCE RELATIONS FOR MATRIX ELEMENTS IN RELATIVISTIC QUANTUM MECHANICS

***A.C. Ilarraza-Lomelí*[*]*, M.N. Valdés-Martínez*[†]*, A.L. Salas-Brito*[‡]**
Laboratorio de Sistemas Dinámicos, Departamento de Ciencias Básicas,
Universidad Autónoma Metropolitana, Unidad Azcapotzalco,
Apartado Postal 21-267, Coyoacán 04000 D. F., México,
***R.P. Martínez-y-Romero*[§]**
Facultad de Ciencias, Universidad Nacional Autónoma de México,
Apartado Postal 50-542, Coyoacán 04510, D. F., México
***H.N. Núñez-Yépez*[¶]**
Departamento de F'sica, Universidad Autónoma Metropolitana,
Unidad Iztapalapa, Apartado Postal 55-534, Iztapalapa 09340 D. F., México.

Abstract

We review recent attempts aimed at the recursive calculations of radial matrix elements in relativistic quantum mechanics. We first discuss how to obtain sum rules relating matrix elements of f and of βf with matrix elements of their first and second derivatives —f is an arbitrary radial function and β is the standard Dirac matrix. Such elements are assumed to be taken between radial energy eigenfunctions corresponding to two different radial potentials $V_1(r)$ and $V_2(r)$ in the unshifted case. That is, the validity of our obtained relations requires that both potentials attain a minimum at the same radial position. To obtain directly usable results we then insert in our general formulas specific expressions for the two radial potentials and a definite form of the radial

[*]E-mail address: aquiles_ilarraza@hotmail.com
[†]E-mail address: ody123@prodigy.net.mx
[‡]Corresponding author. E-mail: asb@correo.azc.uam.mx
[§]E-mail address: rodolfo@dirac.fciencias.unam.mx
[¶]E-mail address: nyhn@xanum.uam.mx

function to obtain recurrence relations between the relativistic matrix elements of the mentioned function. We give specific examples of such procedure obtaining both new and previously reported recurrence relations between matrix elements of powers of the radial coordinate, r^λ and βr^λ, where λ can in general be a complex number, between radial hydrogenic states. We also obtain recurrence relations for two center matrix elements of r^λ in potential energy functions of the form $V_i(r) = a_i r^{\alpha_i}, i = 1, 2$, where a_i and α_i are constants.

1 Introduction

At present, the consideration of relativistic phenomena in atomic or molecular physics is significative not only for understanding properties of the heavy elements (*i.e.* those with $Z > 80$) since the heavier the element the more relativistic the electron motion [1, 2], or for understanding the problem of electron motion in two Coulomb centers (the so-called Z_1eZ_2 problem) [3, 4, 5, 6, 7], but also for explaining the results of experiments using synchroton radiation which attain a precision typically less than the expected values of relativistic corrections [8, 9, 10, 11, 12, 13, 14]. All of this in spite of the "variational collapse" or the spurious states that may plague relativistic atomic calculations [15, 16, 17]. Also, there is the recognition of the in uence of relativistic phenomena on the structure of molecules and on their interactions with electromagnetic fields [3, 18, 19].

To interpret phenomena like the just mentioned, the calculation of a great number of matrix elements of the operators of the problem could be required. This calls for matrix elements of a function of the radial coordinate, r, between states of the radially symmetric system [20, 21, 22, 23, 24, 26, 27, 28, 29, 30, 31, 32, 33, 34, 35, 37, 38]. We should mention that some results for radial matrix elements in nonradial potentials have been also recently obtained [39]. Any technique making less cumbersome the evaluation of a series of such matrix elements is thus welcome. In nonrelativistic quantum mechanics the task of calculating matrix elements can be simplified using diverse approaches [26, 27, 40, 41, 42, 43, 44, 45, 46], even the convertion of recurrence relations into differential equations [47]. As it has been shown in last years [33, 34, 48, 49, 50, 51, 52, 53, 54], this is also possible in relativistic quantum mechanics.

In this work we review the progress made by means of hypervirial-like formulas and other operator manipulations for obtaining sum rules of relativistic matrix elements of a radial function, $f(r)$, between eigenstates of a particle moving in the field of two radial potentials, as was initially proposed in [48]. Notice, however, that this is not an standard review since it is formulated not only to review the method and the results obtained in the last few years in this area. We decided instead to offer a sort of generalized approach first presenting new results for evaluating relativistic two-center sum rules and then specializing them for obtaining both previous and new one-center results. We regard this approach to be best suited for people who wants to pursue the method for obtaining their own results.

So, let $V_1(r)$ and $V_2(r)$ be two local radial potentials, let $f(r)$ be an *unspecified* radial function, and let $|1\rangle$, and $|2\rangle$ be energy eigenstates of the systems having, respectively, V_1

and V_2 as their potential energy function—see equations (2) below. We apply the term *sum rule* to any relation of the general form

$$\sum_{i=0}^{2} a_i \langle 2|A_i(V_1, V_2, f, r) f^{(i)}(r)|1\rangle = 0, \quad \text{or} \quad \sum_{i=0}^{2} b_i \langle 2|B_i(V_1, V_2, f, r) \beta f^{(i)}(r)|1\rangle = 0, \tag{1}$$

where the A_i and the B_i are given functions depending on the two local potentials introduced above and on their first and second r-derivatives, on the radial function $f(r)$, and on the radial coordinate r; additionally, $f^{(i)}$ stands for the i-th r-derivative of f, and the a_i and the b_i are constants, that is, r-independent coefficients that nevertheless depend on the energies, the angular momenta and other features of the relativistic eigenstates involved in (1). (The order of the r-derivatives of f or of V appearing in (1) can be greater than two for a general sum-rule, but for the purposes of this work such value is enough). This use of the term sum-rule should be contrasted with the meaning in, for example, [55, 56]. Our rules may be regarded as somewhat analogous to the Blanchard sum rule of non-relativistic quantum mechanics [20, 27]. However, with the assumption of specific forms for both the two potentials and for the radial function f, we shall be able to obtain instances of recurrence relations between relativistic radial matrix elements of a radial function, getting, in particular, some of our already published previous results [33, 48, 51, 52, 53]. Besides, we also manage to recover the relativistic virial theorem as applied to the hydrogen atom [26, 25, 57].

We pay particular attention to the case where both of the potentials are Coulomb's —where most of the work has been done [33, 49, 51, 52]. The importance of Coulomb sum rules stems from the role of hydrogenic states for physicochemical calculation and diverse approximation schemes. They are also of importance in electron capture and atomic collisions at relativistic speeds [58, 59]. We recover certain relations we have discovered for simplifying the calculation of relativistic one-center and two-center matrix elements [33, 48, 51, 52, 53]—though the two-center result recovered corresponds to just a preliminary step in such a direction. In the relativistic approach we need to take into account the presence of the Dirac's β matrix. This feature is natural since the β matrix has an eigenvalue $+1$ for the positive energy eigenstates of the Hamiltonian and -1 for the negative ones, producing some differences in the treatment of both matrix elements. The scheme thus produces, in general, two series of sum rules, one for matrix elements without the β matrix and other one for matrix elements with the β matrix explicitly appearing—though both series might not be in practice disentangled. To get the sum rules we employ the procedure, inspired on a nonrelativistic hypervirial method [27], already used to deduce recurrence relations for the radial matrix elements of $f(r)$ and $\beta f(r)$ —β is a standard Dirac matrix [60]— and the matrix elements of its first and second radial derivatives taken between relativistic energy eigenstates corresponding to two different local potential energy functions. Such radial potentials are assumed to behave as the time component of a 4-vector. Other assumptions we will be using are that the nucleus is point-like and fixed in space and that the potential acts locally. In this paper we use atomic units throughout: $\hbar = 1$, $m_e = 1$, $q_e = 1$.

2 The Relativistic Method

It is the purpose of this section to review the hypervirial-like technique and the operator calculations employed for obtaining sum rules relating matrix elements of an arbitrary radial function, f, in relativistic quantum mechanics. We address here the two-center case. The sum rules become explicit recurrence relations between matrix elements of succesive powers of r when the specific form $f(r) = r^\lambda$ (λ a constant) is substituted for the arbitrary function f. We pinpoint that our sum rules and corresponding recurrence relations are obtained only in the so-called *unshifted* case, where the position of equilibria of the two potentials are coincident. That is, we discuss only the restricted ("unshifted") two-center problem. As a sort of consistency test, we have managed to obtain the relativistic virial theorem for Coulomb systems from our sum rules [49, 51].

2.1 The Unshifted Two-Center Problem

Let us consider then two radial Dirac Hamiltonians with two different radial potentials (each behaving as the temporal component of a fourvector) $V_1(r)$ and $V_2(r)$. We further assume that these potentials have the same equilibrium position which, furthermore, is coincident with the origin of coordinates. That is, the recurrence relations correspond to the so-called unshifted case [41]. The main difficulty for not dealing yet with the general (shifted) case is the angular couplings introduced by the relative displacement of one of the potentials respect the other in the otherwise purely radial interactions.

The relativistic radial Hamiltonians can be written as [51, 61]

$$
\begin{aligned}
H_1 &= c\alpha_r[p_r - i\beta\epsilon_1(j_1 + 1/2)/r] + M_1\beta c^2 + V_1(r), \\
H_2 &= c\alpha_r[p_r - i\beta\epsilon_2(j_2 + 1/2)/r] + M_2\beta c^2 + V_2(r),
\end{aligned}
\tag{2}
$$

where c is the speed of light and the masses M_1 and M_2 are measured in atomic units. Note that $\kappa = -\epsilon(j + 1/2)$ where κ is the often used [60] eigenvalue of the operator $\Lambda = \beta(1 + \mathbf{\Sigma} \cdot \mathbf{L}$, where $\mathbf{\Sigma} = \text{diag}(\sigma, \sigma)$. The eigenstates of these radial Hamiltonians correspond to definite values of the total angular momentum (orbital plus spin) $\mathbf{J} = \mathbf{L} + \mathbf{S}$ and of the quantum number $\epsilon = (-1)^{j+l-1/2}$, where, respectively, j is the total, l the orbital, and s the spin angular momentum quantum numbers, the radial momentum operator p_r is

$$
p_r = -\frac{i}{r}\left(1 + r\frac{d}{dr}\right), \tag{3}
$$

and the Dirac matrices α y β appropriate for the radial problems at hand are given by

$$
\alpha_r = \begin{pmatrix} 0 & -1 \\ -1 & 0 \end{pmatrix}, \qquad \beta = \begin{pmatrix} 1 & 0 \\ 0 & -1 \end{pmatrix}. \tag{4}
$$

Each Dirac equation can be written as $H_k\psi_k(r) = E_{n_kj_ks_k}\psi_{n_kj_ks_k}(r)$ where the energy eigenvalues $E_{n_kj_ks_k} \equiv E_k$ and the corresponding eigenfunctions $\psi_{n_kj_ks_k}(r) \equiv \psi_k(r)$ are assumed known.

Let us take the difference between the radial Hamiltonians H_1 and H_2 in (2), to obtain

$$H_1 = H_2 + ic\alpha_r\beta\frac{\Delta^-}{2r} - c^2\beta M^- - (V_2(r) - V_1(r)). \tag{5}$$

where $M^\pm \equiv M_2 \pm M_1$,

$$\Delta^\pm \equiv \epsilon_2(2j_2+1) \pm \epsilon_1(2j_1+1) \tag{6}$$

and, in general, if X is any symbol we will be using

$$X^\pm \equiv X_2 \pm X_1. \tag{7}$$

On employing (2) again, we can directly evaluate the commutator

$$[H_1, f(r)] = -ic\alpha_r\frac{df(r)}{dr} \tag{8}$$

where $f(r)$ is an arbitrary radial function and $[H, f(r)]$ stands for the commutator of H and $f(r)$. We calculate this commutator again, but now using equation (5), to get the alternative form

$$[H_1, f(r)] = H_2f(r) - f(r)H_1 + \left(ic\alpha_r\beta\frac{\Delta^-}{2r} - c^2\beta M^- - V^-\right)f(r). \tag{9}$$

It is now simple to obtain, from equations (8) and (9), the relation

$$(E_2 - E_1)\langle 2|f|1\rangle = \langle 2|\left(c^2\beta M^- + V^-\right)f|1\rangle - ic\langle 2|\alpha_r\left(f' + \beta f\frac{\Delta^-}{2r}\right)|1\rangle; \tag{10}$$

where we have additionally taken matrix elements between the eigenstates of the problems, we use the notation $\langle 1| \equiv \langle n_1\, j_1\, \epsilon_1|$ and $|2\rangle \equiv |n_2\, j_2\, \epsilon_2\rangle$. Equation (10) may lead to sum rules connecting relativistic matrix elements of radial functions between radial eigenstates [48, 51, 52] and generalizes a nonrelativistic one useful for similar purposes [27]. Equation (10) is an exact relation for the calculation of any $f(r)$ matrix elements of f between eigenstates of two potentials in relativistic quantum mechanics. Albeit exact, equation (10) is not entirely convenient due to the presence of the operator $\alpha_r\beta$. To write up this procedure, it is convenient to deal directly with operator relations and not with the matrix elements themselves. The matrix elements will be evaluated at the end of the calculations.

2.2 The First Sum Rule

Let us establish that

$$H_2 f - f H_1 = \left(c^2\beta M^- + V^-\right) f - ic\alpha_r \left(f' + \beta f \frac{\Delta^-}{2r}\right), \tag{11}$$

notice that equation (10) above can be obtained from (11) just by taking matrix elements. The following result is also easily established

$$H_2 f + f H_1 = \left(c^2\beta M^+ + V^+\right) f - ic\alpha_r \left(2f\frac{d}{dr} + f' + \frac{2f}{r} + \beta f \frac{\Delta^+}{2r}\right). \tag{12}$$

Then, it can be seen that

$$-ic\left(H_2\alpha_r f + \alpha_r f H_1\right) = ic\alpha_r \left(c^2\beta M^- - V^+\right) f - c^2 \left(2f\frac{d}{dr} + f' + \frac{2f}{r} - \beta f \frac{\Delta^-}{2r}\right), \tag{13}$$

and that

$$H_2 f V^- - f V^- H_1 = \left(c^2\beta M^- + V^-\right) V^- f - ic\alpha_r \left(V^- f' + \frac{dV^-}{dr} f + \beta f V^- \frac{\Delta^-}{2r}\right). \tag{14}$$

The next relation is also readily apparent

$$-ic\left[H_2\alpha_r\beta\frac{f}{r} + \alpha_r\beta\frac{f}{r}H_1\right] = -ic\alpha_r \left(\beta V^+ - c^2 M^-\right)\frac{f}{r} - c^2\left[\beta\left(\frac{f'}{r} - \frac{f}{r^2}\right) - \frac{\Delta^+}{2r}\frac{f}{r}\right]. \tag{15}$$

Let us define $\psi(r) \equiv H_2 f(r) - f(r) H_1$, and evaluate

$$\begin{aligned} H_2 \quad \psi - \psi H_1 &= c^2\beta\frac{\Delta^+}{2r} f' + c^2\left(\frac{\Delta^-}{2r}\right)^2 f + \left(c^2\beta M^- + V^-\right)^2 f \\ &- \quad c^2 f'' - c^2\beta\frac{\Delta^-}{2r}\left(2f\frac{d}{dr} + f' + \frac{f}{r}\right) \\ &- \quad ic\alpha_r \left[\left(f' + \beta f\frac{\Delta^-}{2r}\right)\left(V^- - c^2\beta M^+\right) + c^2\beta M^-\left(2f\frac{d}{dr} + f' + \frac{2f}{r}\right)\right. \\ &+ \quad \left. V^- f' + \frac{dV^-}{dr} f + c^2 M^- \frac{\Delta^+}{2r} f + V^- \frac{\Delta^-}{2r}\beta f\right]. \end{aligned} \tag{16}$$

Given these last expressions [equations (11)–(16)], it is relatively simple to obtain, from (11), the following expression

$$-ic\alpha_r\left(f' + \beta f\frac{\Delta^-}{2r}\right) = (H_2 f - f H_1) - \left(c^2\beta M^- + V^-\right) f, \tag{17}$$

and, from (12), we obtain

$$-ic\alpha_r\beta\left(2f\frac{d}{dr} + f' + \frac{2f}{r}\right) = (H_2\beta f + \beta f H_1) - \left(c^2\beta M^+ + V^+\right)\beta f + ic\alpha_r\frac{\Delta^+}{2r}f. \tag{18}$$

From equation (13), we get

$$-c^2\beta\frac{\Delta^-}{2r}\left(2f\frac{d}{dr} + f' + \frac{2f}{r}\right) = -ic\left(H_2\alpha_r\beta\frac{\Delta^-}{2r}f + \alpha_r\beta\frac{\Delta^-}{2r}fH_1\right)$$
$$- \; ic\alpha_r\beta\frac{\Delta^-}{2r}\left(c^2\beta M^- - V^+\right) f - c^2 f\left(\frac{\Delta^-}{2r}\right)^2. \tag{19}$$

From equation (14), we obtain

$$-ic\alpha_r\left(V^- f' + \frac{dV^-}{dr}f\right) = (H_2 f V^- - f V^- H_1) - \left(c^2\beta M^- + V^-\right) V^- f$$
$$+ic\alpha_r V^-\frac{\Delta^-}{2r}\beta f. \tag{20}$$

Substituting equations (17), (18), (19), and (20), into equation (16) and substituting the terms which include the operator $ic\alpha_r$ with the help of equation (15), we thus get

$$-ic\left(H_2\alpha_r\beta\frac{\Delta^-}{2r}f + \alpha_r\beta\frac{\Delta^-}{2r}fH_1\right) - \; ic\alpha_r\tfrac{\Delta^-}{2r}\left(c^2\beta M^- - V^+\right)\beta f =$$
$$- \; c^2\tfrac{\Delta^-}{2r}\left[\beta\left(f' - \tfrac{f}{r}\right) - \tfrac{\Delta^+}{2r}f\right]. \tag{21}$$

Using again equation (21), it yields

$$H_2\psi - \psi H_1 = -c^2\left(f'' - \beta f'\frac{\Delta^+}{2r}\right) + c^2\frac{\Delta^-}{r^2}\beta f + \left(M^-\right)^2 c^4 f$$
$$- \; c^2 M^+\left(H_2\beta f - \beta f H_1\right) + c^2 M^+ V^-\beta f + c^2 M^-\left(H_2\beta f + \beta f H_1\right)$$
$$- \; c^2 M^- V^+\beta f + V^-\left[\, 2\left(H_2 f - f H_1\right) - V^- f\,\right] - c^2\frac{\Delta^-}{2r}\left(\beta f' - \frac{\Delta^+}{2r}f\right). \tag{22}$$

Evaluating the matrix elements between the Dirac eigenstates $\langle 2|$ and $|1\rangle$ and rearranging, we finally obtain the relation

$$
\begin{aligned}
a_0\langle 2|f|1\rangle + a_2\langle 2|\tfrac{f}{r^2}|1\rangle - 2E^-\langle 2|V^- f|1\rangle + \langle 2|(V^-)^2 f|1\rangle + c^2\langle 2|f''|1\rangle = \\
b_0\langle 2|\beta f|1\rangle + b_1\langle 2|\beta\tfrac{f}{r^2}|1\rangle - c^2M^-\langle 2|V^+\beta f|1\rangle + c^2M^+\langle 2|V^-\beta f|1\rangle \\
+b_4\langle 2|\beta\tfrac{f'}{r}|1\rangle,
\end{aligned} \tag{23}
$$

where

$$
\begin{aligned}
a_0 &= \left(E^-\right)^2 - \left(c^2M^-\right)^2 \\
a_2 &= -\frac{c^2}{4}\Delta^-\Delta^+ \\
b_0 &= c^2\left(M^-E^+ - M^+E^-\right) \\
b_1 &= c^2\Delta^- \\
b_4 &= \frac{c^2}{2}\left(\Delta^+ - \Delta^-\right)
\end{aligned} \tag{24}
$$

This is a sum rule relating matrix elements of an arbitrary radial function $f(r)$ between eigenstates of two different but unshifted potentials as a function of the eigenenergies in relativistic quantum mechanics.

2.3 The Second Sum Rule

As we mentioned in the Introduction, for the purpose of computing matrix elements, more relations are convenient. To obtain a second equation, let us evaluate

$$
\begin{aligned}
& H_2 f V^- + f V^- H_1 = \left(c^2\beta M^+ + V^+\right) V^- f \\
- & \imath c\alpha_r\left(2V^- f\frac{d}{dr} + V^- f' + \frac{dV^-}{dr} f + 2V^-\frac{f}{r} + \beta V^- f\frac{\Delta^+}{2r}\right).
\end{aligned} \tag{25}
$$

Using again the definition $\psi(r) \equiv H_2 f(r) - f(r) H_1$, we can write

$$
\begin{aligned}
& H_2\psi + \psi H_1 = c^2 f\frac{\Delta^-\Delta^+}{4r^2} - \imath c\alpha_r\left(f' + \beta f\frac{\Delta^-}{2r}\right)\left(V^+ - c^2\beta M^-\right) \\
+ & \left(c^2\beta M^+ + V^+\right)\left(c^2\beta M^- + V^-\right) f - c^2\left(2f'\frac{d}{dr} + f'' + 2\frac{f'}{r} - \beta f\frac{\Delta^-}{2r^2}\right) \\
- & ic\alpha_r\left(2V^- f\frac{d}{dr} + V^- f' + \frac{dV^-}{dr} f + 2V^-\frac{f}{r} + \beta V^- f\frac{\Delta^+}{2r} + c^2\beta M^- f'\right. \\
+ & \left. c^2M^-\frac{\Delta^-}{2r} f\right).
\end{aligned} \tag{26}
$$

The calculations required for obtaining the recurrence relation are similar to that used above. So, from equation (11) we obtain

$$-ic\alpha_r c^2\beta M^-\left(f'+\beta f\frac{\Delta^-}{2r}\right)=c^2\beta M^-\left(H_2 f-fH_1\right)-c^2M^-\left(c^2\beta M^-+V^-\right)\beta f, \tag{27}$$

from (13) we get

$$\begin{aligned}-c^2\left(2f'\frac{d}{dr}+f''+\frac{2f'}{r}\right) &= -ic\left(H_2\alpha_r f'+\alpha_r f'H_1\right)\\ &- ic\alpha_r\left(c^2\beta M^- -V^+\right)f'-c^2\beta f'\frac{\Delta^-}{2r},\end{aligned} \tag{28}$$

and from equation (25), we obtain

$$\begin{aligned}-ic\alpha_r\left(2V^-f\frac{d}{dr}+V^-f'+\tfrac{dV^-}{dr}f+2V^-\tfrac{f}{r}\right) &= \left(H_2fV^-+fV^-H_1\right)\\ &+ ic\alpha_r V^-\beta f\tfrac{\Delta^+}{2r}-\left(c^2\beta M^+ +V^+\right)V^-f.\end{aligned} \tag{29}$$

Using equations (17), (27), (28) and (29), in equation (26),

$$\begin{aligned}H_2\psi+\psi H_1 &= -c^2\beta\frac{\Delta^-}{2r}\left(f'-\frac{f}{r}\right)+c^2\frac{\Delta^+}{2r}\frac{\Delta^-}{2r}f-V^-V^+f\\ &+ M^-M^+c^4f+V^+\left(H_2f-fH_1\right)+V^-\left(H_2f+fH_1\right)\\ &- ic\left(H_2\alpha_r f'+\alpha_r f'H_1\right)-\imath c\alpha_r\left(c^2\beta M^- -V^+\right)f'.\end{aligned} \tag{30}$$

Again, the last two terms of (30) can be obtained from (15). Combining these terms with

$$\begin{aligned}-ic\left(H_2\alpha_r f'+\alpha_r f'H_1\right)- & ic\alpha_r\left(c^2\beta M^- -V^+\right)f'=\\ - & c^2\left(f''-\tfrac{f'}{r}\right)+c^2\tfrac{\Delta^+}{2r}\beta f',\end{aligned} \tag{31}$$

we get

$$\begin{aligned}H_2\psi+\psi H_1 &= -c^2\beta\frac{\Delta^-}{2r}\left(f'-\frac{f}{r}\right)+c^2\frac{\Delta^+}{2r}\frac{\Delta^-}{2r}f-V^-V^+f\\ &+ M^-M^+c^4f+V^+\left(H_2f-fH_1\right)+V^-\left(H_2f+fH_1\right)\\ &- c^2\left(f''-\frac{f'}{r}\right)+c^2\frac{\Delta^+}{2r}\beta f'.\end{aligned} \tag{32}$$

Taking matrix elements, we obtain the sought after sum rule

$$
\begin{aligned}
c_0\langle 2|f|1\rangle \quad &+a_2\langle 2|\tfrac{f}{r^2}|1\rangle - E^+\langle 2|V^- f|1\rangle - E^-\langle 2|V^+ f|1\rangle + \langle 2|V^+V^- f|1\rangle \\
&-c^2\langle 2|\tfrac{f'}{r}|1\rangle + c^2\langle 2|f''|1\rangle = \tfrac{b_1}{2}\langle 2|\beta\tfrac{f}{r^2}|1\rangle + b_4\langle 2|\beta\tfrac{f'}{r}|1\rangle,
\end{aligned}
\tag{33}
$$

where the only new coefficient is

$$
c_0 = E^+E^- - c^4M^+M^-. \tag{34}
$$

2.4 The Third and Fourth Sum Rules

Having stablished the general procedure, we can continue using it to get more sum rules. For example,

$$
e_0\langle 2|f|1\rangle = g_0\langle 2|\beta f|1\rangle - \langle 2|\left(V^+ - V^-\right)\beta f|1\rangle, \tag{35}
$$

where

$$
\begin{aligned}
e_0 &= c^2\left(M^+ - M^-\right) \\
g_0 &= E^+ - E^-,
\end{aligned}
\tag{36}
$$

this equation allows writing the matrix elements of f in terms of those of βf, as substitution of equation (35) into (23) shows

$$
\begin{aligned}
&A_0\langle 2|\beta f|1\rangle + A_1\langle 2|\beta\frac{f}{r^2}|1\rangle + A_2\langle 2|V^-\beta f|1\rangle + A_3\langle 2|\left(V^-\right)^2\beta f|1\rangle \\
&+\langle 2|\left(V^-\right)^3\beta f|1\rangle + A_5\langle 2|V^+\beta f|1\rangle + 2E^-\langle 2|V^-V^+\beta f|1\rangle \\
&-\langle 2|\left(V^-\right)^2 V^+\beta f|1\rangle + a_2\langle 2|\left(V^+ - V^-\right)\beta\frac{f}{r^2}|1\rangle \\
&-c^2\langle 2|\left(V^+ - V^-\right)''\beta f|1\rangle = A_9\langle 2|\beta\frac{f'}{r}|1\rangle + 2c^2\langle 2|\left(V^+ - V^-\right)'\beta f'|1\rangle \\
&-c^2 g_0\langle 2|\beta f''|1\rangle - c^2\langle 2|\left(V^+ - V^-\right)\beta f''|1\rangle.
\end{aligned}
\tag{37}
$$

where

$$
\begin{aligned}
A_0 &= \left(E^-\right)^2\left(E^+ - E^-\right) + c^4E^-\left[\left(M^-\right)^2 + \left(M^+\right)^2\right] - c^4M^+M^-\left(E^+ - E^-\right) \\
A_1 &= -\frac{c^2}{4}\left(E^+ - E^-\right)\Delta^+\Delta^- - c^4\Delta^-\left(M^+ - M^-\right) \\
A_2 &= -2E^-\left(E^+ - E^-\right) + \left(E^-\right)^2 - \left(c^2M^-\right)^2 - c^4M^+\left(M^+ - M^-\right) \\
A_3 &= E^+ - 3E^- \\
A_5 &= c^2M^+M^- - \left(E^-\right)^2 \\
A_9 &= \frac{c^4}{2}\left(M^+ - M^-\right)\left(\Delta^+ - \Delta^-\right).
\end{aligned}
\tag{38}
$$

Equation (37) is the last sum rule connecting relativistic $f(r)$ matrix elements between states of different potentials with those of their derivatives, we are going to explicitly exhibit in this review. Notice that, at difference of the previous relations [equations (23), (33) and (35)], equation (37) relates among themselves just matrix elements of βf and its derivatives times a certain known function of r.

2.5 Sum Rules for Constant f

It is also of interest to explicitly write down the case where $f(r)$ is simply a constant. In this case we first obtain

$$a_0\langle 2|1\rangle + \langle 2|\frac{a_2}{r^2}|1\rangle - 2E^-\langle 2|V^-|1\rangle + \langle 2|(V^-)^2|1\rangle = \\ \langle 2|\left(b_0 + \frac{b_1}{r^2} - c^2M^-V^+ + c^2M^+V^-\right)\beta|1\rangle, \tag{39}$$

then

$$e_0\langle 2|1\rangle = \langle 2|\left(g_0 - V^+ + V^-\right)\beta|1\rangle, \tag{40}$$

next

$$c_0\langle 2|1\rangle + \langle 2|\frac{a_2}{r^2}|1\rangle - E^+\langle 2|V^-|1\rangle - E^-\langle 2|V^+|1\rangle + \langle 2|V^-V^+|1\rangle = \\ \frac{b_1}{2}\langle 2|\frac{\beta}{r^2}|1\rangle, \tag{41}$$

and, combining equations (39) and (40), we get

$$\langle 2|\left(A_0 + \frac{A_1}{r^2} + A_2V^- + A_3(V^-)^2 + (V^-)^3 + A_5V^+ + 2E^-V^-V^+ - \\ (V^-)^2V^+ - \frac{a_2}{r^2}(V^+ - V^-) - c^2\left(V^+ - V^-\right)''\right)\beta|1\rangle = 0; \tag{42}$$

using again equations (39) and (40), we obtain the sum rule

$$\langle 2|c_0g_0 + \frac{1}{r^2}\left[a_2g_0 - \frac{1}{2}b_1e_0 - a_2(V^+ - V^-)\right] - (c_0 + E^+g_0)V^+ + \\ (c_0 - E^+g_0)V^- + 2(E^+ - E^-)V^+V^- + E^-(V^+)^2 + E^+(V^-)^2 + \\ V^+(V^-)^2 - V^-(V^+)^2|1\rangle = 0. \tag{43}$$

2.5.1 The Relativistic Virial Theorem in the Coulomb Case

From (40), assuming both potentials as Coulomb's and both the energy eigenstates, $|1\rangle$ and $|2\rangle$, as coincident, we inmediately get

$$\frac{E^2}{c^2} + Z\left\langle \frac{\beta}{r} \right\rangle = c^2, \tag{44}$$

where we have also used that $c^2\langle\beta\rangle = E$ as is valid for the relativistic hydrogen atom [26]. That is, we have been able to obtain the well-known Coulomb-Dirac relativistic virial theorem [33, 51, 62] from our two-center expressions. See also [26, 63, 64], though some of these works deal with relativistic versions of the virial theorem employing equations different from Dirac's.

2.5.2 A Relation among Inverse Powers of r

Moreover, from the sum rule (43) and under the conditions used for obtaining the virial theorem, we can also inmediately obtain

$$2Z(c_0 + E^- g_0)\left\langle \frac{1}{r} \right\rangle + \left(4c^4 Z^2 E^- + a_2 g_0 - \frac{1}{2} b_1 e_0\right)\left\langle \frac{1}{r^2} \right\rangle + 2a_2 Z\left\langle \frac{1}{r^3} \right\rangle = -c_0 g_0. \tag{45}$$

a neat relation valid for expectation values of inverse powers of r in the relativistic hydrogen atom.

3 Recurrence Relations of Matrix Elements of r^λ between Relativistic Energy Eigenstates

Given the sum rules obtained in the previous section, in this section we want to illustrate how we may obtain relations between relativistic matrix elements of the form $\langle 2|\beta^a r^\lambda|1\rangle$ where $a = 0$ or 1, λ is a possibly complex number, and the $|i\rangle$ are radial relativistic energy eigenstates [60, 65].

3.1 Recursions with Power Potentials

To get the recurrence relations, let us substitute potential energy functions of the form $V_1(r) = k_1 r^{\alpha_1}$ and $V_2(r) = k_2 r^{\alpha_2}$ with α_1 and α_2 constants, and use radial function of the form $f(r) = r^\lambda$, λ possibly complex, in equations (23), (33) and (35). The form selected for the potentials is general enough to encompass the Coulomb, Van der Waals, and terms of the Lennard-Jones ones. Performing the substitutions and after some algebra we first get

$$a_0\langle 2|r^\lambda|1\rangle + \left(a_2 + c^2\lambda(\lambda-1)\right)\langle 2|r^{\lambda-2}|1\rangle - 2E^- k_2\langle 2|r^{\lambda+\alpha_2}|1\rangle +$$

$$+2E^{-}k_1\langle 2|r^{\lambda+\alpha_1}|1\rangle + k_2^2\langle 2|r^{\lambda+2\alpha_2}|1\rangle + k_1^2\langle 2|r^{\lambda+2\alpha_1}|1\rangle -$$
$$-2k_1k_2\langle 2|r^{\lambda+\alpha_2+\alpha_1}|1\rangle = b_0\langle 2|\beta r^{\lambda}|1\rangle + (b_1+b_4\lambda)\langle 2|\beta r^{\lambda-2}|1\rangle +$$
$$+c^2(M^+ - M^-)k_2\langle 2|\beta r^{\lambda+\alpha_2}|1\rangle - c^2(M^+ + M^-)k_1\langle 2|\beta r^{\lambda+\alpha_1}|1\rangle, \quad (46)$$

then,

$$c_0\langle 2|r^{\lambda}|1\rangle + (a_2 + c^2\lambda(\lambda-2))\langle 2|r^{\lambda-2}|1\rangle - (E^+ + E^-)k_2\langle 2|r^{\lambda+\alpha_2}|1\rangle +$$
$$+(E^+ - E^-)k_1\langle 2|r^{\lambda+\alpha_1}|1\rangle + k_2^2\langle 2|r^{\lambda+2\alpha_2}|1\rangle - k_1^2\langle 2|r^{\lambda+2\alpha_1}|1\rangle =$$
$$= \left(\tfrac{b_1}{2} + b_4\lambda\right)\langle 2|\beta r^{\lambda-2}|1\rangle, \quad (47)$$

and the relationship

$$e_0\langle 2|r^{\lambda}|1\rangle = g_0\langle 2|\beta r^{\lambda}|1\rangle - 2k_1\langle 2|\beta r^{\lambda+\alpha_1}|1\rangle, \quad (48)$$

plus the next recurrence relation between βr^a terms only

$$A_0\langle 2|\beta r^{\lambda}|1\rangle + (A_1 - A_9\lambda + c^2 g_0\lambda(\lambda-1))\langle 2|\beta r^{\lambda-2}|1\rangle +$$
$$+(A_2+A_5)k_2\langle 2|\beta r^{\lambda+\alpha_2}|1\rangle + (A_5 - A_2)k_1\langle 2|\beta r^{\lambda+\alpha_1}|1\rangle +$$
$$+(A_3 + 2E^-)k_2^2\langle 2|\beta r^{\lambda+2\alpha_2}|1\rangle + (A_3 - 2E^-)k_1^2\langle 2|\beta r^{\lambda+2\alpha_1}|1\rangle +$$
$$-2k_1^3\langle 2|\beta r^{\lambda+3\alpha_1}|1\rangle - 2A_3k_1k_2\langle 2|\beta r^{\lambda+\alpha_2+\alpha_1}|1\rangle + 6k_2k_1^2\langle 2|\beta r^{\lambda+\alpha_2+2\alpha_1}|1\rangle =$$
$$= 2k_1\left(a_2 - c^2\left(\lambda(\lambda-1) - 2\alpha_1(\lambda+\alpha_1-1)\right)\right)\langle 2|\beta r^{\lambda+\alpha_1-2}|1\rangle. \quad (49)$$

The $|1\rangle$ and $|2\rangle$ states should correspond, respectively, to eigenstates of the V_1 and the V_2 potentials.

3.2 Uncoupling the Recurrence Relations

As it is clear, these general relations can be further specialized if we use the specific exponents α_i associated with certain potentials. For example, to evaluate matrix elements of potential functions for the Z_1eZ_2 problem, we use the above relations with $\alpha_1 = \alpha_2 = -1$ and $M^- = 0, M^+ = 2$ $(M_1 = M_2 = 1)$ or we may choose $\alpha_1 = \alpha_2 = -6$ to evaluate certain relations for the (unshifted) long-range electron-atom interaction potential [55]. Besides, the above equations can be written explicitly as recurrence relations of terms of the form βr^a plus other independent recurrence relations of terms r^b. From the above relations we can obtain explicitly relations between r^{λ} matrix elements only or relations between βr^{λ} matrix elements. That is, the relations (46), (47), (48) and (49) can be uncoupled to obtain

$$D_0\langle 2|r^{\lambda}|1\rangle = D_1\langle 2|r^{\lambda+2}|1\rangle + D_2\langle 2|r^{\lambda+\alpha_1}|1\rangle + D_3\langle 2|r^{\lambda+2+\alpha_1}|1\rangle + D_4\langle 2|r^{\lambda+2+\alpha_2}|1\rangle$$

$$+D_5\langle 2|r^{\lambda+2+2\alpha_1}|1\rangle + D_6\langle 2|r^{\lambda+2+2\alpha_2}|1\rangle + D_7\langle 2|r^{\lambda+2+3\alpha_1}|1\rangle$$

$$+D_8\langle 2|r^{\lambda+2+\alpha_1+2\alpha_2}|1\rangle + D_9\langle 2|r^{\lambda+2+\alpha_1+\alpha_2}|1\rangle, \qquad (50)$$

where the D-coefficients are given by

$$\begin{aligned}
D_0 &= \frac{c^6}{4}(M^+ - M^-)\Big((\Delta^-)^2 + \Delta^-(\Delta^+ - \Delta^-)(2\lambda + 4 + \alpha_1) \\
&\quad + (\Delta^+ - \Delta^-)^2(\lambda+2)(\lambda+2+\alpha_1)\Big) \\
&\quad + \frac{c^4}{2}(E^+ - E^-)\left(\left[\frac{\Delta^+\Delta^-}{4} - \lambda(\lambda+2)\right]\left[\Delta^- + (\Delta^+ - \Delta^-)(\lambda+2+\alpha_1)\right]\right), \\
D_1 &= \frac{c^2}{2}(E^+ - E^-)(E^+E^- - c^4M^+M^-)\left[\Delta^- + (\Delta^+ - \Delta^-)(\lambda+2+\alpha_1)\right], \\
D_2 &= c^4k_1\left(\left[\frac{\Delta^+\Delta^-}{4} - (\lambda+\alpha_1)(\lambda+2\alpha_1)\right]\left[\Delta^- + (\Delta^+ - \Delta^-)(\lambda+2)\right]\right), \\
D_3 &= c^2k_1\left(\frac{(E^+ - E^-)^2}{2}\left[\Delta^- + (\Delta^+ - \Delta^-)(\lambda+2+\alpha_1)\right]\right. \\
&\quad \left. - (E^+E^- - c^4M^+M^-)\left[\Delta^- + (\Delta^+ - \Delta^-)(\lambda+2)\right]\right), \\
D_4 &= \frac{c^2}{2}k_2\left[(E^-)^2 - (E^+)^2\right]\left[\Delta^- + (\Delta^+ - \Delta^-)(\lambda+2+\alpha_1)\right], \\
D_5 &= \frac{c^2}{2}k_1^2(E^- - E^+)\left(3\Delta^- + (\Delta^+ - \Delta^-)(3\lambda+6+\alpha_1)\right), \\
D_6 &= \frac{c^2}{2}k_2^2(E^+ - E^-)\left[\Delta^- + (\Delta^+ - \Delta^-)(\lambda+2+\alpha_1)\right], \\
D_7 &= \frac{c^2}{2}k_1^3\left[\Delta^- + (\Delta^+ - \Delta^-)(\lambda+2)\right], \\
D_8 &= -c^2k_1k_2^2\left[\Delta^- + (\Delta^+ - \Delta^-)(\lambda+2)\right], \\
D_9 &= c^2k_1k_2(E^+ + E^-)\left[\Delta^- + (\Delta^+ - \Delta^-)(\lambda+2)\right].
\end{aligned} \qquad (51)$$

Again, when the two potentials become equal $a = b = -1$, $Z_1 = Z_2 = Z$, and $f(r) = r^\lambda$, *i.e.* the case of two Coulomb potentials with a power function of the radial coordinate, we recover our previous formulas for calculating one-center matrix elements of r^λ [51, 52]. That is, we can get the recurrence relation between matrix elements of βr^a (where $M_1 = M_2 = 1$, $M^- = 0$)

$$\eta_0\langle 2|\beta r^\lambda|1\rangle = \eta_1\langle 2|\beta r^{\lambda-1}|1\rangle + \eta_2\langle 2|\beta r^{\lambda-2}|1\rangle + \eta_3\langle 2|\beta r^{\lambda-3}|1\rangle, \qquad (52)$$

with

$$\begin{aligned}
\eta_0 &= \frac{E^+G_1}{2\lambda G_2} - \frac{2}{\Delta^+}(G_4+\lambda), \\
\eta_1 &= Z\left(G_5 - \frac{G_1}{\lambda G_2}\right), \\
\eta_2 &= \left(\frac{c^2(\lambda-1)}{2\lambda\Delta^+}(G_4+\lambda-2) - \frac{c^2\Delta^-}{8\lambda}\right)\frac{G_3}{G_2}, \\
\eta_3 &= \frac{Zc^2(\lambda-1)}{4\lambda}\frac{G_3G_5}{G_2},
\end{aligned} \tag{53}$$

and the recurrence relation between matrix elements of r^b

$$\nu_0\langle 2|r^\lambda|1\rangle = \nu_1\langle 2|r^{\lambda-1}|1\rangle + \nu_2\langle 2|r^{\lambda-2}|1\rangle + \nu_3\langle 2|r^{\lambda-3}|1\rangle + \nu_4\langle 2|r^{\lambda-4}|1\rangle + \nu_5\langle 2|r^{\lambda-5}|1\rangle, \tag{54}$$

where

$$\begin{aligned}
\nu_0 &= 2(E^+)^2(E^-)^2\frac{G_6}{ZG_7}, \\
\nu_1 &= -8E^+(E^-)^2\frac{G_1+2c^2}{G_7}, \\
\nu_2 &= \frac{2\lambda}{Z}G_2G_9 - \frac{E^+}{2Z}G_1G_8 - \frac{8(E^-)^2G_1}{G_7} - \frac{E^+(E^-)^2G_3G_6}{2Z(\lambda-1)G_1}, \\
\nu_3 &= G_1G_8 + \frac{1}{(\lambda-1)\Delta^+}\left(4E^+(\lambda-1)(\lambda-2)G_1 + (E^-)^2\Delta^+G_3\right), \\
\nu_4 &= \frac{c^2(\lambda-1)}{2Z}G_3G_9 - \frac{8Z}{\Delta^+}(\lambda-2)G_1 + \frac{c^2\Delta^+}{8Z}G_3G_8, \\
\nu_5 &= c^2(\lambda-2)G_3.
\end{aligned} \tag{55}$$

To write equations (53) and (55), we have found it convenient to define

$$\begin{aligned}
G_1 &= \Delta^-E^- - 4c^2\lambda, \\
G_2 &= (E^-)^2 - 4c^4, \\
G_3 &= 4\lambda^2 - (\Delta^-)^2, \\
G_4 &= E^+\frac{\Delta^+ + \Delta^-}{4c^2}, \\
G_5 &= \frac{\Delta^+ + \Delta^-}{\Delta^+c^2}, \\
G_6 &= G_1 + 4c^2, \\
G_7 &= c^2(\lambda-1)\Delta^+, \\
G_8 &= \frac{(4\lambda(\lambda-1)E^+ - \Delta^+G_1)\,G_6}{(\lambda-1)G_1\Delta^+} - \frac{4c^2\Delta^+}{\Delta^+ + \Delta^-},
\end{aligned}$$

$$G_9 = \frac{\lambda E^+ G_6}{(\lambda-1)G_1} - \frac{G_6}{\Delta^+} - E^+ - \frac{4c^2(\lambda-2)}{\Delta^+ + \Delta^-}. \tag{56}$$

As we have exhibited in [51], the exponent, λ, of the radial coordinate used in the previous equations can even be complex. In the case of the Coulomb potential, the restriction can be stated as $\omega_1 + \omega_2 + 1 > |\lambda|$, where the $\omega_i \equiv \sqrt{(j_i+1/2)^2 - Z_i^2/c^2}$, $i = 1, 2$, are real numbers. In any case, the restriction is nothing more than guaranteeing that any integral vanishes faster than $1/r$ as $r \to \infty$.

4 Conclusions

We have reviewed how to obtain sum rules and recurrence relations for the calculation of matrix elements of a radial function between states of two different radial potentials sharing a common origin. We have exhibited that hypervirial-like formulas and other operator identities suffice to obtain relations between relativistic matrix elements of an arbitrary radial function between eigenstates corresponding to two different potentials sharing a single equilibrium point. The obtained sum rules are given in the most general case of an arbitrary function taken between any non necessarily diagonal radial eigenstates of the two radial potentials. Our relations are of interest in, for example, the Z_1eZ_2 problem or for calculations related with the long range e-atom interaction. The rules have, as particular cases, recursion relations between one-potential integrals or, in other particular cases, between overlap and one center integrals in Dirac relativistic quantum mechanics and lead to explicit recursion relations when specific forms of the radial functions are used and particular potentials are substituted. Calculations of this sort are exemplified in subsections 3.1 and 3.2 where we are able to get sum rules relating matrix elements of succesive powers $r^{\lambda-a}$ and matrix elements of $\beta r^{\lambda-b}$ between relativistic eigenstates of power potentials and, in particular, of the hydrogen atom that can be useful for studying radiative transitions in Rydberg atoms, in analysing atomic photorecombination and photoionization processes, for evaluating relativistic corrections to ionic oscillator strengths. The potentials could be regarded as describing the electronic configurations in processes involving electronic rearrangements in atomic transitions. For example, any transition to an autoionizing state studied in the central field approximation where the electron motion is solution of the Dirac equation with an effective central potential created by a k or a $k-x$ electron ion core. Or in any other atomic processes involving highly excited electrons which need to be studied using multichannel spectroscopy or quantum deffect theory [34, 66, 67, 68]. The recursions can be also useful for studying radiative transitions in Rydberg atoms, in analysing atomic photorecombination and photoionization processes, for calculating relativistic corrections to ionic oscillator strengths, or in analysing impact ionization or vibrational transitions in molecules — albeit in the last two cases in a crude manner — [11, 69, 70, 71].

We expect the obtained recursions, together with the previous relations we have obtained [48, 51, 52], to be useful in atomic or molecular physics calculations as they may simplify calculation of matrix elements in the range of applicability of Dirac's relativistic quantum mechanics [55, 72, 73]. For most uses of the relations we first have to set

$M_1 = M_2$, *i.e.* $M^- = 0$ and $M^+ = 2$—the particles are electrons. For the Z_1eZ_2 problem other settings may be necessary. Besides, the sum rules or the recurrence relations that follow from tem can be useful for calculation of diverse physical quantities [50], for example, for the relativistic calculation of magnetic or electric splittings in multiple charged ions [74], or for calculating the quadrupole moment of atomic hydrogen [75, 76], or the dipole moment of the electron[82]. Of course that we may use also relations discovered by other people, *e. g.* the interesting work reported in [49, 50]. We have illustrated the utility behind our relationships by deriving the virial theorem for the relativistic Coulomb problem from them.

From a more practical angle, little can be done for the analytical evaluation of these integrals beyond the Coulomb and a few similarly exactly solvable potentials. However, there are relativistic theoretical techniques [40, 81] and relativistic [3, 83] and nonrelativistic numerical methods[77] that, after being adapted to relativistic conditions, can provide the crucial starting results needed for the systematic use of the recurrence relations obtained here [78]. Our results can be also useful in the so-called perturbation theory of relativistic corrections, in relativistic quantum defect calculations, and for the relativistic extension of the calculations of exchange integrals using Slater orbitals with Coulomb-Dirac wave functions [3, 34, 36, 37, 50, 72, 79, 80].

We should mention that in this review we have only considered some possible uses of the approach and of the sum rules that may follow from it. It should be clear that their applications can be extended.

Acknowledgements

This work has been partially supported by PAPIIT-UNAM (grant 108302). We acknowledge with thanks the collaboration of C. Cisneros and I. Álvarez. We also want to thank the friendly support of Danna and Dua Castillo-Vásquez, and the cheerful enthusiasm of G. R. Inti, F. A. Maya, P. M. Schwartz (Otello), P. O. M. Yoli, A. S. Ubo, P. L. M. Yagnionak, G. Sieriy, M. Chiornaya, P. A. Koshka, G. D. Abdul, L. S. Micha, P. V. Patricio, G. Sadi, P. L. M. Elf, P. Lobitta, P. R. Kranquen, M. Katze, P. Lobo, and D. Gorbe. This work is for Mileva Sofˇa in her first birthday.

References

[1] P. Pyykkö, *Chem. Rev.* **88** 563 (1988).

[2] L Gagliardi, N. C. Handy, A. G. Ioannou, C.-K. Skylaris, S. Spencer, A. Willets, A. M. Simper, Chem. *Phys. Lett.* **283** 187 (1998).

[3] H. Zhang, O. Kullie, D Kolb, *J. Phys. B: At. Mol. Opt. Phys.* **37** 905 (2004).

[4] P. Briet, H. Hogreve, Few-Body Systems, to appear (2004). DOI: 10.1007/s00601-004-0035-5.

[5] P. Briet, H. Hogreve, *Ann. Phys.* (NY) **306** 159 (2003).

[6] A. A. Balinsky and W. D. Evans, *J. of Comp. Appl. Math.*, **148**, 239 (2002).

[7] O. K. Reity, V. Yu. Lazur, and A. V. Katernoha, *J. Phys. B: At. Mol. Opt. Phys.* **35** 1 (2002).

[8] A. Aguilar, J. S. Thompson, D. Calabrese, A. M. Covington, C. Cisneros, V. T. Davis, M. S. Gulley, M. Halka, D. Hanstorp, J. Sandström, B. M. McLaughlin, and D. J. Pegg, *Phys. Rev. A* **69** 022711 (2004).

[9] J. B. West, *Rad. Phys. Chem.* **70** 275 (2004).

[10] A. M. Covington, A. Aguilar, I. R. Covington, M. F. Gharaibeh, C. A. Shirley, R. F. Phaneauf, I. Álvarez, C. Cisneros, G. Hinojosa, J. D. Bozek, I. Dominguez, M. M. Sant'Anna, A. S. Schlachter, N. Berrah, S. N. Nahar, B. M. McLaughlin, *Phys. Rev. Lett.* **87** 243002 (2001).

[11] S. Schippers, A. Müller, S. Ricz, M. E. Bannister, G. H. Dunn, J. Bosek, A. S. Slachter, G. Hinojosa, C. Cisneros, A. Aguilar, A. M. Covington, M. F. Gharaibeh, and R. F. Phaneauf, *Phys. Rev. Lett.* **89** 193002 (2002).

[12] S. Schippers, A. Müller, S. Ricz, M. E. Bannister, G. H. Dunn, A. S. Slachter, G. Hinojosa, C. Cisneros, A. Aguilar, A. M. Covington, M. F. Gharaibeh, and R. F. Phaneauf, *Phys. Rev. A* **67** 032702 (2003).

[13] A. Müller, R. A. Phaneuf, A. Aguilar, M. F. Gharaibeh, A. S. Schlachter, I. Alvarez, C. Cisneros, G. Hinojosa, and B. M. McLaughlin, *J. Phys. B: At. Mol. Opt. Phys.* **35** L137 (2002).

[14] A. Aguilar, J. B. West, R. A. Phaneuf, R. L. Brooks, F. Folkmann, H. Kjeldsen, J. D. Bozek, A. S. Schlachter, and C. Cisneros, *Phys. Rev. A* **67** 012701 (2003).

[15] J. D. Talman, *Phys. Rev. Lett.* **57** 1091 (1986).

[16] G. L. Malli, *Relativistic and Electron Correlation Effects in Molecules and Solids* (Plenum, New York, 1994).

[17] J. Dolbeault, M. J. Esteban, E. Seré, M. Van Breugel, *Phys. Rev. Lett.* **85** 4020 (2000).

[18] H. M. Quiney, in J. Maruani, C. Minot, R McWeeny, Y. G. Smeyers, S Wilson (Eds.) *New Trends in Quantum Systems in Chemistry and Physics*, Vol. 1, (Kluwer, Dordrecht, 2001) 135.

[19] Gh. E. Draganescu, A. Messina and A. J. Napoli, *J. Opt. B* **4** 240 (2002).

[20] P. Blanchard, *J. Phys. B: At. Mol. Phys.* **7** 1993 (1974).

[21] M. Brack, *Phys. Rev. D* **27** 1950 (1983).

[22] N. Bessis, G. Bessis, and D. Roux, *Phys. Rev. A* **32** 2044 (1985).

[23] M. K. F. Wong and H-Y. Yeh, *Phys. Rev. A* **27** 2300 (1983).

[24] M. K. F. Wong and H-Y. Yeh, *Phys. Rev. A* **27** 2305 (1983).

[25] V. A. Fock, *Z. Phys.* **63** 855 (1930).

[26] O. L. de Lange and R. E. Raab, *Operator Methods in Quantum Mechanics* (Oxford, Clarendon, 1991).

[27] H. N. Núñez-Yépez, J. L. López-Bonilla, A. L. Salas-Brito, *J. Phys. B: At. Mol. Opt. Phys.* **28** L525 (1995).

[28] S. G. Nana Engo, M. G. Kwato Njock, L. C. Owono Owono, G. Lagmago Kamta and O. Motapon, *Phys. Rev. A* **56** 2624 (1997).

[29] H. M. Quiney, H. Skaane and I. P. Grant, *J. Phys. B: At. Mol. Opt. Phys.* **30** L829 (1997).

[30] I. V. Dobrovolska and R. S. Tutik, *Phys. Lett. A* **260** 10 (1999).

[31] J. B. West, *J. Phys. B: At. Mol. Opt. Phys.* **34** R45 (2001).

[32] F. Elboudali and B. Joulakian, *J. Phys. B: At. Mol. Opt. Phys.* **34** 4877 (2001).

[33] R. P. Mart´nez-y-Romero, H. N. Núñez-Yépez, and A. L. Salas-Brito, *J. Phys. B: At. Mol. Opt. Phys.* **33** L367 (2000).

[34] L. C. Owono Owono, M. G. Kwato Njock and B. Oumaro, *Phys. Rev. A* **66** 052503 (2002).

[35] N. M. Avram and Gh. E. Draganescu, *Int. J. Quantum Chem.* **65** 655 (1997).

[36] E. Charro, S. López-Ferrero, and I. Martin, *J. Phys. B: At. Mol. Opt. Phys.* **34** (2001) 4243 (1997).

[37] E. Charro and I. Martin, *J. Phys. B: At. Mol. Opt. Phys.* **35** 3227 (2002).

[38] L. Nagy and L. Végh, *Phys. Rev. A* **46** 284 (1992).

[39] Chang-Yuan Chen, Dong-Shen Sun, Cheng-Lin Liu, *Phys. Lett. A* **317** 80 (2003).

[40] E. Ley-Koo, R. Jáuregui, A. Góngora, and C. F. Bunge, *Phys. Rev. A* **47** 1761 (1993).

[41] F. M. Fernández and E. A. Castro, *Hypervirial Theorems* (Berlin, Springer, 1987).

[42] F. M. Fernández and E. A. Castro, *Algebraic Methods in Quantum Chemistry and Physics* (Boca Ratón, CRC, 1996).

[43] K. Pachucki, M. Puchalski, E. Remiddi, arXiv:physics/0405057 (2004).

[44] J. Morales, *Phys. Rev. A* **36** 4101 (1987).

[45] J. Morales, L. Sandoval, A. Palma, J. L. López-Bonilla, *Chem. Phys. Lett.* **135** 149 (1987).

[46] H. N. Núñez-Yépez, J. L. López-Bonilla, D. Navarrete, and A. L. Salas-Brito, *Int. J. Quantum Chem.* **62** 177 (1997).

[47] P. A. Braun, *Phys. Rev. A* **46** 6108 (1992).

[48] A. C. Ilarraza-Lomel′, M. N. Valdés-Mart′nez, A. L. Salas-Brito, R. P. Mart′nez-y-Romero, and H. N. Núñez-Yépez, *Int. J. Quantum Chem.* **90** 195 (2002).

[49] V. M. Shabaev *J. Phys. B: At. Mol. Opt. Phys.* **24**, 4479 (1991).

[50] D. Andrae, *J. Phys. B: At. Mol. Opt. Phys.* **30** 4435 (1997).

[51] R. P. Mart′nez-y-Romero, H. N. Núñez-Yépez, and A. L. Salas-Brito, *J. Phys. B: At. Mol. Opt. Phys.* **34** 1261 (2001).

[52] R. P. Mart′nez-y-Romero, H. N. Núñez-Yépez, and A. L. Salas-Brito, *J. Phys. B: At. Mol. Opt. Phys.* **35** L71 (2002).

[53] R. P. Mart′nez-y-Romero, H. N. Núñez-Yépez, and A. L. Salas-Brito, *Recent Research Developments in Physics* Vol. 4 (Kirala, Transworld Research Network, 2004), in press.

[54] L. C. Owono Owono, M. L. C. Owona Angue, M. G. Kwato Njock, B. Oumarou, *Phys. Lett. A* (2004) in press.

[55] S. P. Goldman and G. W. F. Drake, *Phys. Rev. A* **25** 2877 (1982).

[56] S. M. Cohen, P. T. Leung, *Phys. Rev. A* **57** 4994 (1998).

[57] F. Rosicky, F. Mark, *J. Phys. B: Atom. Molec. Phys.* **8** 2581 (1975).

[58] G. R. Deco, R. D. Rivarola, *Phys. Rev. A* **39** 5451 (1989).

[59] N. Toshima, J. Eichler, *Phys. Rev. A* **42** 3896 (1990).

[60] I. P. Grant in G. W. F. Drake Editor, *Atomic, Molecular and Optical Physics Handbook*, (American Institute of Physics, Woodbury, 1996) Ch. 32.

[61] R. P. Mart′nez-y-Romero, *Am. J. Phys.* **68** 1050 (2000).

[62] J. N. Bahcall, *Phys. Rev.* **124** 923 (1961).

[63] W. Lucha, F. F. Schöberl, *Phys. Rev. Lett.* **64** 2733 (1990).

[64] Zhen-Fen Luo, Xi-Jun Qiu, *J. Phys. G: Nucl. Part. Phys.* **18** 221 (1992).

[65] H. A. Bethe and E. E. Salpeter *Quantum Mechanics of One- and Two-Electron Atoms* (New York, Academic, 1957).

[66] S. Mannervik, D. DeWitt, L. Engström, J. Lindberg, E. Lindroth, R. Schuch, and W. Zong, *Phys. Rev. Lett.* **81** 313 (1998).

[67] Xiao-Min Ton, Lei Liu, Jia-Ming Li, *Phys. Rev. A* **49** 4641 (1994).

[68] M. Aymar, C. H. Greene, E. Luc-Koenig, *Rev. Mod. Phys.* **68** 1015 (1996).

[69] S. K. Semenov and N. A. Cherepkov, *J. Phys. B: At. Mol. Opt. Phys.* 36 1409 (2003).

[70] P. Weck, O. A. Fojón, J. Hanssen, B. Joulakian, and R. D. Rivarola, *Phys. Rev. A***63** 042709 (2001).

[71] C. Bauche-Arnould, J. Bauche, E. Luc-Koenig, J.-F. Wyart, R. M. More, C. Chenais-Popovics, J.-C. Gauthier, J.-P. Geindre, and N. Tragin, *Phys. Rev. A* **39** 1053 (1989).

[72] J. M. Bang and J. M. Hansteen, *J. Phys. B: At. Mol. Opt. Phys.* **35** 3979 (2002).

[73] R. E. Moss, *Advanced Molecular Quantum Mechanics* (Chapman and Hall, London, 1972)

[74] V. M. Shabaev, *Vestn. Leningrad Univ.* **4** 15 (1984).

[75] I. B. Khriplovich, and A. S. Yelkhoovsky, *Phys. Lett. A* **135** 353 (1988).

[76] I. B. Khriplovich, and A. S. Yelkhoovsky, *Phys. Lett. A* **137** 496 (1989).

[77] G. Gaigalas, arXiv:physics/0405072 May 14 (2004).

[78] Z. Chen, D. Bessis, and A. Z. Msezane, *Phys. Rev. A* **47** 4756 (1993).

[79] J. Kuang and C. D. Lin, *J. Phys. B: At. Mol. Opt. Phys.* **29** L889 (1996).

[80] A. Rutkowski, *Phys. Rev. A* **53** 145 (1996).

[81] M. J. Jamieson, A. Dalgarno, L. Wolniewicz, *Phys. Rev. A* **61** 042705 (2000).

[82] T. A. Isaev, A. N. Petrov, N. S. Mosyagin, A. V. Titov, E. Eliav, U. Kaldor, *Phys. Rev. A* **69** 030501(R) (2004).

[83] O. Kullie, O Düsterhöft, D. Kolb, *Chem. Phys. Lett.* **314** 307 (1999).

In: Trends in Chemical Physics Research
Editor: A.N. Linke, pp. 117-127

ISBN 1-59454-483-2

Chapter 5

THE DIFFERENCE BETWEEN THE HEATS OF CHEMICAL REACTIONS OBTAINED BY THE VAN'T-HOFF EQUATION AND BY CALORIMETRY

I.A. Stepanov [1]
Juglas 1a-20, Riga, LV-1024, Latvia

Abstract

It is shown that the difference between the heat of chemical reaction derived by the Van't-Hoff equation, ΔH^0, and measured by calorimetry, ΔQ^0, is $P\Delta V^0$. An explanation is given.

Keywords: entropy, heat of chemical reactions, the Van't-Hoff equation, chemical thermodynamics, the first law of thermodynamics

1 Introduction

There is a paradox in chemical thermodynamics: the heat of chemical reactions, that of dilution of liquids and that of micelle formation measured by calorimetry and by the Van't-Hoff equation differs significantly [1-3]. The difference is far beyond the error limits. According to thermodynamics, the Van't-Hoff equation must give the same results as calorimetry because the Van't-Hoff equation is derived from the first and the second law of thermodynamics without simplifications. Nobody succeeded to explain this phenomenon. Let us cite [1]: "Historically, the Van't-Hoff isochore has frequently been used as a means of deriving enthalpy change values for reactions of interest from experimental data gathered at a number of temperatures. ... Detailed review of the data in articles for reactive systems where the enthalpy change has been determined both by calorimetric measurement and other

[1] E-mail address: istepanov@email.com.

temperature-change techniques show that there is often a sizeable discrepancy between the two values. ... It is possible to show that improvements in instrumentation and computing power have not overcome earlier difficulties in the estimation of enthalpy values from data taken at several temperatures. Perusal of literature over the last three decades shows that many authors have been aware that obtaining enthalpy values by these techniques is inherently unsatisfactory". In private communication to the author the author of [1] wrote: "There are a number of comprehensive collections of enthalpy data, particularly those authored by R. M. Izatt and J. J. Christensen, which contain all the thermodynamic data for many systems. Further, many of these systems contain data collected by calorimetric determination and data from 1/T estimation methods in the same table. As I examined those it seemed that there was always a consistent difference between the two methods, with the estimates always being lower than the measured numbers".

In the present paper it is shown that

$$\Delta H^0 - \Delta Q^0 = P\Delta V^0 \quad (1)$$

where ΔH^0 is the heat of reaction obtained by the Van't-Hoff equation, and ΔQ^0 is that measured by calorimetry.

2 Results and Discussion

The Van't-Hoff equation is:

$$d/dT \ln K = \Delta H^0/RT^2 \quad (2)$$

Table 1. Dependence of the heat of $N_2O_4=2NO_2$ reaction on the temperature [4]

T,°K	K (atm) [4]	ΔH^0, kJ/mol Eq. (3)	ΔQ^0, kJ/mol [5]	$\Delta H^0-P\Delta V^0$, kJ/mol
273.15	0.01436	-	-	-
293.15	0.09600	59.68	57.30	57.24
303.15	0.2140	59.68	57.25	57.16
323.15	0.9302	59.68	57.15	56.99

In [4] the dependence of $N_2O_4=2NO_2$ reaction equilibrium constant on the temperature was given which was found experimentally (Table1). From Table 1 one sees that the equilibrium constant at 293, 303 and 323 K obeys the equation

$$\ln K = 22.151 - 59677/RT \quad (3)$$

with high accuracy. Values calculated by Eq. (3) differ from experimental ones at 0.1, 0.2 and 0.03%, respectively. The equilibrium constant at 273 K deflects from this relationship. Dependence of K on temperature is a broken line. The points 293, 303 and 323 K are at one segment, 273 K is at the other one. From Eqs. (2) and (3) it is possible to calculate ΔH^0 and

$\Delta Q^0=\Delta H^0 - P\Delta V^0$. It is supposed that gas is ideal. Below it will be shown that it does not lead to big mistakes. These values are given in Table 1. There also the values of the heat of reaction ΔQ^0 are given taken from [5]. One sees a good agreement with Eq. (1). The K in Eq. (3) is the true reaction equilibrium constant which depends on the activities.

In [6] the following reaction was considered:

$$2Zn(gas)+Se_2(gas)=2ZnSe(solid) \tag{4}$$

$$\Delta G_f^0=-723.614+382.97T \text{ kJ/mol}, \qquad 1260<T<1410 \text{ K} \tag{5}$$

As $\ln K=-\Delta G^0/RT$, from Eq. (5) one may find ΔH^0 and $\Delta Q^0=\Delta H^0-P\Delta V^0$. The heat of reaction ΔQ is taken from [7] for T=1300 (Table 2). For T=1400 the data in [7] are absent.

Table 2. The heat of some chemical reactions obtained by the Van't-Hoff equation and by calorimetry

	$2Zn(gas)+Se_2(gas)=2ZnSe(solid)$ [6]		
T,°K	**ΔH^0, kJ/mol Eq. (5)**	**ΔQ^0, kJ/mol [7]**	**$\Delta H^0-P\Delta V^0$, kJ/mol**
1300	-723.614	-690.80	-691.20
	$2HgTe(solid)=2Hg(gas)+Te_2(gas)$ [8]		
T,°K	**ΔH^0, kJ/mol Eq. (7)**	**ΔQ^0, kJ/mol [7]**	**$\Delta H^0-P\Delta V^0$, kJ/mol**
800	348.496	329.50	328.56
900	348.496	325.72	326.06
	$2ZnS(solid)=2Zn(gas)+S_2(gas)$ [9]		
T,°K	**ΔH^0, kJ/mol Eq. (9)**	**ΔQ^0, kJ/mol [7]**	**$\Delta H^0-P\Delta V^0$, kJ/mol**
1100	774.142	753.00	746.72
1200	774.142	750.00	744.22
1300	774.142	746.92	741.74
1400	774.142	743.76	739.24
	$2Te(liquid)=Te_2(gas)$ [10]		
T,°K	**ΔH^0, kJ/mol Eq. (11)**	**ΔQ^0, kJ/mol [7]**	**$\Delta H^0-P\Delta V^0$, kJ/mol**
800	119.76	112.92	113.11
900	116.71	109.41	109.27
1000	114.05	105.97	105.74
1100	114.05	102.59	104.91

In [8] the following reaction was given:

$$2HgTe(solid)=2Hg(gas)+Te_2(gas) \tag{6}$$

$$\ln K=-348496/RT+13.77, \quad 778<T<943 \text{ K} \tag{7}$$

In [9] the following reaction was given:

$$2ZnS(solid)=2Zn(gas)+S_2(gas) \quad (8)$$

$$\ln K=-774142/RT+21.976,\ 1095<T<1435\ K \quad (9)$$

The equilibrium constant K in Eqs. (6) and (8) depends on pressure, not on fugacity. But at a temperature many hundreds grades Kelvin and at the atmospheric pressure one can neglect intermolecular interactions and pressure is very close to fugacity.

In [10] the following reaction was considered:

$$2Te(liquid)=Te_2(gas) \quad (10)$$

$$\ln P=-119.755/RT+11.672469, \quad 722.65<T\leq 800\ K \quad (11)$$

$$-116.706/RT+11.21345, \quad 800<T<921.6\ K$$

$$-114.045/RT+10.8661, \quad 921.6<T<1142\ K.$$

Assuming K=P one can obtain the enthalpy. The heat of these reactions is given in Table 2.

In [11] the following processes were considered:

$$Si(liquid)=Si(gas) \quad (12)$$

$$\log P=\log K=-2.08\cdot 10^4/T+10.84, \quad 1762\leq T\leq 1998K \quad (13)$$

$$2Si(liquid)=Si_2(gas) \quad (14)$$

$$\log P=\log K=-2.46\cdot 10^4/T+10.93, \quad 1842\leq T\leq 1992K \quad (15)$$

The 1st process is evaporation, the 2nd one is a chemical reaction. Their heats are given in Table 3. One sees that the heat of evaporation measured by the Van't-Hoff equation is much closer to the experiment than the heat of the reaction. Evaporation is a physical process, it obeys the traditional formula. The second process obeys Eq. (1).

In [12] the following reaction has been studied:

$$Si(solid)+SiO_2(solid)=2SiO(gas) \quad (16)$$

$$\log P=13.613-1.785\cdot 10^4/T, \quad 1300\leq T\leq 1580K \quad (17)$$

In [14] the following reaction was considered:

$$Al_2Br_6(g)=2AlBr_3(g) \quad (18)$$

$$\ln K=-14077.0/T+147.5, \quad 500\leq T\leq 900K \quad (19)$$

In [15] the following processes were considered:

$$AlBr_3(l) = AlBr_3(g) \tag{20}$$

$$2AlBr_3(l) = Al_2Br_6(g) \tag{21}$$

Table 3. The heat of evaporation and of some chemical reactions obtained by the Van't-Hoff equation and by calorimetry

	Si(liquid)=Si(gas) [11][2]		
T,°K	**ΔH^0, kJ/mol Eq. (13)**	**ΔQ^0, kJ/mol [5, 7]**	**ΔH^0-$P\Delta V^0$, kJ/mol**
1800	397.55	395.78	-
1900	397.55	395.23	-
	2Si(liquid)=Si_2(gas) [11][1]		
T,°K	**ΔH^0, kJ/mol Eq. (15)**	**ΔQ^0, kJ/mol [5, 7]**	**ΔH^0-$P\Delta V^0$, kJ/mol**
1850	470.19	455.57	454.82
1900	470.19	455.05	454.40
1990	470.19	454.11	453.65
	Si(solid)+SiO_2(solid)=2SiO(gas) [12]		
T,°K	**ΔH^0, kJ/mol Eq. (17)**	**ΔQ^0, kJ/mol [13]**	**ΔH^0-$P\Delta V^0$, kJ/mol**
1300	683.10	661.81	661.50
1400	683.10	657.90	659.83
1500	683.10	652.43	658.17
	Al_2Br_6(g)=2$AlBr_3$(g) [14]		
T, K	**ΔH^0, kJ/mol Eq. (19)**	**ΔQ^0, kJ/mol [7]**	**ΔH^0-$P\Delta V^0$, kJ/mol**
500	116.98	113.32	112.83
600	116.98	111.76	111.9
700	116.98	110.17	111.16
800	116.98	108.57	110.33
900	116.98	106.95	109.50
	$AlBr_3$(l)=$AlBr_3$(g) [15]		
T, K	**ΔH^0, kJ/mol**	**ΔQ^0, kJ/mol [7]**	**ΔH^0-$P\Delta V^0$, kJ/mol**
400	85.68	86.19	-
500	82.72	81.72	-
	2$AlBr_3$(l)=Al_2Br_6(g) [15]		
T, K	**ΔH^0, kJ/mol**	**ΔQ^0, kJ/mol [7]**	**ΔH^0-$P\Delta V^0$, kJ/mol**
400	60.06	57.5	56.74
500	53.42	50.12	49.27

The 1st process is a phase transition, the 2nd one is a chemical reaction. Their heats are given in Table 3. One sees that the heat of evaporation measured by the Van't-Hoff equation

[2] Thermodynamic data for Si(liquid) is taken from [7], these for Si(gas) and Si_2(gas) is from [5].

is much closer to the experiment than the heat of the reaction. Evaporation is a physical process, it obeys the traditional formula. The second process obeys Eq. (1).

In [16-21] the following reactions have been studied:

$$2V_2O_5(s)=V_4O_{10}(g)\ [16] \tag{22}$$

$$2SnO_2(s)=2SnO(g)+O_2\ [17] \tag{23}$$

$$2HgO(g)=2Hg(g)+O_2\ [18] \tag{24}$$

$$SiO_2(s)=SiO(g)+O\ [19] \tag{25}$$

$$SiO_2(g)=SiO(g)+O\ [20] \tag{26}$$

$$Hg(g)+Te(l)=HgTe(s)\ [21] \tag{27}$$

$$2Hg(g)+Te_2(g)=2HgTe(s)\ [21] \tag{28}$$

The heat of these reactions is given in Table 4.

Table 4. The heat of some chemical reactions obtained by the Van't-Hoff equation and by calorimetry

Reaction	T, K	ΔH^0, kJ/mol	ΔQ^0, kJ/mol	ΔH^0-$P\Delta V^0$, kJ/mol
$2V_2O_5(s)=V_4O_{10}(g)$ [16]	1300	113.01	105.05 [5]	102.21
	1400	113.01	99.35	101.38
$2SnO_2(s)=2SnO(g)+O_2$ [17]	1300	1184.1	1151.4 [5]	1151.7
	1400	1184.1	1145.5	1149.2
$2HgO(g)=2Hg(g)+O_2$ [18]	500	45.91	39.74 [7]	41.76
	600	45.91	39.90	40.92
	700	45.91	40.05	40.09
	800	45.91	40.21	39.26
$SiO_2(s)=SiO(g)+O$ [19]	1700	1045.0	1022.3 [13]	1016.8
	1800	1045.0	1019.9	1015.1
	1900	1045.0	1015.3	1013.4
$SiO_2(g)=SiO(g)+O$ [20]	1700	458.1	448.40[a]	443.97
	1800	458.1	444.96	443.14
	1900	458.1	441.44	442.31
$Hg(g)+Te(l)=HgTe(s)$ [21]	800	-113.91	-108.30 [7]	-107.26
	900	-113.91	-108.16	-106.43
$2Hg(g)+Te_2(g)=$ $2HgTe(s)$ [21]	800	-345.49	-329.51 [7]	-325.55
	900	-345.49	-325.73	-323.05

One sees that heats of chemical reactions obey Eq. (1). Heats of physical processes obey the traditional formula.

[a] Data for SiO and O are from [13], $H(SiO_2(g))=H(SiO_2(s)$, [13])+ΔH(evaporation, [19])

In [22, 24, 25, 29-31] the following reactions were considered:

$$2Cu_2O=4Cu+O_2 \quad (29)$$

$$2MgO=2Mg(g)+O_2 \quad (30)$$

$$Mo_3Si+SiO_2(\alpha\text{-quartz})=3Mo+2SiO(g) \quad (31)$$

$$\ln K=0.524-803.83/RT, \quad (1512\text{-}1663\ K) \quad (32)$$

$$3Mo_5Si_3+4SiO_2\ (\alpha\text{-quartz})=5Mo_3Si+8SiO(g) \quad (33)$$

$$\ln K=2.052-3064.7/RT, \quad (1408\text{-}1674\ K) \quad (34)$$

$$5MoSi_2+7SiO_2(\alpha\text{-quartz})=Mo_5Si_3+14SiO(g) \quad (35)$$

$$\ln K=3.671-5145.6/RT, \quad 1400\text{-}1500\ K \quad (36)$$

$$2Al_2O_3=Al(g)+AlO(g)+Al_2O(g)+4O[29] \quad (37)$$

$$Al(g)+O=AlO(g)\ [30] \quad (38)$$

$$Al_2O(g)=Al(g)+AlO(g)\ [30] \quad (39)$$

$$2CdTe=2Cd(g)+Te_2(g)\ [31] \quad (40)$$

The heat of these reactions is given in Table 5.

In the present paper it is supposed that PV=RT for one mole. For molecular gases at atmospheric pressure and room temperature PV/RT=1 with a good accuracy [32]. At higher temperatures and lower pressures the accuracy increases. In this paper the biggest part of the reactions are at pressures 1 atmosphere and less, and at temperatures 800-2800 K. In [33] dependence of PV/RT on the temperature and pressure is given for H_2, N_2 and O_2 in the interval 220 K - 570 K. In the whole temperature range PV/RT differs from 1 by less than 10% till pressure about 100 atmospheres.

One sees that heats and enthalpies of chemical reactions obey Eq. (1). The result is important for calculation of heats of chemical processes. An attempt to explain this phenomenon was given in [34-37]. There it is supposed that for endothermic chemical reactions with increasing volume and for exothermic chemical reactions with decreasing volume the law of conservation of energy is

$$\Delta U = \Delta Q + P\Delta V + \Sigma\mu_i\Delta N_i \quad (41)$$

If to derive the Van't-Hoff equation using (41), it will be

$$d/dT\ \ln K = (\Delta Q^0 + P\Delta V^0)/RT^2 \quad (42)$$

Table 5. The heat of some chemical reactions obtained by the Van't-Hoff equation and by calorimetry

Reaction	T, K	ΔH^0, kJ/mol	ΔQ^0, kJ/mol	ΔH^0-$P\Delta V^0$, kJ/mol
$2Cu_2O$=4Cu+O_2 [22]	900	343,58	336.72 [23]	336.10
	1000	343,58	335.38	335.27
	1100	343,58	333.99	334.44
	1200	343,58	332.57	333.61
	1300	343,58	331.64	332.78
2MgO=2Mg(g)+O_2 [24]	1900	1504,2	1462.4 [23]	1456.9
	2000	1504,2	1453.3	1454.3
Mo_3Si+SiO_2=3Mo+2SiO(g) [25]	1600	803,83	776.37 [13, 26]	777.24
$3Mo_5Si_3$+$4SiO_2$=$5Mo_3Si$+ 8SiO(g) [25]	1200	3064,7	2987.1 [13, 26, 27]	2984.9
	1300	3064,7	2966.3	2978.3
	1400	3064,7	2945.4	2971.6
	1500	3064,7	2924.1	2965.0
	1600	3064,7	2902.6	2958.3
$5MoSi_2$+$7SiO_2$=Mo_5Si_3+ 14SiO(g) [25]	1400	5145,6	4984.6 [13, 27, 28]	4982.7
	1500	5145,6	4947.5	4971.1
	1600	5145,6	4909.5	4959.5
$2Al_2O_3$=Al(g)+AlO(g)+ Al_2O(g)+4O [29]	2000	4684.0	4526.1 [7, 23]	4567.7
	2100	4684.0	4519.9	4561.8
	2200	4684.0	4513.5	4556.0
	2300	4684.0	4507.13	4550.2
Al(g)+O=AlO(g) [30]	2000	-544.28	-517.56 [7, 23]	-527.66
	2100	-544.28	-517.18	-526.83
	2200	-544.28	-516.74	-526.0
	2300	-544.28	-516.27	-525.17
	2400	-544.28	-515.75	-524.34
	2500	-544.28	-515.20	-523.51
	2600	-544.28	-513.94	-522.67
	2700	-544.28	-514.11	-521.84
	2800	-544.28	-513.57	-521.01
Al_2O(g)=Al(g)+AlO(g) [30]	2000	564.8	540.86 [7, 23]	548.19
	2100	564.8	541.28	547.35
	2200	564.8	541.76	546.52
	2300	564.8	542.28	545.69
2CdTe=2Cd(g)+Te_2(g) [31]	1100	573.84	550.41 [7]	546.42
	1200	573.84	544.04	543.92
	1300	573.84	537.33	541.43

The enthalpy for SiO_2, SiO and Mo is taken from [13], that for Mo_3Si from [26], that for Mo_5Si_3 from [27], that for $MoSi_2$ is calculated as $\Delta_f H^0(300\ K)+\int C_P dT$, $\Delta_f H^0$ is from [28], C_P is from [13]. The enthalpy for Al(g) is from [7], that for AlO, Al_2O, Al_2O_3, O is from [23].

The first law of thermodynamics for heat exchange is

$$dQ = dH = dU + PdV. \quad (43)$$

Here dH is the change in enthalpy. The author showed that for substances with negative thermal expansion this law reads

$$dQ = dU - PdV \quad (44)$$

[38 - 44].

The following consideration supports the result of this paper. The Van't - Hoff equation (2) was often used for determination of the heat of chemical reactions. It gives correct results for reactions with $\Delta V = 0$. Therefore, it was assumed that for chemical reactions $\Delta Q = \Delta H = \Delta U + P\Delta V$. However, from the traditional thermodynamics it follows that $\Delta Q = \Delta U + P\Delta V - \sum \mu_i \Delta N_i$. One may not neglect the term $\sum \mu_i \Delta N_i$. Thus, the Van't-Hoff equation must not give correct results neither for $\Delta V = 0$ nor for $\Delta V \neq 0$. The theory developed in [34-37] explains this contradiction: in chemical reactions (42) must be used.

References

[1] P. R. Brown, *Proc. 12th IUPAC Conf. on Chem.Thermodynamics*, Snowbird, Utah (1992) pp 238-239.

[2] D. S. Reid, M. A. Quickenden, and F. Franks, *Nature* **224** (1969) 1293.

[3] J. C. Burrows, D. J. Flynn, S. M. Kutay, T. G. Leriche, and D. G. Marangoni, *Langmuir* **11** (1995) 3388.

[4] H. Blend, *J. Chem. Phys.* **53** (1970) 4497.

[5] L. V. Gurvich and I. V. Veitz, *Thermodynamic Properties of Individual Substances*, Hemisphere Pub Co., NY, L. 4th edition (1989 etc).

[6] R. Brebrick and H. Liu, *High. Temp. Mater. Sci.* **35** (1996) 215.

[7] I. Barin, *Thermochemical Data of Pure Substances*, VCH, Weinheim (1989).

[8] R. F. Brebrick and A. J. Strauss, *J. Phys. Chem. Solids* **26** (1965) 989.

[9] W. Hirschwald, G. Neumann, and I. N. Stranski, Z. *Physik. Chem. (NF)* **45** (1965) 170.

[10] R. F. Brebrick, *High Temp. Science* **25** (1988) 187.

[11] T. Tomooka, Y. Shoji, and T. Matsui, *J. Mass Spectrom. Soc. Jpn.* **47** (1999) 49.

[12] O. Kubaschewski and T. G. Chart, *J. Chem. Thermodyn.* **6** (1974) 467.

[13] L. V. Gurvich and V. S. Iorish, IVTANTHERMO - *A Thermodynamic Database and Software System for the Personal Computer*. User's Guide, CRC Press, Inc., Boca Raton, (1993[a]).

[14] A. D. Rusin, *Vestnik MGU. Ser. Khim.* **39** (1998) No 1, 25-29 (www.chem.msu.su:8081/rus/vmgu/welcome.html).

[15] A. D. Rusin, *Vestnik MGU. Ser. Khim.* **39** (1998) No 3, 147-152.

[a] I used data from the Russian version of this database on www.openweb.ru/windows/thermo/index_eng.htm

[16]G. Semenov, K. Frantseva, and E. Shalkova, *Vestnik LGU, Fiz.-Khim.* **16** (1970) No 3, 82.

[17]E. K. Kazenas and D. M. Chizhikov, Davlenie i Sostav Para nad Okislami Khimicheskikh Elementov, *Nauka*, Moscow (1976).

[18]M. V. Gorbachova, *Zh. Fiz. Khim.* **72** (1998) 416-420 [*Russ. J. Phys. Chem.*].

[19]R. H. Lamoreaux, D. L. Hildenbrand and L. Brewer, *J. Phys. Chem. Ref. Data.* **16** (1987) 419-445.

[20]S. Shornikov, I. Archakov and M. Shults, *Russ. J. Gen. Chem.* **70** (2000) 360.

[21]A. Nasar and M. Shamsuddin, *J. Less-Common Metals* **161** (1990) 87-92.

[22]G. B. Barbi, *Gazz. Chim.* **100** (1970) 64.

[23]http://webbook.nist.gov.

[24]E. K. Kazenas, in: Tsvetnye Metally. Nauchnye Poiski i Perspektivy, *Nauka*, Moscow (1975).

[25]T. G. Chart, *Metal Science* **8** (1974) 344-348.

[26]I. Tomaszkiewicz, G. A. Hope, C. M. Beck II, and P. A. G. O'Hare, *J. Chem. Thermod.* **28** (1996) 29-42.

[27]I. Tomaszkiewicz, G. A. Hope, C. M. Beck II, and P. A. G. O'Hare, *J. Chem. Thermod.* **29** (1997) 87-98.

[28]P. A. G. O'Hare, *J. Chem. Thermod.* **25** (1993) 1333-1343.

[29]R. P. Burns, *J. Chem. Phys.* **44** (1966) 3307.

[30]M. Furber, R. D. Srivastava, and O. M. Uy, *J. Chem. Soc. Faraday Trans. pt. 1*, **68** (1972) 249.

[31]J. P. Schwartz, Tse Tung, and R. F. Brebrick, *J. Electrochem. Soc.* **128** (1981) 438-451.

[32]Chemical Encyclopedia, *Sovetskaja Entsiklopedija*, Moscow, vol. 2 (1990) p. 398.

[33]F. Daniels and R. A. Alberty, *Physical Chemistry*, 4th ed., p. 84, Mir, Moscow (1978) Russian translation.

[34]I. A. Stepanov, *Modification of the First Law of Thermodynamics for Chemical Reactions*. DEP VINITI. N37-B96. 03.01-96. Moscow. (Available from VINITI, Moscow, www.viniti.ru).

[35]I. A. Stepanov, *The First Law of Thermodynamics for Chemical Reactions*. DEP VINITI. No 3387-B98. 20.11.98. (Available from VINITI, Moscow, www.viniti.ru).

[36]I. A. Stepanov. Modification of the 1st Law of Thermodynamics for Chemical Reactions.- 7th European Symposium on Thermal Analysis and Calorimetry. Aug. 30 - Sept. 4. Balatonfüred, Hungary. 1998. *Book of Abstracts*. P. 402.

[37]I. A. Stepanov. The 1st Law of Thermodynamics for Chemical Reactions. In: *Trends in Chemical Physics Research*. Ed. A. N. Linke. Nova Science Publishers, Inc. Hauppauge, NY. 2006. In press.

[38]I. A. Stepanov. Thermodynamics of Substances with Negative Thermal Expansion Coefficient. *Computer Modelling & New Technologies*. 2000. V. 4. No 2. 72 - 74. (Available at www.tsi.lv).

[39]I. A. Stepanov. Thermodynamics of Iodine under High Pressure at Low Temperatures. *Physica B*. 2004. V. 349. No 1 - 4. 251 - 253.

[40]I. A. Stepanov, Negative Thermal Expansion of ZrW_2O_8 and HfW_2O_8: Contradiction to a Thermodynamic Relation. In: *Progress in Chemical Physics Research*, Ed. A. N. Linke. Nova Science Publishers, Inc., Hauppauge, NY (2005). 255-262.

[41]I. A. Stepanov. Paradoxes of Thermal Expansion. – In: *Progress in Chemical Physics Research*. Ed. A. N. Linke. Nova Science Publishers, Inc. Hauppauge, NY. 2005, 263-269.

[42]I. A. Stepanov. $(\partial S/\partial V)_U$ is negative for substances with negative thermal expansion. In: *Trends in Chemical Physics Research.* Ed. A. N. Linke. Nova Science Publishers, Inc., Hauppauge, NY, 2005. 120-125.

[43]I. A. Stepanov. Supercooled Water: Contradiction to a Thermodynamic Relation. In: *Trends in Chemical Physics Research.* Ed. A. N. Linke. Nova Science Publishers, Inc. Hauppauge, NY. 2006. In press.

[44]I. A. Stepanov. Calculation of Phase Transitions in Ice which do not Obey the Clapeyron Equation. – Submitted.

In: Trends in Chemical Physics Research
Editor: A.N. Linke, pp. 129-134 ISBN 1-59454-483-2

Chapter 6

$(\partial S/\partial V)_U$ IS NEGATIVE FOR SUBSTANCES WITH NEGATIVE THERMAL EXPANSION

I.A. Stepanov [1]
Institute of Physical Chemistry, University of Freiburg
Albertstrasse 23a, Freiburg, 79104 Germany

Abstract

According to the first law of thermodynamics, $(\partial S/\partial V)_U$ is always positive. According to other equations of thermodynamics, for heat exchange it is positive if the thermal expansion coefficient is positive, and negative if the thermal expansion is negative. A contradiction in thermodynamic equations is shown. It is also shown that this result is successfully used for calculation of phase diagrams for substances with negative thermal expansion (ice, iodine) and for calculation of anomalous dependence of the heat capacity of ZrW_2O_8, HfW_2O_8, water and other substances on pressure. This contradiction is explained.

PACS: 05.70. -a, 05.70. Ce
Keywords: entropy, internal energy, negative thermal expansion coefficient, the first law of thermodynamics

Introduction

One can write

$$\alpha = (1/V)(\partial V/\partial T)_P = -(1/V)(\partial V/\partial P)_T(\partial P/\partial T)_V = \beta(\partial S/\partial V)_T \qquad (1)$$

where α is the thermal expansion coefficient, V is the volume, T is the temperature, S is the entropy and β is the isothermal compressibility $-(1/V)(\partial V/\partial P)_T$. Therefore,

[1] E-mail address: istepanov@email.com, Tel.: 3717-617136, Fax: 3717-820113; Permanent address: I. A. Stepanov, Juglas 1a-20, Riga, LV-1024, Latvia

$$(\partial S/\partial V)_T = \alpha/\beta. \tag{2}$$

If $\alpha < 0$ then $(\partial S/\partial V)_T < 0$.
On the other hand,

$$\delta Q \geq TdS = dU + PdV \tag{3}$$

and

$$(\partial S/\partial V)_U = P/T > 0 \tag{4}$$

for all α.

However, $(\partial U/\partial T)_X \neq 0$ where X = P, V, S, H, F and G [1], H is the enthalpy, F is the Helmholtz energy and G is the Gibbs energy. Therefore, almost always, if U = const then T = const. If U = const then the right sides of Eqs. (2) and (4) must have the same sign. So, there is a certain contradiction in thermodynamic equations.

One has to stress that here heat exchange is considered: one introduces the quantity of heat δQ in substance, and U and T vary. For this case if U changes then T changes, and vice versa, thus if U is a constant then T is a constant, and vice versa. One does not have to confuse this problem with compression of substance without introducing heat. In that case, of course, $(\partial U/\partial V)_T \neq 0$.

In the present paper this paradox is explained.

Theory

According to [1], $(\partial U/\partial T)_P = C_P - \alpha PV$, $(\partial U/\partial T)_V = C_V$, $(\partial U/\partial T)_S = C_V\beta P/\alpha T$,

$$(\partial U/\partial T)_H = ((C_P - \alpha PV) - C_V\beta P)/(1 - \alpha T) \tag{5}$$

$$(\partial U/\partial T)_F = (\beta SP - \alpha TS + C_V\beta P)/\beta P \tag{6}$$

$$(\partial U/\partial T)_G = C_P - \alpha PV - (\alpha T - \beta P)S \tag{7}$$

So, usually, dU/dT is not equal to 0. For the biggest part of the processes $dU/dT \neq 0$ thus if U = const then T = const and a contradiction does exist.

For example,

$$dU = (\partial U/\partial P)_V dP + (\partial U/\partial V)_P dV = \beta/\alpha \cdot C_V dP + (C_P - \alpha PV)/\alpha V \cdot dV \tag{8}$$

[1]. From this, if $dU/dT = 0$ then

$$dP/dV = (\alpha PV - C_P)/\beta C_V V. \tag{9}$$

For every process which does not satisfy Eq. (9), $dU/dT \neq 0$ and if U = const then T = const.

For water at 273 K, $\alpha = -68.05 \cdot 10^{-6}$ K^{-1}, $C_P = 4217.6$ J(kg/K), $C_P - C_V \approx 2.5$ J(kg/K) and $\beta = 5.09 \cdot 10^{-10}$ Pa^{-1} [2]. Substituting these values in Eq. (9) one gets $dP/dV = 1/\beta V \sim 10^{12}$ $Pa \cdot kg/m^3$ with the highest accuracy. But for heat exchange at constant pressure, dP = 0 thus Eq. (9) is not true. Only for such very seldom case when $dP/dV \sim 10^{12}$ $Pa \cdot kg/m^3$, U is a constant when $T \neq$ const.

Another interesting example. Recall that

$$(\partial U/\partial H)_P = (C_P - \alpha PV)/C_P \tag{10}$$

and

$$(\partial U/\partial P)_H = (C_V\beta PV - (C_P - \alpha PV)V)/C_P \tag{11}$$

[1]. For water with the highest accuracy $(\partial U/\partial H)_P = 1$ and $(\partial U/\partial P)_H = -V$. Therefore, if $dU/dT = 0$ then

$$dP/dH = 1/V = 10^3 \ Pa \cdot kg/m^3 \tag{12}$$

For heat exchange at constant pressure, $dH > 0$ but $dP = 0$, thus Eq. (12) is not true.

Next, recall that

$$dT = (\partial T/\partial P)_V dP + (\partial T/\partial V)_P dV = \beta/\alpha dP + 1/(\alpha V)dV \tag{13}$$

[1]. If $dT/dU = 0$ then from (13) one can find $dP/dV = -1/\beta V \sim -10^{12}$ $Pa \cdot kg/m^3$. Only for processes with such steep dependence P(V), dT = const when $dU \neq$ const.

So, it is proven that for condensed matter in almost all cases, if U is a constant then T is a constant, and vice versa. It means, that for those cases

$$(\partial S/\partial V)_T = \alpha/\beta = (\partial S/\partial V)_U \tag{14}$$

Using (3), (4) and (14) one has to come to conclusion that if $\alpha < 0$ then $TdS = dU - PdV$. Really, from a well-known thermodynamical formula [1]

$$(\partial x/\partial y)_z = (\partial x/\partial y)_w + (\partial x/\partial w)_y(\partial w/\partial y)_z \tag{15}$$

it follows that

$$(\partial S/\partial V)_U = (\partial S/\partial V)_T - (\partial U/\partial V)_T/T. \tag{16}$$

If U is a constant then $(\partial U/\partial V)_T = 0$ and (14) holds.

According to [1], $(\partial U/\partial V)_T = (\alpha T - \beta P)/\beta$. If to introduce this into (16), one obtains

$$(\partial S/\partial V)_U = (\partial S/\partial V)_T - (\alpha/\beta - P/T) \quad (17)$$

This equation has only one solution: $\alpha/\beta = P/T$. It is clear if to introduce it into the brackets of (17) and compare with (2) and (4). If to suppose that $\alpha/\beta = P/T + \Delta$ and to introduce it to (17), then (17) turns to identity and is not informative at all.

Else, if in the left part of (16) U = const then in the right part of (16) U = const. If in calorimeter U is a constant and a scientist wants to calculate $(\partial S/\partial V)_U$ according to (16), he has to assume that U = const in both parts of (16).

Results and Discussion

So, one can make a conclusion: in this paper it is proven that for negative α, $(\partial S/\partial V)_U$ is always negative and equals -P/T. One can apply this result to calculation of phase diagrams and $(\partial C_P/\partial P)_T$.

The Clapeyron equation

$$dT/dP = T(V_2 - V_1)/\Delta H \quad (18)$$

is derived from the first and second laws of thermodynamics without simplification and must give correct results. However, for substances with negative thermal expansion coefficient it often gives wrong results deflecting greatly from the experimental values [3, 4]. Even the sign of dT/dP is often wrong. If to derive the Clapeyron equation assuming that $(\partial S/\partial V)_U = -P/T$ for $\alpha < 0$, then it gives correct results for phase transitions in ice [3]. The new equation will have the form

$$dT/dP = T(\text{sign}\alpha_2 V_2 - \text{sign}\alpha_1 V_1)/\Delta H \quad (19)$$

Here α_1 and α_2 are thermal expansion coefficients of the two phases. This modified Clapeyron equation was successfully applied to phase transitions in iodine under high pressure [4].

There is an equation:

$$(\partial C_P/\partial P)_T = -TV((\partial\alpha/\partial T)_P + \alpha^2) \quad (20)$$

It is also derived from the first and second laws of thermodynamics without simplification [11]. However, for substances with negative thermal expansion (ZrW_2O_8, HfW_2O_8, water etc.), it gives wrong results and the following equation is true:

$$(\partial C_P/\partial P)_T = TV((\partial\alpha/\partial T)_P + \alpha^2) \quad (21)$$

[5, 6] (notice the change of sign with respect to (20)). One can obtain this equation if in the derivation of (20) to suppose that $(\partial S/\partial V)_U < 0$ for $\alpha < 0$.

The idea that $(\partial S/\partial V)_U = -P/T$ for $\alpha < 0$ was firstly declared in [2]. (There is a misprint in Eq. (11) in [2]. The correct equation is: for $\alpha < 0$, $C_V = C_P k_S/k_T$).

The only reason why it is still supposed that $(\partial S/\partial V)_U > 0$ for $\alpha < 0$ is that the Clapeyron equation gives the correct result for melting of ice. Let us scrutinize melting of ice I_h. Water has negative expansion till $P \approx 14$ MPa [7]. At this pressure ice melts at $T \approx 272$ K. In [8] a few dT/dP values are given in the interval 272 - 273 K, however, they are calculated by interpolation and not found experimentally. The data in [8] are based on experimental values [9, 10]. In [10] there are no experimental points in this interval but in [9] there is only one: 4.9 MPa and T = 272.8 K. It obeys (18): dT/dP = -0.074 K/MPa. Equation (19) gives dT/dP = -1.71 K/MPa. Nevertheless, one may not confirm that (19) is not true for melting of ice. One can strongly suggest that the result that water has negative expansion at 272.8 K was obtained by an indirect measurement. It means that another value was measured and then the thermal expansion coefficient was calculated from it assuming that $(\partial S/\partial V)_U > 0$.

The following consideration supports the conclusion made in this paper. There is the Mayer's relation [11]:

$$C_P - C_V = T\alpha^2 V/\beta \tag{22}$$

It is derived from the first and second laws of thermodynamics without simplification. If to derive it assuming that $(\partial S/\partial V)_U = -P/T$ for $\alpha < 0$, then it will look like

$$C_P - C_V = - T\alpha^2 V/\beta < 0 \tag{23}$$

[2]. Let us consider thermal expansion of a solid. In the first approximation, V is constant. Then, in the first approximation,

$$C_P = (\partial H/\partial T)_V = C_V + \alpha V/\beta \tag{24}$$

[1]. From (24), at $\alpha < 0$, $C_P < C_V$ and (23) is true, not (22).

In the second approximation, P = const. Let us suppose that in that case $C_P > C_V$. There is the Rolle's lemma: If f(x) is continuous in the interval $a \le x \le b$, and if $f(a) > 0$ and $f(b) < 0$ then there exists at least one value of x (say x = c) such that f(c) = 0, where $a < c < b$. As heat capacity is a continuous function, from the Rolle's lemma it follows that there exists one value of V (say $V_c = V_0 + \Delta V$, V_0 is the constant V in (24)) such that $C_P(V_c) - C_V = 0$. It is a contradiction because from (22), at $\alpha < 0$, $C_P(V) - C_V \neq 0$.

Else, if P = const, for $V_0 - \Delta V < V < V_0 + \Delta V$ and $\alpha < 0$, due to continuity of heat capacity, $C_P - C_V < 0$. It is another contradiction to (22). So, it is proven that for negative thermal expansion (23) is true.

An important result is obtained in [12]. There it is found that the difference between the heats of chemical reactions measured by the Van't-Hoff equation, ΔH^0, and obtained by calorimetry, ΔQ^0, is $P\Delta V^0$. The Van't-Hoff equation must give the same results as calorimetry because it is derived from the first and the second laws of thermodynamics without simplifications. But the difference between both methods is far beyond the error limits. Nobody succeeded to explain this phenomenon. In [12] it is supposed that for endothermic chemical reactions with increasing volume and for exothermic chemical reactions with decreasing volume the law of conservation of energy is

$$\Delta U = \Delta Q + P\Delta V + \Sigma\mu_i\Delta N_i \tag{25}$$

(notice the change of sign in front of P and that $(\partial S/\partial V)_{U,Ni} = -P/T < 0$), here μ_i are chemical potentials and ΔN_i are changes in the number of moles of the substance i. If to derive the Van't-Hoff equation using (25) then it will be

$$d/dT \ln K = \Delta H^0/RT^2 = (\Delta Q^0 + P\Delta V^0)/RT^2. \tag{26}$$

References

[1] Karapetyants M. H., *Himicheskaja Termodinamika.* (Vysshaja Shkola, Moscow) 1974, pp. 108-110. [English translation: Karapetyants M. Kh. *Chemical Thermodynamics.* (Translated by G. Leib), (Mir, Moscow) 1978].

[2] Stepanov I. A., *Computer Modelling & New Technologies*, **4** (2) (2000) 72-74. (Available at www.tsi.lv).

[3] Stepanov I. A., http://preprint.chemweb.com/physchem/0306007. A much larger version of this paper is submitted to publication.

[4] Stepanov I. A., *Physica B*, **349** (2004) 251-253.

[5] Stepanov I. A., Negative Thermal Expansion of ZrW_2O_8 and HfW_2O_8: Contradiction to a Thermodynamic Relation. In: *Progress in Chemical Physics Research*, Ed. A. N. Linke. Nova Science Publishers, Inc. Hauppauge, NY, 2005. 255-262.

[6] Stepanov I. A., Supercooled Water: Contradiction to a Thermodynamic Relation. In: *Trends in Chemical Physics Research*, Ed. A. N. Linke. Nova Science Publishers, Inc. Hauppauge, NY, 2006, accepted.

[7] http://webbook.nist.gov.

[8] *CRC Handbook of Chemistry and Physics*, 80th Ed. CRC Press, Inc, Boca Raton, Florida. 1999-2000.

[9] Henderson S. J. and Speedy R. J., *J. Phys. Chem.* **91,** 3069 – 3072 (1987).

[10]Bridgman P. W. *Proc. Am. Acad. Arts Sci.* **47,** 441 (1912).

[11]Guemez J., Fiolhais C. and Fiolhais M. (1999) *Am. J. Phys.* **67**, 1100-1104.

[12]Stepanov I. A., The Difference Between the Heats of Chemical Reactions Obtained by the Van't-Hoff Equation and by Calorimetry. In: *Trends in Chemical Physics Research*, Ed. A. N. Linke. Nova Science Publishers, Inc. Hauppauge, NY, 2005. 110-119.

In: Trends in Chemical Physics Research
Editor: A.N. Linke, pp. 135-154
ISBN 1-59454-483-2

Chapter 7

Inner-Sphere Reorganization Energies of Outer-Sphere Self-Exchange Reactions Utilizing Density Functional Theory

Robert G. Endres
Center for Computational Sciences and Computer Science &
Mathematics Division, Oak Ridge National Laboratory,
Oak Ridge, TN 37831-6164, USA

Abstract

Ab initio density functional theory is increasingly applied to accurately estimate parameters for the Marcus theory of electron transfer. In particular, the conceptually simple outer-sphere self-exchange of transition-metal ions in aqueous solution has been a frequent choice. Although transition metal ions only exist in trace quantities in natural aquatic environments, their redox cycling in the oceans and atmosphere is geo- and biochemically important. The experimental determination of their reaction rates is furthermore extremely tedious, partly because different transition-metal ions can have rates that vary by several orders of magnitude. A convenient theoretical approach consists of separating the various contributions to the electron transfer rate into a pre-exponential and an exponential, Arrhenius-type factor, $\exp(-\Delta G^*/kT)$, with ΔG^* being the important activation free energy. This article reviews contributions from density functional theory to the study of outer-sphere self-exchange reactions. The calculation of activation energies (and of related reorganization energies) is our main focus. A variety of approaches will be discussed, such as the George-Griffith-Marcus method, the Klimkāns and Larsson method, the polynomial curve-fitting method, as well as the Hessian matrix method.

1 Introduction

Transition metal ions and their redox chemistry are of fundamental importance in bio- and geochemistry. The catalytic properties of metal ions play a crucial role in molecular evolution scenarios [1], and make up the redox-active sites of today's metalloproteins [2]. They also carry a structural role in protein enzyms, in particular in the common zinc-finger proteins [3]. Although transition metal ions occur only in trace-quantities in aqueous environments ($< 0.1\mu$M), they are essential micronutrients for phytoplankton in surface seawater and even constrain global carbon fixation [4]. They also play a crucial role in aqueous-phase atmospheric chemistry, and participate in a variety of redox reactions in natural waters [5]. Knowing their electron transfer rates is important, even if they are slow. Rate limiting reaction steps control the overall rate, and are important for specification and transport of contaminants in the environment [6]. Some oxidation states (e.g. Cr(VI)) are thought to be carcinogenic and their release in the environment constitutes a potential health risks [7].

The subject of this article is the self-exchange electron transfer reactions of selected transition metal ions in aqueous solution. Self-exchange reactions have been studied experimentally [8, 9] and theoretically [10, 11] for about 50 years. There is an inherent two-fold simplicity [12]. Since reactants and products are identical, these reactions do not have a thermodynamic driving force. Second, no chemical bonds are broken or formed. On the theoretical side, self-exchange reactions were initially examined in terms of the empirical ligand field theory. Later, calculations were performed using semi-empirical quantum chemical electronic structure schemes and the fully *ab initio* Hartree Fock (HF) theory (UHF for unrestricted open-shell calculations) [13]. In the latter approach, electronic correlations are often included at the Møller-Plesset second-order perturbation theory level (MP2 or UMP2). Only during the last decade has the popular *ab initio* density functional theory (DFT) been increasingly applied, mainly in order to obtain ground state and spin-excited state energies, nuclear activation and inner-sphere reorganization energies, as well as metal-ligand stretching frequencies [15, 14].

This review is structured as follows. In section 2 we give an overview of some experimentally determined rates and their interpretation in terms of the Marcus theory of electron transfer. In section 3 we talk in detail about outer-sphere electron transfer. Here we define certain concepts such as free energy surfaces, activation and reorganization energies. In section 4 we review computational methods of inner-sphere activation and reorganization energies based on *ab initio* DFT electronic structure calculations. We discuss the George-Griffith-Marcus method (Sec. 4.1), the Klimkāns and Larsson method (Sec. 4.2), the polynomial curve-fitting method (Sec. 4.3), and the Hessian matrix method (Sec. 4.4). We finally conclude with a summary in section 5.

2 Experiments and their Interpretation

Basic facts about the ion hydration chemistry are well established from X-ray scattering and extended X-ray absorption fine structure (EXAFS) studies, as well as from vibrational spectra (infra-red and Raman spectroscopy). A large number of divalent and trivalent transition metal ions $M^{2+/3+}$ exit as monomeric hexaaqua- and hexaammine species with an octahedrally coordinated first solvation shell of six solvent molecules or ligands. Among the ions shown in table 1, Cr^{2+} and Mn^{2+} are strongly tetragonally distorted through the Jahn-Teller effect, while Co^{2+} has a tendency to be tedrahedrally coordinated (for an excellent review see Ref. [18]).

Table 1: **Selected properties of some octahedrally coordinated aqua ions** (data compiled from Refs. [16, 17, 18, 6], newer data was preferred). Shown are the first row transition metal ions V, Cr, Mn, Fe, and Co, and the second row ion Ru, their metal-ligand bond length difference $\Delta r = r_{2+} - r_{3+}$, their total activation free energy ΔG^*, the ionic strength μ, and their rate constant k. The listed hexaammine complexes are also in aqueous solution. Alternative quasi-experimental reorganization energies can be found in Ref. [19].

Redox couple	d-orbitals	Δr $[A]$ [1]	$\Delta G^*[kcal/mol]$	μ [M]	$k[(Ms)^{-1}]$
$V(H_2O)_6^{2+/3+}$	t_{2g}^3/t_{2g}^2	0.13	16.1	2	1.0×10^{-2}
$Cr(H_2O)_6^{2+/3+}$	$t_{2g}^3e_g^1/t_{2g}^3$	0.09_{eq} [2] 0.36_{ax}	20.5	1	2.0×10^{-5}
$Mn(H_2O)_6^{2+/3+}$	$t_{2g}^3e_g^2/t_{2g}^3e_g^1$	0.18	19.0	3	3.0×10^{-4}
$Fe(H_2O)_6^{2+/3+}$	$t_{2g}^4e_g^2/t_{2g}^3e_g^2$	0.14	13.0	0.6	4.2
$Co(H_2O)_6^{2+/3+}$	$t_{2g}^5e_g^2/t_{2g}^6$	0.21	13.3 [3] - 21.7 [4]	0.5 [5], 3 [6]	3.3
$Co(NH_3)_6^{2+/3+}$	$t_{2g}^5e_g^2/t_{2g}^6$	0.22	23.2	1.0	$10^{-7} - 10^{-5}$ [7]
$Ru(H_2O)_6^{2+/3+}$	t_{2g}^6/t_{2g}^5	0.09	10.5	1.0	$20 - 10^2$
$Ru(NH_3)_6^{2+/3+}$	t_{2g}^6/t_{2g}^5	0.04	7.4	0.1	3.2×10^3

The experimental self-exchange rates of a limited number of redox couples are shown in table 1. These rates cover a range of 10 orders of magnitude, and have been subject to a lot of speculation. The experiments are generally tedious and error-prone with resulting rates having large error bars of about ± 2 orders of magnitude [18]. Forgotten competing side reactions lead to an underestimation of the rate [21]. Often, fast self-exchange rates can only be estimated from cross-reactions between different metal species. Slower reactions can be measured using isotope labeling in such methods as NMR. The rates from table 1 follow certain trends which are correlated with the change in metal-ligand bond length Δr, and the electronic configuration. If the electron transfer involves changes in the occupation of anti-bonding e_g d-orbitals pointing toward the ligands, like for the first row transition metal ions, the metal-ligand changes are generally larger. This implies that a large reorganization energy is necessary to convert electron donors to acceptors and vica versa, leading to a small rate. On the other hand, the second row redox couple $Ru^{2+/3+}$ is all low-spin leading to relatively large rates. This rate is further enhanced by the more extended 4d orbitals.

The experimental rates are generally rationalized in terms of the classical theory of electron transfer developed by Marcus in 1956 [22, 23] and by Hush [24, 25]. Also early on, Levich and Dogonadze formulated a quantum mechanical theory for condensed systems [26, 27, 28]. The theory of electron transfer in solution was later extended to a semi-quantum mechanical [29] and a fully quantum mechanical approach [30]. Although Marcus theory is subject of the next section, the main idea is as follows. The rate constant is factorized in a pre-exponential and an exponential factor. The former depends on the electronic donor-acceptor coupling, while the latter, the nuclear Frank-Condon factor, can be expressed as $exp(-\Delta G^*/kT)$ where ΔG^* is the activation free energy [8], k is the Boltzmann constant, and T is the temperature. This free energy barrier is due to a reorganization of the environment necessary to conserve energy during electron transfer. The variations of the ΔG^* values from table 1 are mainly due to changes in metal-ligand bond lengths Δr of the first solvation shell, and can be estimated from experimental stretching frequencies (see Sec. 4.1) or from the experimental rates and the Marcus formula Eq. 8 (see below). Activation energies of self-exchange reactions can also be roughly estimated from reorganization energies of bimolecular cross-reactions (using the Marcus cross relation) [16, 31]. If both reactants are stable (half-lives greater than $10s$), reorganization energies can be measured using electron spin resonance (ESR) line-broadening or scanning tunneling microscopy (STM) [32, 31]. For less stable molecules, electrochemical techniques can be applied [33, 34], but have led to an inconsistent picture [31]. Useful for unstable redox couples in solution has been an approach which directly allows the determination of total reorganization energies. This approach is based on photoemission experiments with photon energies in the range of 6 - 11 eV [35].

Although the Marcus theory has contributed a lot of valuable insights into the physical principles of electron transfer, the quantitative calculation of actual rates is hindered by uncertainties of the actual reaction mechanism to be modeled. The Marcus theory is mainly a theory of outer-sphere electron transfer meaning that both first solvation shells stay intact during the encounter-complex formation. Deviations from this ideal scenario are expected to occur, e.g. for $Co(H_2O)_6^{2+/3+}$. A favorable inner-sphere pathway due to partial destruction of the first solvation shells could involve a bridging water ligand [36]. This notion is supported by the fact that $Co(H_2O)_6^{3+}$ is strongly oxidizing [36], and that water ligand exchange is fast (10^6 times faster than ammonium exchange [18]). Furthermore, it is possible that the reaction involves thermally excited spin-states, rather the ground states. This is particularly important when the reaction via ground states is spin-forbidden like for $Co(H_2O)_6^{2+/3+}$ and $Co(NH_3)_6^{2+/3+}$ [37, 38, 39, 40, 20].

3 Theory of Outer-Sphere Electron Transfer

The key aspects of electron transfer theory are conveniently introduced by figure 1. It shows simplified free energy surfaces (FESs) of the reactants plus their environment (R) as well as of the products plus their environment (P) as a function of a nuclear reaction coordinate. In order to go from the reactants' FES (R) to the products' surface (P), a transition state has to be passed requiring the earlier mentioned activation energy ΔG^*. In the diabatic limit, when there is no electronic coupling between donor and acceptor ($H_{DA} = 0$) the R-FES and P-FES intersect and no electron transfer is possible. In contrast, for finite electronic coupling, the R-FES and P-FES split at the transition state by an amount $2H_{DA}$. In the

[8] For brevity, from now on we refer to the "activation free energy" simply as the "activation energy". However, in order to assure correctness we use G for Gibbs free energies and E for potential energies.

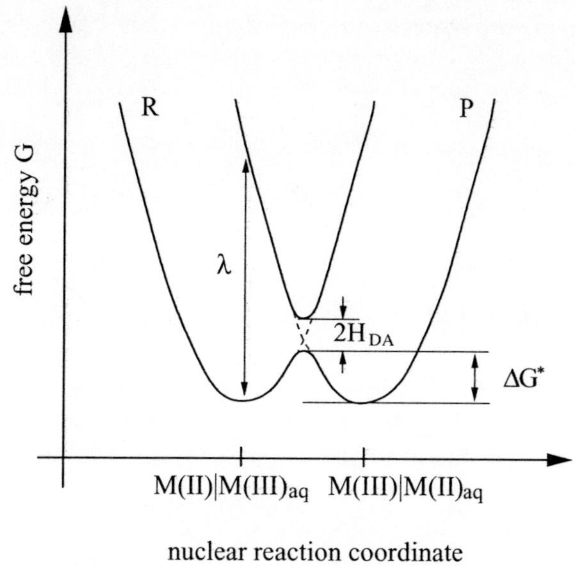

Figure 1: Schematic profiles of the many-dimensional Gibbs free energy surfaces of the reactants plus environment (R) and of the products plus environment (P). If the system has N degrees of freedom, then the reaction path, as well as the intersection of the two surfaces are $(N-1)$ dimensional surfaces.

adiabatic limit (typically $H_{DA} > kT \approx 200cm^{-1}$ at room temperature [39, 41]), electron transfer occurs always when the transition state is reached. Also included in Fig. 1 is the reorganization energy λ defined as the deformation energy of P-FES at the nuclear equilibrium configuration of the R-FES. Equivalently, λ is defined as the vertical (optical) transition energy to go from R-FES to P-FES.

As for the case of electron transfer between di- (II) and trivalent (III) metal ions (M) in solution, the self-exchange reaction with rate constant k_{et} can generally occur over a wide range of ion separations R with varying solvent coordination numbers and ways of encounter. The overall observable, weighted rate constant k_{obs} can be expressed in terms the pair distribution function $g(R)$ as [16]

$$k_{obs} = \int 4\pi R^2 g(R) k_{et}(R) dR \quad (1)$$

$$\approx K_{pre}(\bar{R}) k_{et}(\bar{R}), \quad (2)$$

where $\bar{R}$ is the value of separation at which the integrand has a maximum. K_{equ} is the pree-quilibrium constant of forming the encounter complex [16], and re ects the electrostatic work required to bring the two reactants together by a diffusive process. Eq. 2 can be further separated into the following five steps [42]:

$$M(II) + M(III) \xrightarrow{K_{pre}} M(II)|M(III) \tag{3}$$
$$M(II)|M(III) \xrightarrow{\nu_n \exp(-\Delta G^*/kT)} [M(II)|M(III)]^* \tag{4}$$
$$[M(II)|M(III)]^* \xrightarrow{\Gamma_n \kappa_{el}} [M(III)|M(II)]^* \tag{5}$$
$$[M(III)|M(II)]^* \xrightarrow{(fast)} M(III)|M(II) \tag{6}$$
$$M(III)|M(II) \xrightarrow{(fast)} M(III) + M(II) \tag{7}$$

After formation of the encounter complex in Eq. 3, Eq. 4 denotes activation to the transition state with the nuclear attempt frequency ν_n (s^{-1}) driven by thermal uctuations. Eq. 5 is the actual electron transfer within the complex to form the successor complex where $\Gamma_n \geq 1$ and $\kappa_{el} \leq 1$ are the nuclear tunneling factor and electronic transmission coefficient, respectively. At room temperature, we have typically $\Gamma \approx 1$ [43]. The size of electronic transmission factor depends on the electronic donor-acceptor coupling H_{DA}. For the classical adiabatic Marcus theory, $\kappa_{el} \approx 1$, while for the diabatic situation, $\kappa_{el} << 1$ [39, 44]. This is in particular true for spin-forbidden processes [37]. Finally, Eq. 6 is the nuclear relaxation, and Eq. 7 is the dissociation of the complex into separated products. Both steps are assumed to occur fast. In summary, we can write the rates in the separable semiclassical approximation as [29]

$$k_{obs} = K_{pre}\, \nu_n\, \Gamma_n\, \kappa_{el}\, e^{-\Delta G^*/kT}. \tag{8}$$

For conceptual convenience only, the activation barrier is generally regarded to have an inner-sphere and an outer-sphere contribution. The inner-sphere contribution ΔG_i^* stems from changes of the metal-ligand bonds, while the outer-sphere contribution ΔG_o^* is attributed to polarization of solvation layers further outside the complex. Marcus found a simple expression of the activation energy just in terms of the driving force $\Delta G^o(\leq 0)$ and the reorganization energy $\lambda(> 0)$. Assuming that the FESs are harmonic along the reaction path (as illustrated in Fig. 1), one can write

$$\Delta G^* = \frac{(\lambda + \Delta G^o)^2}{4\lambda} \tag{9}$$
$$\rightarrow \lambda/4 \quad \text{at zero driving force.} \tag{10}$$

The second line corresponds to homogeneous self-exchange reactions. Neglecting interactions between the inner-sphere complex and the outer-sphere environment, one often writes $\Delta G^* = \Delta G_i^* + \Delta G_o^*$, and analogously $\lambda = \lambda_i + \lambda_o$. This separation implies further $\Delta E_i^* = \lambda_i/4$ and $\Delta G_o^* = \lambda_o/4$, where inner-sphere entropic contributions are generally neglected. This was demonstrated to be true for the case where inner-sphere vibrations are harmonic and the activated complex and the reactants have the same frequencies [45, 16]. Deviations from the $\Delta E_i^* = \lambda_i/4$ formula can generally be expected due to anharmonicities. The outer-sphere reorganization energy λ_o is typically calculated with the classical Born dielectric continuum model [22, 23, 16].

In the following we neglect entropy contributions and use solely the potential energy and corresponding potential energy surfaces (PESs) to calculate inner-sphere reorganization energies.

4 Computational Methods of Inner-Sphere Activation

Considering the above mentioned difficulties of determining reorganization energies and hence activation energies from experiments, quantum chemical calculations seem the only possibility [31]. In particular, inner-sphere reorganization energies, which correspond to the total reorganization energy of gas-phase electron transfer, are relatively accessible with DFT electronic-structure codes. These tools have proved to be extremely useful and computationally efficient for systems with a relatively large number of atoms that are difficult to treat with *ab initio* Hartree-Fock (HF) theory [47, 46]. Furthermore, DFT automatically includes certain electron correlations, which are expected to play an important role [47, 46]. In what follows we review a variety of computational schemes.

4.1 George-Grif th-Marcus Method

Traditionally, the inner-sphere contribution is obtained with the classical formalism by George and Griffith [48], as well as by Marcus [49]. This method was originally dependent on vibrational spectroscopic data, which was not abundantly available, but can equally be used with computational data.

In the harmonic approximation, the inner-sphere reorganization and activation energies are

$$\lambda_i = 6\left\{\frac{1}{2}\left[(k_{2+}+k_{3+})\Delta r^2\right]\right\} \tag{11}$$

$$\Delta E_i^*(r^*) = 6\left\{\frac{1}{2}k_{2+}(r_{2+}-r^*)^2+\frac{1}{2}k_{3+}(r_{3+}-r^*)^2\right\}. \tag{12}$$

The transition state bond length r^* can be obtained by minimizing the activation energy Eq. 12, which yields [46]

$$r^* = \frac{k_{2+}r_{2+}+k_{3+}r_{3+}}{k_{2+}+k_{3+}}. \tag{13}$$

With this value for r^*, the inner-sphere activation energy Eq. 12 can be written in terms of a reduced force constant $\bar{k} = 2k_{2+}k_{3+}/(k_{2+}+k_{3+})$, *viz.* [50, 16]

$$\Delta E_i^* = \frac{3}{2}\bar{k}\Delta r^2. \tag{14}$$

Even though the vibrations of each ion are treated as harmonic, the PES along the reaction path is not parabolic. Hence, one can see deviations from $\Delta E_i^* = \lambda_i/4$. Using Eqs. 11 and 14, one obtains

$$\lambda_i = \Delta E_i^*\left[4+\frac{(k_{2+}-k_{3+})^2}{k_{2+}k_{3+}}\right]. \tag{15}$$

Analogue discrepancies have been obtained numerically using more realistic anharmonic and Morse potentials instead. The relation $\Delta E_i^* \approx \lambda_i/3$ has been obtained [19]. However, it was shown that when the solvent is treated explicitly as a collection of dipoles, the projection of the many-dimensional free energy surfaces along the reaction path is indeed parabolic (see review [12]).

In order to calculate inner-sphere activation energies with Eq. 14, Δr and Hooke's force constants $k_{2+/3+}$ are needed. Originally, estimates for Δr were obtained from diffraction data, and the $k_{2+/3+}$ were derived from [16, 51]

$$k_q^{expt.} = 4\pi^2 c^2 \omega_q^2 (\mu/N_A), \quad (q = 2+ \ or \ 3+), \tag{16}$$

where μ is the reduced mass (generally set equal to the molecular weight of a single ligand [29]), c is the speed of light in vacuum, N_A is the Avogadro number, and ω_q is the experimentally obtainable symmetric stretching frequency of the ion with charge q. In the absence of experimental data, the frequencies ω_q are often standard output of DFT packages using the frozen phonon approximation [52, 53, 54] or linear response schemes [55]. Alternatively, k_q can be obtained by hand from the relation

$$k_q^{theor.} = \left. \frac{\partial^2 E_q(r)}{\partial r^2} \right|_{r=r_q}, \tag{17}$$

where $E_q(r)$ is the potential energy of the complex as a function of the ion-ligand bond length. This effectively involves plotting and fitting $E_q(r)$ to a parabola.

In order to compute force constants of the bonds, methods like locked-dipole orientation (LDO), improved average dipole orientation (IADO), the semiclassical perturbed rotation state (PRS) model [51, 56] and the intermediate neglect of differential overlap (INDO/2-MO) [57] were employed. Recently, calculations on $M(NH_3)_6^{2+/3+}$ $(M = Mn, Fe, Co)$ were carried out with the Amsterdam density functional (ADF) program package [41] using a triple-zeta basis set with polarization orbitals. Force constants from DFT were generally in better agreement with experimental values compared to results from LDO, IADO, and INDO/2-MO. A similar trend was observed for the change in metal-ligand bond length Δr. INDO/2-MO gave too small values, while UHF with correlation-correction at the UMP2 level gave too large values [40, 58].

4.2 Klimkāns and Larsson Method

In general, the vibrations of the first solvation shell are not harmonic as assumed in the previous section. Klimkāns and Larsson put forward a simple procedure of calculating inner-sphere reorganization energies, which was originally applied to organic π systems like benzene [59]. As shown in Fig. 2 the inner-sphere reorganization energy is the sum of contributions from the individual reactants, and can formally be written in several ways

$$\lambda_i = \lambda_{2+} + \lambda_{3+} \tag{18}$$

$$= (E_2 - E_3) + (E_4 - E_1) \tag{19}$$

$$= IP - |EA|. \tag{20}$$

Obviously, the total reorganization energy can be obtained by first calculating the energies E_1 and E_3 of the independently relaxed complexes M(II) and M(III), respectively. Subsequently, by removing an electron from M(II) and by adding an electron to M(III), one obtains the respective energies E_2 and E_4. This procedure delivers a value of λ_i via Eq. 19. Consequently, this results in a small energy difference of the relatively large ionization potential (IP) and the electron affinity (EA) viz. Eq. 20. Such an approach is inherently inaccurate [31]. In order to assure computational accuracy, it is advisable to use a single complex and to move in a closed loop, e.g. 1-4 and back to 1. The reorganization energies λ_{2+} and λ_{3+} can thus be obtained directly by following the PESs downward according to the Hellman-Feynman forces [20].

Inner-sphere reorganization energies obtained from the method by Klimkāns and Larsson are accurate and include effects from Jahn-Teller distortions and anharmonicity to all orders. These effects can be substantial. For instance, inner-sphere reorganization energies calculated with the harmonic George-Griffith-Marcus method can lead to underestimations of approximately 20% [46].

The described approach appeals due to its simplicity and has been applied to re-examine electron transfer rates of the redox couples $M(OH_2)_6^{2+/3+}$ for M=Fe, Co, V, Mn, and Cr [6]. The B3LYP hybrid DFT method with a 6-311+G basis set from the Gaussian94 package was used. The theoretical inner-sphere reorganization energies were plotted against the corresponding quasi-experimental values, and performed better than previously published theoretical results. These were based on the George-Griffith-Marcus method [60] and the approach by Klimkāns and Larsson with energies based on the restricted open-shell Hartree-Fock method and a Slater-type orbital basis set [61]. Both former results, however, neglected Jahn-Teller distortions by assuming that all 6 metal-ligand bonds have the same length making a direct comparison difficult. Despite presumably accurate inner-sphere reorganization energies from density functional theory in Ref. [6], activation energies were obtained from the rule $\Delta E_i^* \approx \lambda_i/4$. As mentioned before, this can only be regarded as a crude approximation.

4.3 Polynomial Curve-Fitting Method

Despite the advantages of the previous method by Klimkāns and Larsson, the difficulty still exists of how to actually go from the reorganization energy to the activation energy which is a necessary ingredient of Marcus theory.

If all metal-ligand bonds have the same length r, then a potential remedy is as follows. For each specie, with charge q, the calculated potential energy curve can be fitted to an adequate functional form, e.g. a higher-order polynomial [15, 58, 46, 20]

$$E_q(r) = c_{q,0} + c_{q,1}r + c_{q,2}r^2 + c_{q,3}r^3 + c_{q,4}r^4 \ldots, \quad (q = 2,3+) \tag{21}$$

with fitting parameters $c_{q,i}$. Energy conservation for a radiation-less transition of an electron requires that the potential energy of the reactants $E_R = E_{2+}(r_{2+}) + E_{3+}(r_{3+})$ is equal to the potential energy of the products $E_P = E_{3+}(r_{3+}) + E_{2+}(r_{2+})$, i.e. $E_R = E_P$ with

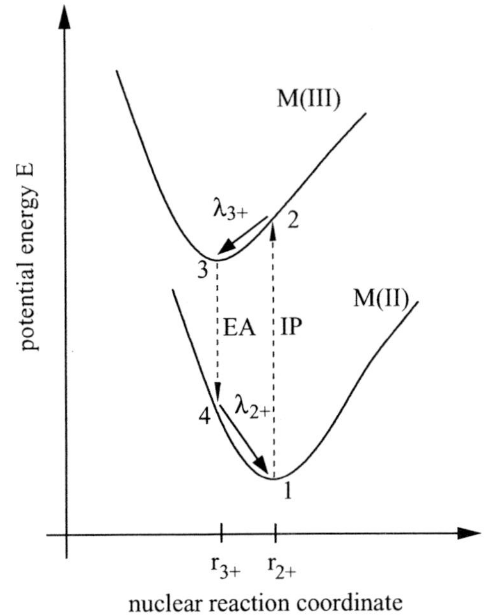

Figure 2: Gas-phase potential energy profiles of the inner-sphere reduced, M(II), and oxidized, M(III), complexes. λ_{2+} and λ_{3+} denote the reorganization energies of M(II) and M(III), respectively. *IP* illustrates the ionization potential, while *EA* describes the electron affinity. (adapted from Ref. [59])

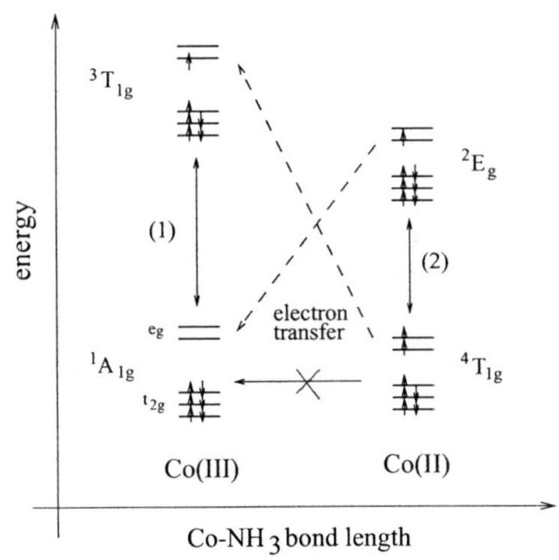

Figure 3: The hexaamminecobalt problem: The ground state electron transfer is only allowed by weak spin-orbit coupling and hence is very diabatic [37], although experiments indicate an adiabatic reaction [21]. While electron transfer via excited states is principally spin-allowed, the vertical excitation energy $^1A_{1g} \rightarrow {}^3T_{1g}$ of Co(III) is too large ($\sim 13,700\text{cm}^{-1}$ [62]) ruling out significant thermal population (path 1). A possible reaction intermediate is the excited state ^2E_g of Co(II) (path 2) [38, 40, 20].

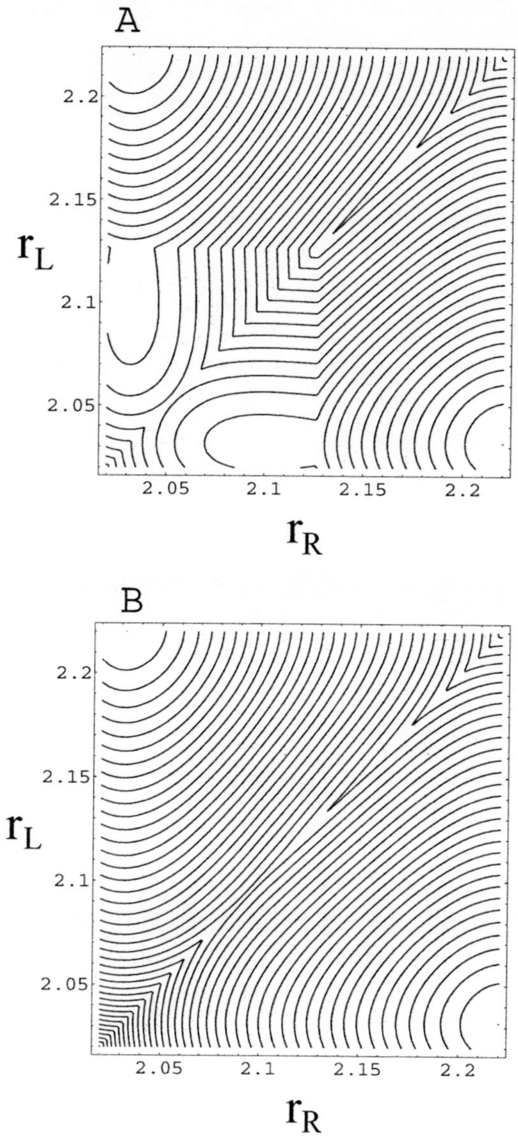

Figure 4: Contour plot of the PES of the total 2 Co-system. The average Co-NH_3 bond lengths of the "left" and "right" complex are denoted by respective r_L and r_R (see text). A) Example of a spin-allowed reaction involving the ground state Co(III, S=0) and the excited state Co(II, S=1/2 $\hbar$) near the transition state region. B) PES of conventional spin-forbidden reaction originating from the ground states Co(III, S=0) and Co(II, S=3/2 $\hbar$). For details see Ref. [20]. Preprinted with permission, © 2003 American Institute of Physics.

$r_{2+} = r_{3+} = r^*$. Analogue to Eq. 12, the activation energy reads then

$$\Delta E_i^*(r^*) = [E_{2+}(r^*) - E_{2+}(r_{2+})] + [E_{3+}(r^*) - E_{3+}(r_{3+})]. \quad (22)$$

Minimization of the activation energy yields the value r^* to be used in Eq. 22. Alternatively, one can determine r^* graphically by plotting E_P or E_R as a function of r^* and locating the saddle point [20].

While this method assumes that the first solvation shells have octahedral symmetry, tetragonal Jahn-Teller distortions are common to the high-spin complexes in Table 1 [18]. The fitting method may, however, still be applicable, if one argues that in order to maximize the Franck-Condon overlap, a octahedral uctuation of the Jahn-Teller unstable specie may be required. An important example is the redox couple $Co(NH_3)_6^{2+/3+}$, where high-spin Co(II) has a significant Jahn-Teller stabilization energy of $\sim 2,000 cm^{-1}$ [38], while low-spin Co(III) is not distorted. As summarized in Fig. 3, the many decades old question is why the theoretically spin-forbidden hexaamminecobalt self-exchange has a rather large rate of $\sim 10^{-6}(Ms)^{-1}$ [21] and, hence, seems to be adiabatic instead [37, 16, 38, 39, 64, 40, 63, 20]. As mentioned in Sec. 2, one possible solution is a thermally excited and spin allowed pathway most likely involving low-spin Co(II), although the actual energy gap, $^4T_{1g} \rightarrow {}^2E_g$, could not been determined experimentally and its calculation has been the subject of some debate [37, 38, 40, 20].

Some of the earlier computational tools, which were applied to the hexaamminecobalt self-exchange, consisted of the semi-empirical INDO, the uncorrelated UHF [38], and the correlated UMP2 [40] method. While HF-based methods likely overstabilized the high-spin groundstate of Co(II) [38], the energy mimimum obtained from INDO of the excited state PES indicates its thermal accessibility. Recent DFT calculations with the SIESTA code and a basis set of double-zeta plus polarization orbitals are strong in favor of this adiabatic pathway [20]. Figure 4 shows the diabatic ground state PES obtained from the sum of the isolated contributions (neglecting interactions), i.e.

$$E_R(r_L, r_R) = E_{Co(II)}(r_L) + E_{Co(III)}(r_R) \quad \text{for upper left triangle} \quad (23)$$

$$E_P(r_L, r_R) = E_{Co(III)}(r_L) + E_{Co(II)}(r_R) \quad \text{for lower right triangle.} \quad (24)$$

The subscripts R and P are arbitarily assigned to distinguish reactants and products. Similarly, the radial parameters r_L and r_R distinguish the bond lengths of the two involved species. While contour plot A shows a lowering of the transition state region by the intersecting excited state, plot B shows the spin-forbidden ground state process. In contrast, process A is spin-allowed and adiabatic. The total PES of the complexed redox couple (and not from the sum of the isolated species) has recently been calculated with a computationally efficient model [65]. While not *ab initio*, it emphazises the multi-Slater determinant character of the involved wavefunctions, and addresses questions of configurational correlation effects beyond conventional DFT.

It is worth mentioning that besides fitting to energy values from *ab initio* calculations, the fitting parameters can also be constrained through experiments. Hence, in certain cases, this method can be used to obtain quasi-experimental total reorganization and activation energies. The potential energy of a single complex (parameter q omitted) can either be modelled by the Morse function

$$E(r) = D(1 - exp(-br))^2 \tag{25}$$

or as an anharmonic oscillator

$$E(r) = k/2r^2 + g/6r^3. \tag{26}$$

If k is known from experimental vibrational spectroscopic frequencies via Eq. 16, and D is obtained from the ion hydration energy by substracting its vaporization heat, the missing parameters b and g may be deduced through [66, 19, 61]

$$b = \left(\frac{k}{2D}\right)^{1/2} \tag{27}$$

$$g = \frac{3k^{3/2}}{(2D)^{1/2}} \tag{28}$$

This method correctly incorporates anharmonicities and outer-sphere solvation effects. A comparison with the George-Griffith-Marcus method and the Born model indicates that the latter overestimates reorganization and activation barriers significantly [19]. Obviously, the described method depends very sensitively on the choice of used parameters.

4.4 Hessian Matrix Method

All previous *ab initio* calculations were based on isolated complexes. However, the donor and acceptor compounds undergo their reorganization in the presence of each other. This is, of course, important, when donor and acceptor sites are covalently connected with each other leading to the problem of intramolecular reorganization energies of large molecules. In principle, the following method can be applied to such molecules or complexes, but such studies have not been conducted on the redox couples at hand. Extended non-covalently bonded complexes are generally difficult to accurately treat with HF or DFT [67], and hybrid approaches will likely become necessary [68, 69]. The method is nevertheless mentioned here for completeness.

In general, the total energy can be expanded in a vector Taylor series around the fully optimized total equilibrium structure. If $\mathbf{r} = \{r_i\}$ is the vibrational degrees of freedom around the reference configuration $\mathbf{r_0}$, then we get [70]

$$E(\mathbf{r}) = \sum_{n=0}^{\infty} \frac{1}{n!}[(\mathbf{r} - \mathbf{r_0}) \cdot \nabla]^n E(\mathbf{r_0}) \tag{29}$$

$$\approx E(\mathbf{r_0}) + \Delta\mathbf{r}^T\mathbf{G} + 1/2\Delta\mathbf{r}^T\mathbf{H}\Delta\mathbf{r}, \tag{30}$$

where superscript T indicates transposition of the vector with the internal coordiantes of the molecule, $\Delta\mathbf{r} = \mathbf{r} - \mathbf{r_0}$ is the difference vector, and the first and second derivatives are the usual gradient and Hessian matrices

$$\mathbf{G}_i(\mathbf{r_0}) = \left.\frac{E(\mathbf{r})}{r_i}\right|_{\mathbf{r}=\mathbf{r_0}} \tag{31}$$

$$\mathbf{H}_{ij}(\mathbf{r_0}) = \left.\frac{{}^{E}(\mathbf{r})}{r_i\, r_j}\right|_{\mathbf{r}=\mathbf{r_0}}. \tag{32}$$

Since the method employs geometry-optimized structures at the *ab initio* level, the gradient vanishes. The Hessian matrix for the optimized structure is termed the force constant matrix and reads

$$\mathbf{F} = \mathbf{H}(\mathbf{r} = \mathbf{r_0}). \tag{33}$$

For practical applications, the fully optimized molecular equlibrium structures of the reactants and products have to be determined, e.g configurations $\mathbf{r}_0^{\mathbf{R}}$ and $\mathbf{r}_0^{\mathbf{P}}$ for the respective redox couples M(II)|M(III) and M(III)|M(II) (see Fig. 1), along with their force constant matrices. Using changes in the internal coordinates, the projected force constants for the reactants and products can be obtained

$$k^{R/P} = \frac{\Delta\mathbf{r}^T \mathbf{F} \Delta\mathbf{r}}{|\Delta\mathbf{r}|^2}. \tag{34}$$

This projects both force constants along the reaction coordinate. The inner-sphere reorganzition energy for one particular mode in the harmonic approximation is given via

$$\lambda_i = \frac{k^R k^P}{k^R + k^P} |\Delta\mathbf{r}|^2. \tag{35}$$

There are some benefits attached to this approach [70]. Unlike the George-Griffith-Marcus method, the Hessian matrix method includes the coupling between the different vibrational modes. These off-diagonal terms in the Hessian are generally non-zero, and were found to be important to include in certain cases. Another bonus is that the projected force constants are related to the generalized reaction coordinates and describe the overall motion of the molecular complex. Furthermore, the larger mass associated with these configurational changes make nuclear tunneling corrections less improtant to include.

Sample studies using this approach were on the alkyl and benzyl radicals [70], a variety of aromatic compounds [31], and metal-benzine systems [41]. At least for the case of pure aromatic hydrocarbons [31], reorganziation energies could be reduced and brought closer to experimental data compared to results from the Klimkāns and Larsson method [59].

5 Summary

While inner-sphere reorganziation energies, and hence activation energies, are not easily accessible to experiments, theoretical approaches from quantum chemical have become powerful tools. In many cases, they can guide time demanding experimental investigations, or even replace them in some cases. In particular, density functional theory is increasingly applied due to its computational efficiency, even to tricky transition metal complexes. Clearly, certain defronciencies must be further remedied. Among these are multi-configurational correlations, excited states and optical gaps, as well as, generally missing, non-local van der Waals interactions.

In this article, several theoretical and computational methods have been discussed. These are the elementary George-Griffith-Marcus method using the harmonic approximation, the Klimkāns and Larsson method, which corrects for anharmonicities, as well as the polynomial curve-fitting method, which additionally achieves the actual location of the transition state, and hence, calculation of the inner-sphere activation energy in certain cases. Last but not least, the Hessian matrix method emphasizes the coupling between the vibrational modes, but again returns to the harmonic approximation. While these methods have been applied to various charge transfer systems at various levels of computational sophistication, in this article the emphasize was on self-exchange reactions of solvated metal ions. These conceptually simple systems largely shaped the development of Marcus theory about 50 years ago. Its universal applicability, however, encompasses many other exciting fields of charge transfer. Among these are are thermal and light-induced charge transfer in polymers, light-harvesting and photosynthetic complexes, redox active proteins, hole hopping in DNA, and many others.

Acknowledgement

The author is grateful to Erin Ingrassia and Don Nicholson for proofreading and valueable comments. Financial support was from the Division of Mathematical Information and Computational Sciences, U.S. Department of Energy, under Contract DE-AC05-00OR22725 with UT-Battelle LLC.

References

[1] Williams, R.J.P., & Frausto Da Silva, J.J.R. (2003). Evolution was chemically constrained. *J. theor. Biol.*, **220**, 323-343.

[2] Rul′sek, L., & Vondrásek, J. (1998). Coordination geometries of selected transition metal ion (Co^{2+}, Ni^{2+}, Cu^{2+}, Zn^{2+}, Cd^{2+}, and Hg^{2+}) in metalloproteins. *J. inorg. Biochem.*, **71**, 115-127.

[3] Frausto Da Silva, J.J.R., & Williams, R.J.P. (2001). *The biological chemistry of the elements*. 2nd Edn. Oxford: Oxford University Press.

[4] Morel, F.M.M., & Price, N.M. (2003). The biogeochemical cycles of trace metals in the oceans. *Science*, **300**, 944-947.

[5] Siefert, R.L., Johansen, A.M., Hoffmann, M.R., & Pehkonen, S.O. (1998). Measurements of trace metal (Fe, Cu, Mn, Cr) oxidation states in fog and stratus clouds. *J. Air & Waste Manage. Assoc.*, **48**, 128-143.

[6] Rosso, K.M., & Rustad, J.R. (2000). Ab initio calculations of homogeneous outer sphere electron transfer rates: application to $M(OH_2)_6^{3+/2+}$ redox couples. *J. Phys. Chem. A*, **104**, 6718-6725.

[7] Seigneur, C., & Constantinou, E. (1995). Chemical kinetic mechanism for atmospheric chromium. *Envirn. Sci. Technol.*, **29**, 222-231.

[8] Adamson, A.W. (1951). The kinetics of the manganous-permanganate reaction. *J. Phys. Chem.* **55**, 293-303.

[9] Anderson A., & Bonner, N.A. (1954). The exchange reaction between chromous and chromic ions in perchloric acid solution. *J. Am. Chem. Soc.* **76**, 3826-3830.

[10] Libby, W.F. (1952). Theory of electron exchange reactions in aqueous solution. *J. Phys. Chem.* **56**, 863-868.

[11] Marcus, R.J., Zwolinski, B.J., & Eyring, H. (1954). The electron tunnelling hypothesis for electron ex-change reactions. *J. Phys. Chem.* **58**, 432-437.

[12] Marcus, R.A. (1997). Electron transfer reactions in chemistry. Theory and expperiment. *J. Electroanal. Chem.* **438**, 251-259.

[13] Newton, M.D. (2002). Thermal and optical electron transfer involving transition metal complexes: insights from theory and computation. *Coord. Chem. Rev.* **238-239**, 167-185.

[14] Amini, A., & Harriman, A. (2003). Computational methods for electron-transfer systems. *J. Photochem. Photobiol. C* **4**, 155-177.

[15] Rul´sek, L., & Havlas, Z. (2000). Ab initio calculations of $[Co\,Y_{6-n}X_n]^{2+}$ complexes. *J. Chem. Phys.* **112**, 149-157.

[16] Sutin, N. (1983). Theory of electron transfer reactions: insights and hindsights. In S.J. Lippard (Ed.), *Prog. Inorg. Chem.* **30** (pp. 441-498). New York: John Wiley & Sons.

[17] Sutin, N. (1986). Theory of electron transfer reactions. In J.J. Zuckermann (Ed.), *Electron transfer and electrochemical reactions; Photochemical and other energized reactions* **15** (p. 16). New York: VCH.

[18] Richens, D.T. (1997). *The chemistry of aqua ions*. New York: John Wiley and Sons.

[19] Bu, Y., & Deng, C. (1997). Hydration model for the energy barrier in self-exchange electron transfer reactions in solution. *J. Phys. Chem. A* **101**, 1198-1205.

[20] Endres, R.G., LaBute, M.X., & Cox, D.L. (2003). Theory of adiabatic hexaamminecobalt self-exchange. *J. Chem. Phys.* **118**, 8706-8714.

[21] Hammershøi, A., Geselowitz, D., & Taube, H. (1984). Redetermination of the heaxaamminecobalt(III/II) electron-self-exchange rate. *Inorg. Chem.* **23**, 979-982.

[22] Marcus, R.A. (1956). On the theory of oxidation-reduction reactions involving electron transfer. part I. *J. Chem. Phys.* **24**, 966-978.

[23] Marcus, R.A. (1957). On the theory of oxidation-reduction reactions involving electron transfer. part II. applications to data on the rates of isotopic exchange reactions. *J. Chem. Phys.* **26**, 867-871.

[24] Hush, N.S. (1958). Adiabatic rate processes at electrodes. I. Energy-charge relationships. *J. Chem. Phys.* **28**, 962-972.

[25] Hush, N.S. (1961). Adiabatic theory of outer-sphere electron-transfer reactions in solution. *Trans. Faraday Soc.* **57**, 557-580.

[26] Levich, V.G., & Dogonadze, R.R. (1959). *Dokl. Akad. Nauk* **124**, 123.

[27] Levich, V.G., & Dogonadze, R.R. (1960). *Dokl. Akad. Nauk* **133**, 158.

[28] Dogonadze, R.R. (1960). *Dokl. Akad. Nauk* **133**, 1368.

[29] Brunschwig, B.S., Logan J., Newton, M.D., & Sutin, N. (1980). A semiclassical treatment of electron-exchange reactions. Application to the hexaaquo iron (II) - hexaaquo iron (III) system. *J. Am. Chem. Soc.* **102**, 5798-5809.

[30] Kestner, N.R., Logan J., & Jortner, J. (1974). Thermal electron transfer reactions in polar solvents. *J. Phys. Chem.* **78**, 2148-2166.

[31] Jakobsen, S., Mikkelsen, K.V., & Pedersen, S.U. (1996). Calculation of intramolecular reorganization energies for electron-transfer reactions involving organic systems. *J. Phys. Chem.* **100**, 7411-7417.

[32] Tao, N.J. (1996). Probing potential-tuned resonant tunneling through redox molecules with scanning tunneling microscopy. *Phys. Rev. Lett.* **76**, 4066-4069.

[33] Grampp, G., Shohoji, M.C.B.L., Herold, B.J., & Steeken, S. (1990). Solvent induced intramolecular electron exchange kinetics in 1,3-isodisubstituted aromatic radical-anions. 2. Activation parameters. *Ber. Bunsen-Ges. Phys. Chem.* **94**, 1507-1511.

[34] Eberson, L., & Shaik, S.S. (1990). Electron-transfer reactions of radical anions: do they follow outer- or inner-sphere mechanisms? *J. Am. Chem. Soc.* **112**, 4484-4489.

[35] Delahay, P., & Dziedzic, A. (1984). Inner-sphere reorganziation in optical electron transfer. *J. Chem. Phys.* **80**, 5793-5798.

[36] Endicott, J.F., Durham, B., & Kumar, K. (1982). Examination of the intrinsic barrier to electron transfer in hexaaquocobalt(III): evidence for very slow outer-sphere self-exchange resulting from contributions of franck-condon and electronic terms. *Inorg. Chem.* **21**, 2437-2444.

[37] Buhks, E., Bixon, M., Jortner, J., & Navon, G. (1979). The hexaaminecobalt electron-exchange reaction. *Inorg. Chem.* **18**, 2014-2018.

[38] Larsson, S., Stahl, K., & Zerner, M.C. (1986). Hexaammine cobalt electron-self-exchange reaction. *Inorg. Chem.* **25**, 3033-3037.

[39] Newton, M.D. (1986). Comparison of electron-transfer matrix elements for transition-metal complexes: t_{2g} vs. e_g transfer and NH_3 vs. H_2O ligands. *J. Phys. Chem.* **90**, 3734-3739.

[40] Newton, M.D. (1991). The $Co(NH_3)_6^{2+/3+}$ exchange reaction: ground-state versus thermally excited pathways. *J. Phys. Chem.* **95**, 30-38.

[41] Zhengyu, Z., Jian, X., Chuansong, Z., Xingming, Z., Dongmei, D., & Kezhong, Z. (1999). Theoretical study of inner-sphere reorganization energy of the electron transfer reactions between $M - C_6H_6$ and $M^+ - C_6H_6$ complexes in the gas-phase: an ab initio computation. *J. Mol. Struct. (Theochem)* **469**, 1-6.

[42] Bu, Y. (2001). Distance dependence analysis of the electron transfer reactivity for the metastable $Fe^{2+}OH_2/Fe^{3+}OH_2$ system: an ab initio investigation. *J. Mol. Struct. (Theochem)*, **540**, 193-209.

[43] Marcus, R.A., & Sutin, N. (1985). Electron Transfers in Chemistry and Biology. *Biochim. Biophys. Acta* **811**, 265-322.

[44] Newton, M.D. (1988). Electronic structure analysis of electron-transfer matrix elements for transition-metal redox pairs. *J. Phys. Chem.* **92**, 3049-3056.

[45] Marcus, R.A. (1960). Exchange reactions and electron transfer reactions including isotopic exchange. *Disc. Faraday Soc.* **29**, 21-31.

[46] Zhang, D., Liu, Y., Hu, H., & Liu, C. (2002). Inner-sphere reorganization for redox pairs $M(NH_3)_6^{2+/3+}$ (M=Mn, Fe, and Co): models and calculations. *Int. J. Quant. Chem.* **86**, 468-477.

[47] Ziegler, T. (1991). Approximate density functional theory as a practical tool in molecular energetics and dynamics. *Chem. Rev.* **91**, 651-667.

[48] George, P. & Griffith, J.S. (1959). in *"The enzyms,"* Vol. 1, Boyer, P.D., Lardy, H., & Myrback, K., Academic Press, New York, Chapter 8, p.347.

[49] Marcus, R.A. (1963). On the Theory of Oxidation-Reduction Reactions Involving Electron Transfer. V. Comparison and Properties of Electrochemical and Chemical Rate Constants *J. Phys. Chem.* **67**, 853-857.

[50] Brunschwig, B.S., Creutz, C., Macartney, D., & Sutin, N. (1982). The role of inner-sphere configuration changes in electron-exchange reactions of metal-complexes. *Faraday Discuss. Chem. Soc.* **74**, 113-127.

[51] Tunuli, M.S., & Khan, S.U.M. (1985). Application of ion-dipole capture models to inner-sphere reorganization energy of ions in solution. *J. Phys. Chem.* **89**, 4667-4669.

[52] Kunc, K., & Dacosta, P.G. (1985). Real-space convergence of the force series in the lattice-dynamics of germanium. *Phys. Rev. B* **32**, 2010-2021.

[53] Srivastava, G.P., & Kunc, K. (1988). Phonon dispersion in the (110) direction: a testing ground for phenomenological models of germanium. *J. Phys. C* **21**, 5087-5106.

[54] Wei, S., & Chou, M.Y. (1992). *Ab initio* calculation of force constants and full phonon dispersion. *Phys. Rev. Lett.* **69**, 2799-2802.

[55] Baroni, S., Giannozzi, P., & and Testa, A. (1987). Green's-function approach to linear response in solids. *Phys. Rev. Lett.* **58**, 1861-1864.

[56] Tunuli, M.S., & Khan, S.U.M. (1987). Ion-dipole capture force field and its role in electron-transfer reactivity. A formalism for inner-sphere reorganization energy. *J. Phys. Chem.* **91**, 3474-3478.

[57] Zhou, Z., & Khan, S.U.M. (1989). Inner-sphere reorganization energy of ions in solution: a molecular orbital calculation. *J. Phys. Chem.* **93**, 5292-5295.

[58] Zhang, D., Zhunag, B., Feng, D., & Liu, C. (2000). Theoretical study of self-exchange electron-transfer reactions for $M(H_2O)_6^{2+/3+}$ (M=V, Cr, Mn, and Fe) systems. *Int. J. Quant. Chem.* **78**, 32-41.

[59] Klimkāns, A., & Larsson, S. (1994). Reorganization energies in benzene, naphthalene, and anthracene. *Chem. Phys.* **189**, 25-31.

[60] Delahay, P. (1982). Correlation between thermal electron-transfer in solution and photo-electron emission. *Chem. Phys. Lett.* **87**, 607-611.

[61] Bu, Y.X., Ding, Y.J., He, F.X., Jiang, L.F., & Song, X.Y. (1997). Nonempirical ab initio studies on inner-sphere reorganziation energies of $M^{2+}(H_2O)_6/M^{3+}(H_2O)_6$ redox couples at valence basis level. *Int. J. Quant. Chem.* **61**, 117-126.

[62] Wilson, R.B., & Solomon, E.I. (1980). Spectroscopic studies of photochemically important transition metal excited states. 2. the ${}^1T_{1g}$, ${}^3T_{1g}$, ${}^5T_{2g}$ excited states of hexaamminecobalt(III). *J. Am. Chem. Soc.* **102**, 4085-4095.

[63] Formosinho, S., & Arnaut, L.G. (1994). Internal reorganziation effects on electron self-exchange reactions. *J. Mol. Struct. (Theochem)* **130**, 105-124.

[64] Geselowitz, D.A. (1988). Is hexaamminecobalt(III/II) self-exchange spin forbidden? Analogy to photochemical processes. *Inorg. Chim. Acta* **154**, 225-228.

[65] LaBute, M.X., Endres, R.G., & Cox, D.L. (2004). An Anderson impurity model for efficient sampling of adiabatic potential energy surfaces of transition metal complexes. Submitted to *J. Chem. Phys.*. E-print: physics/0406043.

[66] Bu, Y. (1994). Experiment scale of reorganization free energy for diatomic molecules in the gaseous phase in electron-transfer processes. 1. The method of modifying the oscillator potential. *J. Phys. Chem.* **98**, 2290-2293.

[67] Kohn, W., Meir, Y., & Makarov, D.E. (1998). Van der Waals energies in density functional theory. *Phys. Rev. Lett.* **80**, 4153-5156.

[68] Baerends, E.J. & Gritsenko, O.V. (1998). A quantum chemical view of density functional theory. *J. Phys. Chem. A* **101**, 5383-5403.

[69] Kafafi, S.A. Novel density functional methodology for the computation of accurate electronic and thermodynamic properties of molecular systems and improved long-range behavior. *J. Phys. Chem. A* **102**, 10404-10413.

[70] Mikkelsen, K.V., Pedersen, S.U., Lund, H., & Swanstrøm, P. (1991). A new and rigorous method of calculating intramolecular reorganization enegies for electron-transfer reactions: applied for self-exchange reactions involving alkyl and benzyl radicals. *J. Phys. Chem.* **95**, 8892-8899.

In: Trends in Chemical Physics Research
Editor: A.N. Linke, pp. 155-176

ISBN 1-59454-483-2

Chapter 8

TWISTED LASER BEAMS AND THEIR OPTICAL INTERACTIONS WITH CHIRAL MATTER

***D.L. Andrews*[†], *L.C. Dávila Romero*[‡] *and M. Babiker*[§]**

[†]School of Chemical Sciences, University of East Anglia, Norwich, NR4 7TJ, U.K.
[‡]School of Physics and Astronomy, University of St Andrews, North Haugh, St Andrews, KY16 9SS, U.K.
[§]Department of Physics, University of York, Heslington, York YO10 5DD, U.K.

Abstract

Recently, optical beams with a new and distinctive type of helicity have become the subject of considerable interest. While circularly polarised light is associated with electromagnetic fields whose locus takes a helical form, for these optically engineered phase-bearing or 'twisted' laser beams it is the wave-front surface that is helical. Consequently, a new and intriguing manifestation of optical handedness is available for characterising optical interactions with matter. Many of the interactions that have been studied with such beams utilise or exhibit their intrinsic 'orbital' angular momentum. To date, however, such studies have mostly focused on interactions with achiral (generally mirror-symmetric) matter. It is timely to assess what new features, if any, can be expected when such beams are used to interrogate a chiral (molecularly right- or left-handed) system. In this review we introduce and analyse the concepts that determine whether optically twisted beams should offer a controllable specificity in their interactions with chiral matter, associated with orbital angular momentum transfer. By explicit analysis of the grounds for chiroptical behaviour, and with a specific focus on circular dichroism, we establish the principle that the helicity of optical vortices cannot directly engage through any parametric or non-parametric optical process with the chirality of a molecular system, up to the level of electric quadrupole interactions. These conclusions have been corroborated by recent experimental results.

1 Introduction

Of the many notable advances marked by its designation as World Year of Physics, 2005 is the hundredth anniversary of the photon concept. As the quantum of electromagnetic radiation – the gauge boson for electromagnetic coupling – the photon has properties that are generally well known and which can be found listed in any table of elementary particles. Amongst these properties it is usual to include the unit spin angular momentum – whose manifestation in circular polarisations is also very well known and corresponds to projections $\pm\hbar$, according to the circular sense, along the direction of propagation. [1] Circularly polarised light is intrinsically handed, its left- or right-handed helicity associated with the locus of the electromagnetic field vectors. Less well known is the fact that such features are usually interpreted with implicit reference to a well-collimated non-divergent beam with a planar wavefront, reflecting common experimental practice. When other kinds of light beam are engineered, novel features can emerge that have no counterpart in conventional optics. Just over ten years ago, [2] recognition first emerged that light beams with a helical wavefront could display other kinds of angular momentum, now generally called *orbital*. Specifically, if such a *twisted beam* has an integer number l twists within its wavelength, an orbital angular momentum of $l\hbar$ is associated with each photon – quite distinct from any spin angular momentum determined by the polarisation. Such beams are now readily producible in the laboratory. [3-5] This new field is rife with synonyms: the radiation itself is often described as comprising *optical vortices*, and the associated technology has also been termed an *optical spanner* (though this term has also been applied to beams whose wave-front is mechanically rotated). Amongst the many issues that arise in the construction, application and representation of twisted beams is the question of how their handedness may or may not engage with chiral matter. [6]

Molecular chirality signifies a structural handedness that changes under spatial inversion or a combination of inversion and rotation, equivalent to the usually stated criterion of a lack of any improper axes of rotation. The optical manifestations of such chirality known as *optical activity* are well attested, [7] ranging from the optical rotation widely used to characterise the enantiomeric (left- and right-handed optical isomeric) forms of sugars, to the circular differential Raman scattering that has emerged as a means of exhibiting the chirality of protein folding. [8] Traditionally, chiral optics engages circularly polarised light – even in the case of optical rotation, the phenomenon is commonly interpreted through a description of the plane polarised state as a superposition of circular polarisations with opposite handedness. It is for this reason that the interactions of circularly polarised light with matter are enantiomerically specific. Whereas twisted beam have until now been studied primarily in their interactions with achiral matter, extensive interest has arisen in the intriguing possibilities represented by their interactions with chiral systems, such as some recent work with liquid crystals. [9] Interested parties range from those researching in the theory of optical angular momentum to others interested in possible applications to molecular motors, nano-manipulation and biotechnology. It is in this broad context that we assess what new features, if any, can be expected if such beams are used to interrogate a system whose optical response is associated with enantiomerically specific molecules.

In the following, we introduce salient quantum electrodynamical principles for subsequent discussion of issues relating to the interplay of quantised spin and orbital angular momentum – the latter with particular reference to twisted beams. The theoretical basis for our discussion is molecular quantum electrodynamics – the only theory in which the photon concept has full legitimacy. [10,11] To avoid unnecessary distraction by lengthy detail, the arguments are pursued with only the most salient equations explicitly presented. Section 2 introduces the principles of molecular chirality at a fundamental level appropriate for subsequent discussion, in section 3, of issues relating to the interplay of quantised spin and orbital angular momentum – the latter with particular reference to twisted beams. The key results are summarised in section 4.

2 Foundation Principles

Considerable intricacies of symmetry can arise when a twisted laser beam interacts with chiral matter. Specifically, it is necessary to construct a theory that can accommodate as independent quantities the handedness of the molecules, the circularity of the optical polarisation, and the sense of twist in the laser beam – see Figure 1. In anticipation of the complexity that ensues, we shall therefore first determine a theoretical framework amenable to the issues of interest, to exact in generalised form the criteria for manifestations of chiroptical interactions. From this framework, in which conventional chirality can be understood in terms of molecular symmetry and multipole properties, we can extend the arguments to address light with a twisted wavefront. For simplicity, materials with a unique enantiomeric specificity are assumed – signifying a chirality that is intrinsic and common to all molecular components (or chromophores) involved in the optical response. Results for systems of this kind will also apply to single-molecule studies. Longer-range translation/rotation order can also produce chirality, as in twisted nematic crystals, but such mesoscopic chirality cannot directly engender enantiomerically specific interactions except under particular conditions. The only exception is where optical waves probe two or more electronically distinct, dissymmetrically oriented but intrinsically achiral molecules or chromophores. Then chirality can be manifest as in the two-group model (see below). Here we are concerned with intrinsically chiral centres rather than chiral arrays.

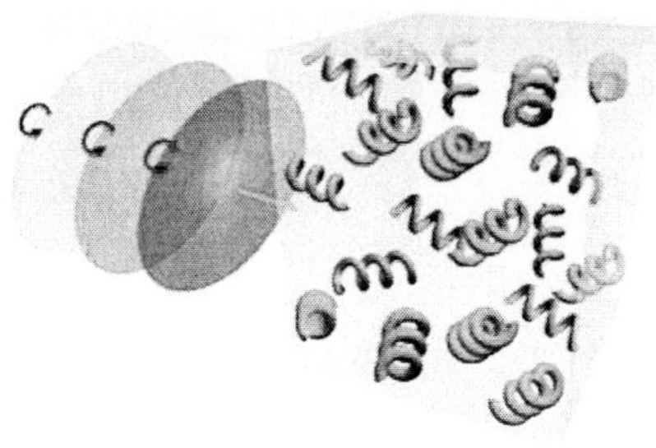

(a)

(b)

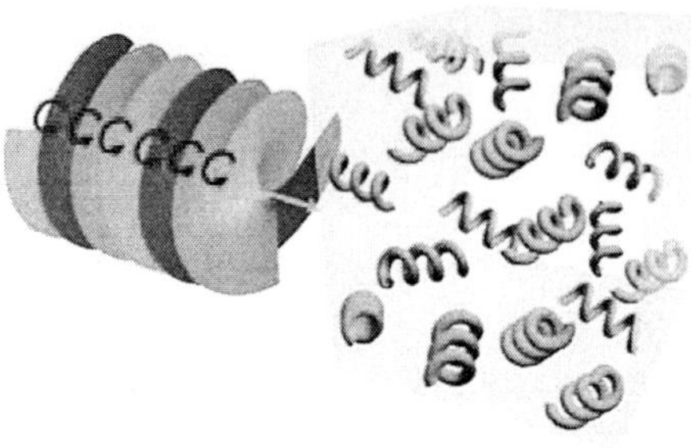

(c)

Fig. 1. Interaction of left circularly polarised radiation (represented by red arrows) with left-hand (randomly oriented) chiral molecules: a) plane waves; b) left-handed Laguerre-Gaussian beam, $l = 3$ and c) right-handed Laguerre-Gaussian beam $l = -3$. In all three cases the radiation is represented by surfaces of constant phase.

Chiroptical interactions can be distinguished by their electromagnetic origins. For molecular systems in their usual singlet electronic ground state (where we exclude free electron spin effects), chiroptical interactions are linked at a fundamental level with the spatial variation of the electric and magnetic fields associated with the input of optical radiation. This variation over space can be understood to engage chirality either through its coupling with dissymmetrically placed, neighbouring chromophore groups (Kirkwood's two-group coupled-dipole model, [12] now of limited application) or more generally through the coupling of its associated electric and magnetic fields with individual groups. As chirality signifies a local breaking of parity selection rules, *i.e.* parity is no longer a good quantum number, it thereby permits the interference of electric and magnetic interactions – or, for example, the interference of multipolar electric interactions having different parity. Even in the two-group case, the paired electric dipole interactions of the system equate to electric and magnetic interactions of the single entity which the two groups comprise. Thus, for convenience, the term 'chiral centre' is used in the following to denote either chromophore or molecule. To understand how this works, it will be necessary to unveil the structure of the quantum amplitude, specified as the amplitude for the optical interaction of a single intrinsically chiral species.

2.1 Hamiltonian and Field Expansions

In order to properly represent the quantum features of the radiation we shall develop the theory using the framework of quantum electrodynamics. The Hamiltonian for any chiroptical interaction comprises the unperturbed operators for the radiation and for N chiral centres, here differentiated by a label ξ, and also a light-matter interaction Hamiltonian;

$$H = H_{\text{rad}} + \sum_{\xi}^{N} H_{\text{centre}}(\xi) + \sum_{\xi}^{N} H_{\text{int}}(\xi) \ . \tag{1}$$

As an energy operator, each component of the Hamiltonian is necessarily of even parity with respect to space inversion and also even with respect to time reversal. The first two terms of (1) determine the basis in which the different states of the system can be described, *i.e.* a direct product of eigenstates of the radiation and the matter. The interaction term can be expressed either in minimal coupling form (cast in terms of coupling with the vector potential of the radiation field) or the more familiar multipolar formulation, directly expressed in terms of electric and magnetic fields. [13] These two (and also other less well-known) options lead to identical results for real processes, that is those subject to overall energy conservation; [14, 15] for convenience in the following we develop theory using the multipolar form.

The multipolar interaction Hamiltonian divides into three parts: one is a linear coupling of the molecular polarisation field (accommodating all electric multipoles En) with the transverse electric field $\mathbf{e}(\mathbf{r})$ of the radiation; another entails linear coupling of the molecular magnetisation field (all magnetic multipoles Mn) with the magnetic field $\mathbf{b}(\mathbf{r})$ of the radiation. The third is a coupling of the molecular diamagnetisation field, quadratically with the magnetic radiation field:

$$H_{\text{int}} = -\int \mathbf{p}(\mathbf{r})\cdot\mathbf{e}^{\perp}(\mathbf{r})\, d^3\mathbf{r} - \int \mathbf{m}(\mathbf{r})\cdot\mathbf{b}(\mathbf{r})\, d^3\mathbf{r} + \tfrac{1}{2}\int O_{ij}(\mathbf{r},\mathbf{r}')\, b_i(\mathbf{r})\, b_j(\mathbf{r}')\, d^3\mathbf{r}\, d^3\mathbf{r}'. \quad (2)$$

In the Schrödinger representation the time-dependence of these operators is suppressed. The first term in equation (2) depends on the molecular polarisation field $\mathbf{p}(\mathbf{r})$ while the second depends on the magnetisation field $\mathbf{m}(\mathbf{r})$; the last term, which involves the diamagnetisation field $O_{ij}(\mathbf{r},\mathbf{r}')$, is associated with the variation of current density when the fields are switched on. The latter term is typically an order of magnitude smaller than the first two (by a factor of the order α^2, with α the fine structure constant) and for present purposes it need concern us no further – indeed there appears to be no suggestion of its engagement in any form of molecular chirality.

The interaction Hamiltonian is considered a perturbation of the system, allowing transition from one eigenstate of the unperturbed Hamiltonian, $H_{\text{rad}} + H_{\text{centres}}$, to another. By expanding the parent polarisation and magnetisation fields as Taylor series about the molecular positions, the terms comprising equation (2) can be expanded in multipolar orders – principal amongst which (retaining terms up to the order of α) are the electric-dipole $\boldsymbol{\mu}(\xi)$, electric-quadrupole $q_{ij}(\xi)$ and the magnetic-dipole $\mathbf{m}(\xi)$. The leading terms of (2) are thus cast as follows;

$$H_{\text{int}} = -\sum_{\xi}\boldsymbol{\mu}(\xi)\cdot\mathbf{e}^{\perp}(\mathbf{R}_{\xi}) + \sum_{\xi} q_{ij}(\xi)\nabla_i e_j^{\perp}(\mathbf{R}_{\xi}) + \ldots - \sum_{\xi}\mathbf{m}(\xi)\cdot\mathbf{b}(\mathbf{R}_{\xi}) + \ldots \quad (3)$$

The parity under time and space symmetry of the electric dipole, electric quadrupole and magnetic dipole are given in Table 1.

Table 1: Space and time parity of electromagnetic quantities

Operator	Time parity	Space parity
Hamiltonian	+1	+1
Electric field	+1	−1
Electric field of LG mode (k,λ,l,p)	+1	$(-1)^{l+1}$
Electric dipole	+1	−1
Magnetic field	−1	+1
Magnetic field of LG mode (k,λ,l,p)	−1	$(-1)^{l}$
Magnetic dipole	−1	+1
Wave-vector	−1	−1

The electric and magnetic fields that feature in (3) are solutions of Maxwell's equations within the QED framework. In free space, and within the Coulomb gauge, the simplest solutions are given by the plane waves;

$$\mathbf{e}(\mathbf{r})=\sum_{\mathbf{k},\lambda}\left\{\mathrm{i}\left(\frac{\hbar ck}{2\varepsilon_0 V}\right)^{\frac{1}{2}}\boldsymbol{\varepsilon}^{(\lambda)}(\mathbf{k})\mathrm{a}^{(\lambda)}(\mathbf{k})\exp[\mathrm{i}\mathbf{k}\cdot\mathbf{r}]+h.c.\right\} \tag{4a}$$

$$\mathbf{b}(\mathbf{r})=\sum_{\mathbf{k},\lambda}\left\{\mathrm{i}\left(\frac{\hbar k}{2\varepsilon_0 cV}\right)^{\frac{1}{2}}\left(\hat{\mathbf{k}}\times\boldsymbol{\varepsilon}^{(\lambda)}(\mathbf{k})\right)\mathrm{a}^{(\lambda)}(\mathbf{k})\exp[\mathrm{i}\mathbf{k}\cdot\mathbf{r}]+h.c.\right\} \tag{4b}$$

Here V is the quantisation volume, $\mathbf{k}$ is the wave-vector and λ is a polarisation label; the latter two quantities together represent four independent degrees of freedom. In the classical interpretation, the terms represented explicitly in equation (4) signify the positive frequency analytic signals [16]; corresponding negative frequency components are conveyed by the hermitian conjugate terms (in which the phase factor has the opposite sign).

For optically engineered radiation with orbital angular momentum, the electromagnetic fields are commonly cast in terms of Laguerre-Gaussian (LG) modes. From an analytical point of view, these modes constitute another complete set, also arising as paraxial solutions of the source-free Maxwell equations. For example the LG_0^1 mode is a phase quadrature superposition of HG_{10} and HG_{01} Hermite-Gaussian modes. [2] Explicitly, the electric and magnetic field expansions for a beam propagating in the $\mathbf{z}$ direction emerge in the following form as functions of cylindrical coordinates (r, z, ϕ);

$$\mathbf{e}^{\perp}(\mathbf{r})=\sum_{\mathbf{k},\lambda,l,p}\left\{\mathrm{i}\left(\frac{\hbar ck}{2\varepsilon_0 A_{lp}V}\right)^{\frac{1}{2}}\boldsymbol{\varepsilon}_{lp}^{(\lambda)}(k\hat{\mathbf{z}})f_{lp}(r)\mathrm{a}_{lp}^{(\lambda)}(k\hat{\mathbf{z}})\exp\left[\mathrm{i}(kz-l\phi)\right]+h.c.\right\}, \tag{5a}$$

$$\mathbf{b}(\mathbf{r}) = \sum_{\mathbf{k},\lambda,l,p} \left\{ \mathrm{i}\left(\frac{\hbar k}{2\varepsilon_0 c A_{lp} V}\right)^{\frac{1}{2}} \left(\hat{\mathbf{k}} \times \boldsymbol{\varepsilon}_{lp}^{(\lambda)}(k\hat{\mathbf{z}})\right) f_{lp}(r)\, \mathrm{a}_{lp}^{(\lambda)}(k\hat{\mathbf{z}}) \exp\left[\mathrm{i}(kz - l\phi)\right] + h.c. \right\}, \quad \text{(5b)}$$

where $\mathrm{a}_{lp}^{(\lambda)}$ is the photon annihilation operator and $\boldsymbol{\varepsilon}_{lp}^{(\lambda)}$ is the wave polarisation vector, which can describe linear, circular or any other basis polarisation. The four degrees of freedom manifest in the summation variables are the magnitude of the wave-vector (its direction assigned to a specified direction z), the polarisation λ and two other labels, l and p, that designate the order of the associated Laguerre polynomial [3,17] – see Appendix 1. Also, l denotes the orbital angular momentum quantum number, for which a positive sign denotes left helicity and a negative sign, right. Here we note that the radial functions $f_{lp}(r)$ are independent on the sign of l; consequently all information of the handness of the optical vortex is in the phase factor $\exp[-\mathrm{i}l\phi]$, a characteristic that is crucial when analysing the interactions of such beams with chiral centres.

Returning to the expression given by equation (3), the electric dipoles $\boldsymbol{\mu}(\mathbf{r})$ (E1) have an odd signature for space parity and even for time; electric quadrupoles $\mathbf{q}$ (E2) are even in both space and time, while the magnetic dipoles $\mathbf{m}(\mathbf{r})$ (M1) are even in space and odd in time. Similarly, independent of which mode expansion is used, the electric field $\mathbf{e}(\mathbf{r})$ of the overall radiation field is space-odd and time-even; conversely the magnetic field $\mathbf{b}(\mathbf{r})$ is of even parity under space inversion and of odd parity under time inversion – the latter temporal features readily apparent in the interaction picture representation.[11] Note, however, that the symmetry assertions for the electric and magnetic fields of the radiation are true *only* for the whole; they do *not* necessarily apply when only one or a finite number of modes is considered (independent of the mode expansion used). These and other fundamental symmetry properties are summarised in Table 1. Cognisance of these features proves extremely significant in developing and identifying non-zero terms in the quantum amplitude for optical processes.

The polarisations of the photon can be described by the vectors $\boldsymbol{\varepsilon}^{(\lambda)}(\mathbf{k})$. For circular polarisations the explicit expressions are;

$$\boldsymbol{\varepsilon}^{(\pm)}(\mathbf{k}) = \tfrac{1}{\sqrt{2}}(\hat{\mathbf{x}} \pm \mathrm{i}\hat{\mathbf{y}}) \quad \text{(6)}$$

where the positive (negative) superscript denotes left-(right-) circular polarisation. It can be seen that $\boldsymbol{\varepsilon}^{(\pm)}(\mathbf{k}) = \overline{\boldsymbol{\varepsilon}}^{(\mp)}(\mathbf{k})$ and $\boldsymbol{\varepsilon}^{(\pm)}(-\mathbf{k}) = -\overline{\boldsymbol{\varepsilon}}^{(\mp)}(\mathbf{k})$. With the wave-vector $\mathbf{k}$, $\boldsymbol{\varepsilon}^{(+)}(\mathbf{k})$ and $\boldsymbol{\varepsilon}^{(-)}(\mathbf{k})$ form a right triad.

2.2 Perturbation Theory and Quantum Amplitudes

The probability of a given chiroptical process is determined by its quantum amplitude. For one molecule ξ we have;

$$M_{fi}^{\xi} \equiv \langle f|U(t,t_0)|i\rangle = \langle f|\left\{1+\sum_{n=1}^{\infty}\left(\frac{1}{i\hbar}\right)^n \int_{t_0}^{t} dt_1 \int_{t_0}^{t_1} dt_2 \ldots \int_{t_0}^{t_n} dt_n \, V_I(t_1)\, V_I(t_2)\ldots V_I(t_n)\right\}|i\rangle \quad (7)$$

where the operator $V_I(t)$ is the interaction Hamiltonian H_{int} in the interaction picture. Depending on the number of photons involved in the process the relevant term of the expansion is considered – for example for second harmonic generation, SHG, where two photons are annihilated and one created in each fundamental interaction, the term $n=3$ is the one that gives the first contribution. Generally, for an n-photon process we have;

$$M_{fi}^{\xi(n)} \equiv \langle f|U(t,t_0)|i\rangle = \langle f|H_{\text{int}}\left(\frac{1}{(H_0-E_i)}H_{\text{int}}\right)^{n-1}|i\rangle \quad (8)$$

The quantum amplitude M_{fi} for a specific optical interaction in a single chiral centre is constructed from time-dependent perturbation theory and entails a linear combination of scalar terms, each of which is the inner product of two rank r tensors: $\mathbf{S}^{(r)}$, a radiation tensor, and $\mathbf{T}^{(r)}$, a molecular tensor, both detailed below; [18]

$$M_{fi}^{\xi} = \exp\left(i\Delta\mathbf{k}\cdot\mathbf{R}_{\xi}\right)\sum_{e,m=0}^{n}\mathbf{S}^{(r)}_{e;\,m;\,n-e-m}\otimes\mathbf{T}^{(r)}_{e;\,m;\,n-e-m} \quad (9)$$

Here, $\Delta\mathbf{k}$ is the mismatch between the wave-vector sum of all input and the sum of all output photons (if any) involved in the process at a chiral centre ξ located at $\mathbf{R}_{\xi}$. To identify the radiation and molecular tensors, $\mathbf{S}^{(r)} \equiv S_{i_1 i_2 \ldots i_r}$ and $\mathbf{T}^{(r)} = T_{i_1 i_2 \ldots i_r}$ respectively, three labels are used: $(e;m;q)$, corresponding to the number of electric-dipole, magnetic-dipole and electric quadrupole interactions, respectively. The sum of these equals the number of photon interactions involved in the process, $e+m+q=n$.

The rank (r) of the tensors $\mathbf{S}^{(r)}$ and $\mathbf{T}^{(r)}$ is determined by the labels $(e;m;q)$ through the simple relation $r=e+m+2q$, which signifies that the rank is between n and $2n$. For plane wave photons, the structure of the radiation tensor $\mathbf{S}^{(r)}$ is that of an outer product of the beam polarisation vectors $\boldsymbol{\varepsilon}^{(\lambda)}(\mathbf{k})$, taking the complex conjugate for emitted photons – and also wave-vector components, in the case where quadrupole (or higher) multipoles are involved. This tensor also carries factors relating to other beam parameters – which, in the signal derived from $\left|M_{fi}^{\xi(n)}\right|^2$ (see below) will manifest the dependence on beam intensities and degrees of coherence. If the LG mode expansion is considered, the radiation tensor $\mathbf{S}^{(r)}$ will also include the complex profile structure of the radiation, given by $f_{lp}(r)$ and its derivatives (see also Appendix 1). The molecular tensor $\mathbf{T}^{(r)}$ can be written in a form that entails a product of one or more molecular transition integrals, as determined by the number of photons involved. In every case the origin of such forms is a single transition integral, expressible as a Dirac bracket with the appropriate number of electromagnetic interactions coupling the initial state to the final state. The simple case of single-photon absorption serves to illustrate the point. Here one interaction couples the two molecular states, and that interaction embodies the entire family of multipolar contributions. In order for the transition integral not to vanish identically, the triple product of the group theoretic representations for the initial and final state wavefunctions with that of the interaction must contain the totally symmetric representation. It is here that the implications of molecular chirality emerge.

Before considering chiral systems in more detail, it is instructive to reflect on the case of a centrosymmetric molecular system, where parity is a good quantum number and all wavefunctions have definite parity. Here, the only non-zero contributions to the transition integral can be those whose parity equates to the product signature of the initial and final state molecular wavefunctions. For example if electric dipole coupling is to give a non-vanishing contribution (i.e. that coupling is *allowed*) then, for that same transition, electric quadrupole and magnetic dipole contributions would indeed vanish (they would be forbidden) by virtue of their opposite spatial parity. Thus, for centrosymmetric species, all multipolar contributions to any given quantum amplitude will have the same spatial parity. Clearly for chiral systems, where parity is not a good quantum number, no such rule applies and the amplitude may entail contributions of both positive and negative parity signature.

2.3 System Response

Not only molecular symmetry determines the nature and extent of any chiroptical response; macroscopic symmetry is important too. In particular, when the system of interest is a fluid or other microscopically disordered medium, the isotropic symmetry of the bulk comes into play. In general, experimental measurements result from the optical interactions of more than one chiral centre, and the observable results from a quantum amplitude comprising an ensemble sum of contributions from all relevant centres. In calculating the observable associated with a particular optical interaction it is usual to apply the Fermi Rule;

$$\Gamma = \frac{d}{dt} P_{TOTAL} = \frac{2\pi}{\hbar} \left| M_{fi} \right|^2 \rho_f \,. \tag{10}$$

Here the transition matrix represents all the system, *i.e.* all the molecules present in the sample and therefore we have:

$$M_{fi} = \sum_{\xi=1}^{N} M_{fi}^{\xi} \tag{11}$$

and hence,

$$\Gamma \propto \left| M_{fi} \right|^2 = \sum_{\xi,\xi'}^{N} M_{fi}^{\xi} \bar{M}_{fi}^{\xi'} \tag{12}$$

a result that emerges from the modulus square of the quantum amplitude. For chiral species this therefore includes, in addition to diagonal terms that are all of even parity in both radiation and molecular parameters, various terms associated with the quantum interference of couplings with opposite parity, correspondingly of odd parity for both the radiation and molecule.

As is apparent from the $\exp\left(i\Delta\mathbf{k}\cdot\mathbf{R}_{\xi}\right)$ factor in (9), each quantum amplitude contribution generally has a different phase associated with spatial variance in the registration of light at the corresponding chiral centre. Most optical processes are incoherent and *non-parametric* (*i.e.* $\Delta\mathbf{k}\neq 0$), such that the interference of quantum amplitudes from different centres gives a null contribution when those centres are isotropically distributed – a principle that holds generally true except in certain nanomaterials.[1][19,20,21] For such processes, no role is played by any orientational order characterising *long*-range chirality. In the following we focus specifically on chiroptical observables associated with non-parametric processes in which the ensemble signal, by virtue of the ergodic theorem, represents the time-averaged response of a single chiral centre. The two reasons for studying such optical processes are; (i) it is processes of this category that are involved in the most common manifestations of chirality (circular dichroism, differential Rayleigh and Raman scattering, circularly polarised luminescence etc.); (ii) where optical vortices are concerned, it is known that parametric processes generally entail conservation of orbital angular momentum by the radiation field, such that chirality is not engaged. [17]

For a non-parametric interaction the transition rate of a single chiral centre is a sum of amplitude contribution products, featuring amongst which are quantum interferences of a particular class, *i.e.* products that are odd in parity for both the matter and radiation – terms that can only arise in the case of chiral systems. Handedness is apparent in two respects. If the space inversion operator (signifying a change to the opposite enantiomeric form) is applied to all properties of the molecule – and in particular its multipole moments – but not the radiation (signifying retention of its circularity), then these interference terms change sign. The same is true if the radiation changes handedness, (on space inversion) but the molecule retains its enantiomeric form; then again, these specific interference terms change sign. Clearly all contributions to the signal are invariant to inversion of the whole system. This is indeed the reason why chiral interactions must involve handed radiation, usually circular (elliptical polarisations are also permissible). Plane polarisations are invariant under space inversion; consequently, applying space inversion to a molecular system interacting with plane polarised light has the same effect as applying it to both the molecule and the radiation; clearly no chiral specificity can emerge. Optical rotation, as noted earlier, is an exception since it is a parametric process.

For the interference terms that support chiral selectivity; it remains to discover any further conditions that need to be satisfied to ensure that they are not identically zero. To this end we focus on the issue of rotational averaging. For any component of a non-parametric signal – and in particular the odd-parity quantum interference terms – rotational averaging effects a disentanglement of the radiation and molecular fields, $\mathbf{S}^{(n)}$ and $\mathbf{T}^{(n)}$ respectively. When rotational averaging is implemented the result is a product of scalars (or, for tensors of

[1] Amplitudes of different centres effectively add within a range $(\Delta\mathbf{k})^{-1}$, eliciting coherent response in mesoscopically disordered systems comprising structured nanoparticles.

rank four or more, a linear combination of such scalars), one scalar for the radiation and one for the molecule (see Appendix 2). Each scalar is derived by contracting the tensor, $\mathbf{S}^{(n)}$ or $\mathbf{T}^{(n)}$, with an isotropic tensor of the same rank – in the language of spherical tensors this signifies the angular momentum $J = 0$ contribution. For example in the E1-M1 interference term for photon absorption by a chiral molecule the molecular tensor $\mathbf{T}^{(2)} = [\boldsymbol{\mu}][\overline{\mathbf{m}}]$, on contraction with the isotropic tensor of rank 2 (the Kronecker delta), yields the scalar $T = (\boldsymbol{\mu} \cdot \overline{\mathbf{m}})$ – signifying that the electric and magnetic transition moments must not be orthogonal (in transitions between non-degenerate states, $\overline{\mathbf{m}}$ will be antiparallel to $\mathbf{m}$). Equally $\mathbf{S}^{(2)} = [\mathbf{e}][\overline{\mathbf{b}}]$ yields the scalar $S = (\mathbf{e} \cdot \overline{\mathbf{b}})$; this determines that chiral resolution vanishes when plane polarised light is employed, for then the field polarisations are real and the orthogonality of the electric and magnetic vectors gives $(\mathbf{e} \cdot \overline{\mathbf{b}}) = 0$. But for any circular (or even for elliptical) polarisation, $\mathbf{S}$ is non-zero and, since it also changes sign on reversal of circularity;

$$S = (\mathbf{e} \cdot \overline{\mathbf{b}}) \xrightarrow{I} ((-\mathbf{e}) \cdot \overline{\mathbf{b}}) = -S \tag{13}$$

chiral specificity is manifest. Since T takes opposite signs for chiral molecules of opposite enantiomeric form,

$$T = (\boldsymbol{\mu} \cdot \overline{\mathbf{m}}) \xrightarrow{I} ((-\boldsymbol{\mu}) \cdot \overline{\mathbf{m}}) = -T \tag{14}$$

the rate of absorption of left-handed circularly polarised radiation, for example, is different for left- and right-handed enantiomers – though only marginally, because of the relative weakness of the salient interference terms compared to the dominant (usually electric dipole) diagonal contributions to the signal. Equally, each enantiomer exhibits a slightly different rate of absorption for left and for right-handed circular input. This is the origin of circular dichroism. Similar remarks apply to optically more intricate processes involving more than one photon, where the quantum amplitude itself comprises products of multipolar couplings, each associated with a definite resultant spatial parity, and the key interference terms arise from products of these products with opposite inversion symmetry. Again, for a chirally specific signal to emerge, it is necessary that the resultant radiation and molecular scalars do not vanish identically.

3 The Interactions of Twisted Beams

Having elicited the key principles, we can now address the central issue of how, if at all, a twisted laser beam can engage in a chirally differentiable way with matter. Most of the tools for addressing this issue have been introduced in the last section. The helicity of optical vortices is present in their wave-front structure. As we have seen, the chirality of the electric and magnetic fields is manifest through additional phase factors in the positive and negative frequency components. In particular, from equation (5a) the (un-normalised) positive frequency analytic signal for the electric field of a Laguerre-Gaussian (LG) mode propagating in the z-direction with a wave-vector of magnitude k, is expressible as;

$$\mathbf{e}^{\perp}(\mathbf{r}) = \mathrm{i}\boldsymbol{\varepsilon}_{lp}^{(\lambda)}(k\hat{\mathbf{z}}) f_{lp}(r) \mathrm{a}_{lp}^{(\lambda)}(k\hat{\mathbf{z}}) \exp\left[\mathrm{i}(kz - l\phi)\right] \tag{15}$$

where $\hat{\boldsymbol{\varepsilon}}$ is the polarisation operator. Of the two space-dependent terms in the field phase, the first is the origin of the wave-vector mismatch factor in (9) – here delivered as $z_{\xi}\Delta\mathbf{k}$ – and the second is an azimuthal phase factor, a characteristic of twisted beams. The first component of the phase in (15) changes sign on space inversion (and equally, on complex conjugation) just as the circular polarisation vectors behave. The second phase factor changes by addition of π on space inversion. The magnetic field for a twisted beam has an exactly similar form. [20]

No difference should be expected between the behaviour of right and left forms of any twisted mode if the photons it comprises are plane polarised and the material they interact with is achiral, in accordance with the dictates of parity. This is equivalent to the case of circularly polarised, planar wave-front (*e.g.* Gaussian) modes, interacting with achiral material. However, there are several differences between LG photons and those associated with planar wave-fronts. In the present context a significant difference is that plane waves have no restriction in the direction of propagation, while the symmetry of LG photons designates propagation in one direction, $\hat{\mathbf{z}}$. This characteristic indicates that mirror inversion along a plane containing the $\hat{\mathbf{z}}$ axis becomes the relevant symmetry element, as opposed to space inversion. Arguing symmetry on these grounds is sufficient to establish that there is no differentiation between LG beams of opposite handedness when achiral material is interrogated, since a left-handed optical vortex transforms to a right-handed one.

Now we come to the crux, the interaction of a twisted beam with chiral matter. Consider first the case of a twisted beam comprising plane polarised photons; superficially this appears exactly analogous to the case of circular polarisations in a plane wave. Mirror inversion of either part of the system (molecule or radiation) gives a system different from the original, and so a change in signal signifying chiral specificity might be anticipated. Representing in brackets the circularity of the radiation and the matter respectively, both cases require identity of (*L*, *L*) and (*R*, *R*), and so too (*L*, *R*) and (*R*, *L*). However (*L*, *L*) and (*L*, *R*) differ. More succinctly, denoting either handedness *L*/*R* by a circularity *c* and the reverse by $\overline{c}$, we have $(c,c) = (\overline{c},\overline{c})$ – which may or may not be equal to $(c,\overline{c})$.

For the twisted beam case it is therefore necessary to ascertain whether there is a mechanism, by means of which signals that are *permissibly* different in these terms may *in fact* differ. In the previous section it was seen that in a fluid, the observable is obtained as the ensemble sum of contributions from all relevant centres. As in the case of plane waves, the factor $\exp(iz_{\xi}\Delta k)$ continues to play an important role and represents any net (longitudinal) optical phase shift associated with a process mediated by a chiral centre ξ located at $\mathbf{r}_{\xi} = (r,\phi,z)$. However, another (azimuthal) phase factor of similar form, $\exp(-i\Delta l\phi)$, now appears alongside it[2] and also enters M_{fi}^{ξ} through the radiation tensor $\mathbf{S}^{(n)}$. Here $\Delta l\hbar$ represents any mismatch between the orbital angular momentum sum of all input and the sum of all output photons involved in the process at the chiral centre. First, if the optical process is parametric, then $\Delta l = 0$ and all dependence on the sign of the orbital angular momentum of the radiation is lost. The transition rate, proportional to the modulus square of the rotationally

[2] In separate work we have verified that this form for the azimuthal phase factor is unchanged when higher order multipole couplings are entertained.

averaged quantum amplitude, $\left|\left\langle M_{fi}\right\rangle\right|^2$, has no dependence on the sign of the orbital angular momentum l and therefore is not a chiral discriminator. If, on the other hand, we consider a *non*-parametric process where $\Delta l \neq 0$, the incoherent character means that the rate is proportional to the rotational average of the modulus square of the quantum amplitude, $\left\langle\left|M_{fi}\right|^2\right\rangle$, in which the azimuthal phase factor cancels each and every centre ξ; again the observable transition rate does not involve the orbital angular momentum or its sign. Hence the circularity of the beam is immaterial.

In the case of chiral molecules interacting with *twisted* light comprising *circular* photons, and in view of the observations made above, the problem reduces to exactly that which applies to other beams of circular polarisation – but now there is a mechanism for the associated signals actually to differ. Again the helicity of the wave-front is immaterial so far as chiral specificity is concerned. Consider for example Figure 1(b), in which a left-handed twisted beam comprising left-handed circular photons interacts with a system of left-handed molecular enantiomers. Let us entertain the simplest optical process in which chirality can be manifest, namely circular dichroism. Then, the quantum interference terms responsible for chiral discrimination change sign if either the circularity of the photons or the isomeric form of the molecules is changed; the interference terms are invariant to a change of both. No component of the rate changes if, for example, the beam becomes a right-handed vortex but the sense of the photons and molecular handedness is the same as before. Note that this conclusion is contingent upon the azimuthal phase factors for the electric and magnetic fields being identical; however this is a necessary consequence of Maxwell's equations, as has been shown explicitly for quantised Laguerre-Gaussian modes. [17] This lack of exchange of orbital angular momentum between the optical vortex and the chiral matter has been experimentally demonstrated by the research group at The Laboratory of Chemical and Biological Dynamics at Leuven, Belgium. [22]

3.1 Explicit Calculation for Circular Dichroism

To ground the above analysis with a specific application, let us consider the simplest case, *i.e.* photon absorption, where any enantiomeric rate dependence would signify a type of dichroism. In this case the initial and final state of the system is given by;

$$|i\rangle = \left|A; n\left(k\hat{\mathbf{z}}, \lambda, l, p\right)\right\rangle \tag{16a}$$

$$|f\rangle = \left|B; (n-1)\left(k\hat{\mathbf{z}}, \lambda, l, p\right)\right\rangle \tag{16b}$$

The three leading contributions to the transition matrix are in this case expressible as;

$$\begin{aligned} M_{fi} &= \langle f| - \mu_i e_i^{\perp} - m_i b_i + q_{ij} \nabla_j e_i^{\perp} |i\rangle \\ &= \underbrace{\langle f|\left(-\mu_i e_i^{\perp}\right)|i\rangle}_{M_{fi}^{\mu}} + \underbrace{\langle f|\left(-m_i b_i\right)|i\rangle}_{M_{fi}^{m}} + \underbrace{\langle f|\left(q_{ij} \nabla_j e_i^{\perp}\right)|i\rangle}_{M_{fi}^{q}}, \end{aligned} \tag{17}$$

whose evaluation invokes the result;

$$\nabla_j e_i^{\perp}(\mathbf{r}) = \sum_{k,\lambda,l,p} \left\{ \mathrm{i} \left(\frac{\hbar ck}{2A_{lp}\varepsilon_0 V} \right)^{\frac{1}{2}} e_i^{(\lambda)}(k\hat{\mathbf{z}}) f_{lp}(r) a_{lp}^{(\lambda)}(k\hat{\mathbf{z}}) \left[\mathbf{G}_{lp}(r,\phi) \right]_j \exp\left[\mathrm{i}(kz - l\phi) \right] + h.c. \right\} \quad (18)$$

with

$$\mathbf{G}_{lp}(r,\phi) = \left[\tfrac{\partial_r f_{lp}}{f_{lp}} \cos\phi + \tfrac{\mathrm{i}l}{r} \sin\phi \right] \hat{x} + \left[\tfrac{\partial_r f_{lp}}{f_{lp}} \sin\phi - \tfrac{\mathrm{i}l}{r} \cos\phi \right] \hat{y} + \left[\mathrm{i}k \right] \hat{z} \quad (19)$$

This is worth considering in detail because it can be seen that, when the quadrupole interaction is involved, not only is the handness of l present in the azimuthal factor, but it is also present in the profile factor $\mathbf{G}_{lp}(r,\phi)$. Therefore it is at least conceivable that, when electric-quadrupole interactions are taken into account, the selectivity of OAM is similar to that of the photon spin. Using the notation given in equation (9) we have;

$$\begin{aligned} M_{fi}^{\xi} &= \exp\left(i\Delta\mathbf{k}\cdot\mathbf{R}_{\xi}\right) \sum_{e,m=0}^{1} \mathbf{S}^{(r)}_{e;\,m;\,1-e-m} \otimes \mathbf{T}^{(r)}_{e;\,m;\,1-e-m} \\ &= \exp\left(i\Delta\mathbf{k}\cdot\mathbf{R}_{\xi}\right) \left[\mathbf{S}^{(1)}_{1;\,0;\,0} \otimes \mathbf{T}^{(1)}_{1;\,0;\,0} + \mathbf{S}^{(1)}_{0;1;\,0} \otimes \mathbf{T}^{(1)}_{0;1;\,0} + \mathbf{S}^{(1)}_{0;0;\,1} \otimes \mathbf{T}^{(1)}_{0;0;\,1} \right], \end{aligned} \quad (20)$$

which we can succinctly represent as a sum of the three terms $M_{fi}^{\mu} = \mathbf{S}^{(1)}_{1;\,0;\,0} \otimes \mathbf{T}^{(1)}_{1;\,0;\,0}$; $M_{fi}^{m} = \mathbf{S}^{(1)}_{0;1;\,0} \otimes \mathbf{T}^{(1)}_{0;1;\,0}$ and $M_{fi}^{q} = \mathbf{S}^{(1)}_{0;0;\,1} \otimes \mathbf{T}^{(1)}_{0;0;\,1}$.

Since single-photon absorption is associated with an inelastic interaction it is non-parametric and the rate observable is proportional to $\left\langle \left| M_{fi} \right|^2 \right\rangle$. Thus, if the rate is to depend on the sign of l it can only be through the x and y components of the $\mathbf{G}_{lp}(r,\phi)$ tensor. When the term $\left\langle \left| M_{fi} \right|^2 \right\rangle$ is calculated we obtain;

$$\begin{aligned} \langle \Gamma \rangle^{\text{incoherent}} &= \frac{2\pi}{\hbar} \rho_f N \Big\{ \left\langle M_{fi}^{\mu} \bar{M}_{fi}^{\mu} \right\rangle + \left\langle M_{fi}^{m} \bar{M}_{fi}^{m} \right\rangle + \left\langle M_{fi}^{q} \bar{M}_{fi}^{q} \right\rangle \\ &\quad + 2\,\mathrm{Re}\left[\left\langle M_{fi}^{\mu} \bar{M}_{fi}^{m} \right\rangle \right] + 2\,\mathrm{Re}\left[\left\langle M_{fi}^{\mu} \bar{M}_{fi}^{q} \right\rangle \right] + 2\,\mathrm{Re}\left[\left\langle M_{fi}^{m} \bar{M}_{fi}^{q} \right\rangle \right] \Big\} \end{aligned} \quad (21)$$

Of the six terms in equation (21) we already know that $\left\langle M_{fi}^{\mu} \bar{M}_{fi}^{\mu} \right\rangle$, $\left\langle M_{fi}^{m} \bar{M}_{fi}^{m} \right\rangle$ and $\left\langle M_{fi}^{\mu} \bar{M}_{fi}^{m} \right\rangle$ are independent of the handness of l. The other three terms, those which are connected to the electric-quadrupole interaction, must be analysed individually. We first consider the $\left\langle M_{fi}^{q} \bar{M}_{fi}^{q} \right\rangle$ case. The explicit expression of this term is;

$$\begin{aligned} \left\langle M_{fi}^{q} \bar{M}_{fi}^{q} \right\rangle &= -\left(\frac{\hbar ck}{2A_{lp}\varepsilon_o V} \right) \frac{\left| f_{lp} \right|^2}{30} \times \\ &\times e_{i_1}^{(\lambda)}(\mathbf{k}) \bar{e}_{i_3}^{(\lambda)}(\mathbf{k}) \left[G_{lp}(r,\phi) \right]_{i_2} \left[\bar{G}_{lp}(r,\phi) \right]_{i_4} I^{(4)}_{i_1 i_2 i_3 i_4;\lambda_1\lambda_2\lambda_3\lambda_4} \, q^{ba}_{\lambda_1\lambda_2} \bar{q}^{ba}_{\lambda_3\lambda_4} \end{aligned} \quad (22)$$

where the 4th rank rotational average tensor is; [13]

$$I^{(4)}_{i_1i_2i_3i_4;\lambda_1\lambda_2\lambda_3\lambda_4} = \frac{1}{30}\begin{bmatrix}\delta_{i_1i_2}\delta_{i_3i_4}\\ \delta_{i_1i_3}\delta_{i_2i_4}\\ \delta_{i_1i_4}\delta_{i_2i_3}\end{bmatrix}^{\mathrm{T}}\begin{bmatrix}4 & -1 & -1\\ -1 & 4 & -1\\ -1 & -1 & 4\end{bmatrix}\begin{bmatrix}\delta_{\lambda_1\lambda_2}\delta_{\lambda_3\lambda_4}\\ \delta_{\lambda_1\lambda_3}\delta_{\lambda_2\lambda_4}\\ \delta_{\lambda_1\lambda_4}\delta_{\lambda_2\lambda_3}\end{bmatrix} \tag{23}$$

Therefore $\left\langle M^q_{fi}\bar{M}^q_{fi}\right\rangle$ takes the form;

$$\left\langle M^q_{fi}\bar{M}^q_{fi}\right\rangle = -\left(\frac{\hbar ck}{2A_{lp}\varepsilon_o V}\right)\frac{\left|f_{lp}\right|^2}{30}\times$$

$$\begin{bmatrix}\left[\mathbf{e}^{(\lambda)}\cdot\mathbf{G}_{lp}(r,\phi)\right]\left[\bar{\mathbf{e}}^{(\lambda)}(\mathbf{k})\cdot\bar{\mathbf{G}}_{lp}(r,\phi)\right]\\ \left[\mathbf{e}^{(\lambda)}(\mathbf{k})\cdot\bar{\mathbf{e}}^{(\lambda)}(\mathbf{k})\right]\left[\mathbf{G}_{lp}(r,\phi)\cdot\bar{\mathbf{G}}_{lp}(r,\phi)\right]\\ \left[\mathbf{e}^{(\lambda)}(\mathbf{k})\cdot\bar{\mathbf{G}}_{lp}(r,\phi)\right]\left[\bar{\mathbf{e}}^{(\lambda)}(\mathbf{k})\cdot\mathbf{G}_{lp}(r,\phi)\right]\end{bmatrix}^{\mathrm{T}}\begin{bmatrix}4 & -1 & -1\\ -1 & 4 & -1\\ -1 & -1 & 4\end{bmatrix}\begin{bmatrix}q^{ba}_{\lambda\lambda}\bar{q}^{ba}_{vv}\\ q^{ba}_{\lambda v}\bar{q}^{ba}_{\lambda v}\\ q^{ba}_{\lambda v}\bar{q}^{ba}_{v\lambda}\end{bmatrix} \tag{24}$$

To proceed further we have to consider the different possible polarisations of the absorbed photon. Let us first consider that the photon is linearly polarised; say $\mathbf{e}^{(\lambda)}(\mathbf{k}) = \hat{y}$. Then;

$$\left\langle M^q_{fi}\bar{M}^q_{fi}\right\rangle = -\left(\frac{\hbar ck}{2A_{lp}\varepsilon_o V}\right)\frac{\left|f_{lp}\right|^2}{30}\times\begin{bmatrix}\left|\frac{\partial_r f_{lp}}{f_{lp}}\right|^2\sin^2\phi + \frac{l^2}{r^2}\cos^2\phi\\ \left|\frac{\partial_r f_{lp}}{f_{lp}}\right|^2 + \frac{l^2}{r^2} + k^2\\ \left|\frac{\partial_r f_{lp}}{f_{lp}}\right|^2\sin^2\phi + \frac{l^2}{r^2}\cos^2\phi\end{bmatrix}^{\mathrm{T}}\begin{bmatrix}4 & -1 & -1\\ -1 & 4 & -1\\ -1 & -1 & 4\end{bmatrix}\begin{bmatrix}q^{ba}_{\lambda\lambda}\bar{q}^{ba}_{vv}\\ q^{ba}_{\lambda v}\bar{q}^{ba}_{\lambda v}\\ q^{ba}_{\lambda v}\bar{q}^{ba}_{v\lambda}\end{bmatrix} \tag{25}$$

which is independent of the sign of l. If, instead, we consider the photon to be circularly polarised; i.e. $\mathbf{e}^{(\lambda)}(\mathbf{k}) = \frac{1}{\sqrt{2}}(\hat{x} + \mathrm{i}\hat{y})$, then;

$$\left\langle M^q_{fi}\bar{M}^q_{fi}\right\rangle = -\left(\frac{\hbar ck}{2A_{lp}\varepsilon_o V}\right)\frac{\left|f_{lp}\right|^2}{60}\times\begin{bmatrix}\left|\frac{\partial_r f_{lp}}{f_{lp}}\right|^2 + \frac{l^2}{r^2}\\ 2\left(\left|\frac{\partial_r f_{lp}}{f_{lp}}\right|^2 + \frac{l^2}{r^2} + k^2\right)\\ \left|\frac{\partial_r f_{lp}}{f_{lp}}\right|^2 + \frac{l^2}{r^2}\end{bmatrix}^{\mathrm{T}}\begin{bmatrix}4 & -1 & -1\\ -1 & 4 & -1\\ -1 & -1 & 4\end{bmatrix}\begin{bmatrix}q^{ba}_{\lambda\lambda}\bar{q}^{ba}_{vv}\\ q^{ba}_{\lambda v}\bar{q}^{ba}_{\lambda v}\\ q^{ba}_{\lambda v}\bar{q}^{ba}_{v\lambda}\end{bmatrix} \tag{26}$$

which is also independent of the l sign.

The next term we consider is $\left\langle M^\mu_{fi}\bar{M}^q_{fi}\right\rangle$, which can be written as;

$$\left\langle M^{\mu}_{fi}\bar{M}^{q}_{fi}\right\rangle = -\left(\frac{\hbar ck}{2A_{lp}\varepsilon_o V}\right)\left|f_{lp}\right|^2 e^{(\lambda)}_{i_1}(\mathbf{k})\bar{e}^{(\lambda)}_{i_2}(\mathbf{k})\left[\bar{G}_{lp}(r,\phi)\right]_{i_3} I^{(3)}_{i_1i_2i_3;\lambda_1\lambda_2\lambda_3}\mu^{ba}_{\lambda_1}\bar{q}^{ba}_{\lambda_2\lambda_3}$$
$$= -\left(\frac{\hbar ckn}{2A_{lp}\varepsilon_o V}\right)\frac{\left|f_{lp}\right|^2}{6}\left[\mathbf{e}^{(\lambda)}(\mathbf{k})\cdot\bar{\mathbf{e}}^{(\lambda)}(\mathbf{k})\times\bar{\mathbf{G}}_{lp}(r,\phi)\right]\left[\varepsilon_{\lambda_1\lambda_2\lambda_3}\mu^{ba}_{\lambda_1}\bar{q}^{ba}_{\lambda_2\lambda_3}\right] \quad (27)$$

where we have used $I^{(3)}_{i_1i_2i_3;\lambda_1\lambda_2\lambda_3} = \frac{1}{6}\varepsilon_{i_1i_2i_3}\varepsilon_{\lambda_1\lambda_2\lambda_3}$. In the case of linearly polarised light it can easily be seen that $\left[\mathbf{e}^{(\lambda)}(\mathbf{k})\cdot\mathbf{e}^{(\lambda)}(\mathbf{k})\times\mathbf{G}_{lp}(r,\phi)\right]\equiv 0$. The other possible case, *i.e.* circularly polarised light, results in $\left[\mathbf{e}^{(\lambda)}(\mathbf{k})\cdot\bar{\mathbf{e}}^{(\lambda)}(\mathbf{k})\times\bar{\mathbf{G}}_{lp}(r,\phi)\right]=(-k)$, and therefore $\left\langle M^{\mu}_{fi}\bar{M}^{q}_{fi}\right\rangle$ does not retain any information of the handness of the optical vortex, therefore independent of *l*.

It remains to consider $\left\langle M^{m}_{fi}\bar{M}^{q}_{fi}\right\rangle$, given by;

$$\left\langle M^{\mu}_{fi}\bar{M}^{q}_{fi}\right\rangle = -\left(\frac{\hbar k}{2A_{lp}\varepsilon_o V}\right)\left|f_{lp}\right|^2 \varepsilon_{i_1i_4i_5}\left(\hat{\mathbf{k}}\right)_{i_4} e^{(\lambda)}_{i_5}(\mathbf{k})\bar{e}^{(\lambda)}_{i_2}(\mathbf{k})\left[\bar{G}_{lp}(r,\phi)\right]_{i_3} I^{(3)}_{i_1i_2i_3;\lambda_1\lambda_2\lambda_3} m^{ba}_{\lambda_1}\bar{q}^{ba}_{\lambda_2\lambda_3}$$
$$= \left(\frac{\hbar k}{2A_{lp}\varepsilon_o V}\right)\frac{\left|f_{lp}\right|^2}{6}\left[\bar{G}_{lp}(r,\phi)\right]_z \varepsilon_{\lambda_1\lambda_2\lambda_3} m^{ba}_{\lambda_1}\bar{q}^{ba}_{\lambda_2\lambda_3} \quad (28)$$

As $\left[\bar{G}_{lp}(r,\phi)\right]_z = -ik$, the result is independent of the signature of *l*. Thus it is confirmed, by explicit calculation, that the helicity of a Laguerre-Gaussian beam cannot engage in any enantiomerically specific sense in its absorption by a chiral molecule.

4 Conclusion

Through rigorous analysis of the grounds for chiroptical behaviour involving electric and magnetic dipole interactions, it transpires that the helicity of twisted beams cannot engage through any parametric or non-parametric optical process with the chirality of a molecular system – other than through any circularity of its photons. Thus, the manifestations of orbital angular momentum differ markedly from those associated with photon spin angular momentum. The analysis is comprehensive, within the scope of the assumptions given, with recent experimental verification confirming our theoretical results. [22] It remains to study those optical processes invoking electric quadrupole and other higher multipole interactions; the analysis reported here excludes the possibility of any magnetic dipole contributions to such chiral specificity. In the specific case of electric quadrupole interactions, the analysis carried out for circular dichroism illustrates the difficulties that higher order multipole considerations introduce, where the simplest symmetry arguments are not sufficient to attest enantiomeric specificity. Assessing the possible chiral selectivity of the *mechanical* interactions of such twisted beams is an issue that merits separate attention, and which is the subject of other recent and ongoing work. [21,23]

Appendix 1 LG Functions, Generalised Laguerre Polynomials and their Properties

The positive frequency, un-normalised electric field of an LG mode as expressed by equation (15) can be written as;

$$\mathbf{e}^{\perp}(\mathbf{r}) = \mathrm{i}\boldsymbol{\varepsilon}_{lp}^{(\lambda)}(k\hat{\mathbf{z}}) f_{lp}(r) \mathrm{a}_{lp}^{(\lambda)}(k\hat{\mathbf{z}}) \exp\left[\mathrm{i}\theta_{klp}(k\hat{\mathbf{z}})\right] \quad \text{(A1.1)}$$

The position-dependent polarisation functions $\boldsymbol{\varepsilon}_{lp}^{(\lambda)}(k\hat{\mathbf{z}})$ and $\theta_{klp}(k\hat{\mathbf{z}})$ are, respectively, the mode amplitude distribution function and phase function, which are explicitly given by; [20]

$$\varepsilon_{lp}^{(\lambda)}(k\hat{z}) = \varepsilon_{k00} \frac{C_{lp}}{\left(1+z^2/z_R^2\right)^{1/2}} \left(\frac{\sqrt{2}r}{w(z)}\right)^{|l|} L_p^{|l|}\left(\frac{2r^2}{w^2(z)}\right) \exp\left[-r^2/w^2(z)\right] \quad ,\text{(A1.2)}$$

$$\theta_{klp}(k\hat{z}) = \frac{kr^2 z}{2\left(z^2+z_R^2\right)} + l\phi + (2p+l+1)\tan^{-1}(z/z_R) + kz \ . \quad \text{(A1.3)}$$

Here ε_{k00} is the amplitude for a plane wave of wave-vector $\mathbf{k}$; $C_{lp} = \sqrt{p!/(|l|+p)!}$ is a normalisation factor; $L_p^{|l|}(x)$ is the generalised Laguerre polynomial of indices $|l|$ and p. The width $w(z)$ of the beam is defined in terms of the Rayleigh range, z_R, by $w^2(z) = 2\left(z^2+z_R^2\right)/kz_R$. The generalised Laguerre polynomials $L_p^l(x)$ are well-known functions whose properties can be found in the established literature. [24,25] Different works of reference give explicit expressions which differ by a constant factor. Table A1 gives the expressions as used in our work.

Table A1: Properties of generalized Laguerre Polynomials [24,25].

Explicit expression:	$L_p^l(x) = \sum_{m=0}^{p} (-1)^m \binom{p+l}{p-m} \frac{1}{m} x^m$
Orthogonality relationship	$\int_0^{\infty} e^{-x} x^l L_p^l(x) L_m^l(x)\, dx = \frac{(l+p)!}{p!} \delta_{p,m}$

The products of two identical generalized Laguerre polynomials are also tabulated in the literature, as;

$$\left[L_p^l(x)\right]^2 = \frac{\Gamma(1+l+p)}{p!} \sum_{k=0}^{\infty} \frac{(2p-2k)!\,(2k)!}{\Gamma(1+l-k)\,(n-k)!} L_{2k}^{2l}(2x) \quad \text{(A1.4)}$$

In passing we note that in reference [24] there is a typographical error in the gamma function in the denominator of the expression (A1.4). However, even with correction the above formula is incorrect. We have found a different expression, which also is a linear combination of $L_{2k}^{2l}(2x)$ polynomials;

$$\left[L_p^l(x)\right]^2 = \sum_{n=0}^{p} c_{p,n}^l L_{2n}^{2l}(2x) \tag{A1.5}$$

Here the coefficients c_{np}^l are more complicated and different to those in equation (A1.4);

$$c_{p,n}^l = \begin{cases} \dfrac{(2p)!}{(2p+2l)!} \displaystyle\sum_{t=0}^{2(n+p)} \sum_{m=0}^{2n} \sum_{k=0}^{n} \frac{1}{2^m} d_{2p,t-m}^{2l}\, d_{n,k}^l\, d_{n,m-k}^l\, (2l+t)! & \text{for } 0 \le n \le p, \\ 0 & \text{otherwise,} \end{cases} \tag{A1.6}$$

with;

$$d_{n,k}^l = \begin{cases} \dfrac{(-1)^k}{k!} \dbinom{n+l}{n-k} & \text{for } 0 \le k \le n, \\ 0 & \text{otherwise.} \end{cases}$$

For $p=0$, a trivial example, the results given by both expressions, (A1.4) and (A1.5) are identical. However, for $p=1$ there is a difference by a numerical factor between the two equations. For $p=2$ the difference is more than just a factor, and as the p index increases this difference is accentuated. We have verified that the expression given in (A1.5), obtained by using the properties listed in Table A1, is correct and should be used instead of (A1.4).

Appendix 2 Optical Interaction Tensors and Rotational Averages

For any chiroptical process taking place in a fluid, or otherwise microscopically disordered system, it is required to evaluate:

$$\left\langle \left|M_{fi}\right|^2 \right\rangle = \left\langle \sum_{\xi,\xi'}^{N} M_{fi}^{\xi} \bar{M}_{fi}^{\xi'} \right\rangle \tag{A2.1}$$

see, for example, equation (12). Using the specific expression for the transition matrix given in equation (9) we have;

$$\left\langle \left| M_{fi} \right|^2 \right\rangle = \sum_{\xi}^{N} \sum_{e,m,e',m'=0}^{n} \mathbf{S}^{(r);\xi}_{e;\,m;\,n-e-m} \overline{\mathbf{S}}^{(r);\xi}_{e';\,m';\,n-e'-m'} \otimes \left\langle \mathbf{T}^{(r);\xi}_{e;\,m;\,n-e-m} \overline{\mathbf{T}}^{(r);\xi}_{e';\,m';\,n-e'-m'} \right\rangle +$$

$$\sum_{\xi \neq \xi'}^{N} \sum_{e,m,e',m'=0}^{n} \exp\left[i\Delta\mathbf{k} \cdot \left(\mathbf{R}_{\xi} - \mathbf{R}_{\xi'} \right) \right] \mathbf{S}^{(r);\xi}_{e;\,m;\,n-e-m} \overline{\mathbf{S}}^{(r');\xi'}_{e';\,m';\,n-e'-m'} \otimes \left\langle \mathbf{T}^{(r);\xi}_{e;\,m;\,n-e-m} \right\rangle \left\langle \overline{\mathbf{T}}^{(r);\xi'}_{e';\,m';\,n-e'-m'} \right\rangle$$

$$= N \sum_{e,m,e',m'=0}^{n} \mathbf{S}^{(r);\xi}_{e;\,m;\,n-e-m} \overline{\mathbf{S}}^{(r);\xi}_{e';\,m';\,n-e'-m'} \otimes \left\langle \mathbf{T}^{(r);\xi}_{e;\,m;\,n-e-m} \overline{\mathbf{T}}^{(r);\xi}_{e';\,m';\,n-e'-m'} \right\rangle +$$

$$\sum_{\xi \neq \xi'}^{N} \sum_{e,m,e',m'=0}^{n} \exp\left[i\Delta\mathbf{k} \cdot \left(\mathbf{R}_{\xi} - \mathbf{R}_{\xi'} \right) \right] \mathbf{S}^{(r);\xi}_{e;\,m;\,n-e-m} \overline{\mathbf{S}}^{(r');\xi'}_{e';\,m';\,n-e'-m'} \otimes \left\langle \mathbf{T}^{(r);\xi}_{e;\,m;\,n-e-m} \right\rangle \left\langle \overline{\mathbf{T}}^{(r);\xi'}_{e';\,m';\,n-e'-m'} \right\rangle \quad \text{(A2.2)}$$

The first term on the right of equation (A2.2) signifies an incoherent contribution, and the second term a coherent response. The latter dominates if the process is parametric, and evaluation of the corresponding term requires the implementation of a rotational average over $\mathbf{T}^{(r)}$. However if the process is *non*-parametric the coherent term vanishes (through the interference of the different phase factors for each molecular centre) and the response derives solely from the incoherent term, where the rotational average is effected over $\left|\mathbf{T}^{(r)}\right|^2$, invoking tensors of rank $(r + r')$. In each case the rotational average involves the molecular tensors alone, since it is considered that the radiation components are fixed with respect to a laboratory-anchored cartesian frame. Since $\mathbf{T}^{(r)}$ refers to a molecular property it is convenient to express it with respect to a molecule fixed frame through the relation;

$$\begin{aligned} \left\langle \mathbf{T}^{(r);\xi}_{e;\,m;\,n-e-m} \right\rangle &= \mathbf{T}^{(r);\xi}_{e;\,m;\,n-e-m} I^{(r)} \\ &= \left(T_{\lambda_1 \lambda_2 \ldots \lambda_r} \right)^{(r);\xi}_{e;\,m;\,n-e-m} I^{(r)}_{\lambda_1 \ldots \lambda_r; i_1 \ldots i_r} \end{aligned} \quad \text{(A2.3)}$$

where $I^{(r)}$ is the rotational average tensor of rank $(2r)$. The Latin and Greek indices refer to space-fixed and molecule-fixed frames respectively. Then the rotational average of the molecular tensor is given by;

$$\left\langle \left| M_{fi} \right|^2 \right\rangle = N \sum_{e,m,e',m'=0}^{n} \mathbf{S}^{(r)}_{e;\,m;\,n-e-m} \overline{\mathbf{S}}^{(r')}_{e';\,m';\,n-e'-m'} I^{(r+r')} \mathbf{T}^{(r)}_{e;\,m;\,n-e-m} \overline{\mathbf{T}}^{(r)}_{e';\,m';\,n-e'-m'} +$$

$$\sum_{\xi \neq \xi'}^{N} \sum_{e,m,e',m'=0}^{n} \exp\left[i\Delta\mathbf{k} \cdot \left(\mathbf{R}_{\xi} - \mathbf{R}_{\xi'} \right) \right] \mathbf{S}^{(r);\xi}_{e;\,m;\,n-e-m} I^{(r)} \mathbf{T}^{(r);\xi}_{e;\,m;\,n-e-m} \overline{\mathbf{S}}^{(r');\xi'}_{e';\,m';\,n-e'-m'} I^{(r')} \overline{\mathbf{T}}^{(r);\xi}_{e';\,m';\,n-e'-m'}$$

Thus for a parametric process, $\Delta\mathbf{k} = 0$, we have;

$$\begin{aligned} \left\langle \left| M_{fi} \right|^2 \right\rangle^{coherent} &\approx N^2 \left| \left\langle M^{\xi}_{fi} \right\rangle \right|^2 \\ &\approx N^2 \sum_{e,m,e',m'=0}^{n} \mathbf{S}^{(r);\xi}_{e;\,m;\,n-e-m} I^{(r)} \mathbf{T}^{(r);\xi}_{e;\,m;\,n-e-m} \overline{\mathbf{S}}^{(r');\xi'}_{e';\,m';\,n-e'-m'} I^{(r')} \overline{\mathbf{T}}^{(r);\xi}_{e';\,m';\,n-e'-m'} \end{aligned} \quad \text{(A2.4)}$$

On the other hand if $\Delta\mathbf{k} \neq 0$ then;

$$\left\langle \left| M_{fi} \right|^2 \right\rangle^{incoherent} \square N \left\langle \left| M_{fi}^{\xi} \right|^2 \right\rangle$$

$$\square N \sum_{e,m,e',m'=0}^{n} \mathbf{S}^{(r)}_{e;\,m;\,n-e-m} \overline{\mathbf{S}}^{(r')}_{e';\,m';\,n-e'-m'} I^{(r+r')} \mathbf{T}^{(r)}_{e;\,m;\,n-e-m} \overline{\mathbf{T}}^{(r)}_{e';\,m';\,n-e'-m'} \quad \text{(A2.5)}$$

To evaluate the result in either case, using equation (A2.4) or (A2.5) as appropriate, it is necessary to engage the double tensor $I^{(k)}$, which is explicitly the rotational average of a direction cosine product;

$$I^{(k)} \equiv \left\langle l_{i_1\lambda_1} l_{i_2\lambda_2} \ldots l_{i_k\lambda_k} \right\rangle = \int\!\!\int\!\!\int l_{i_1\lambda_1} l_{i_2\lambda_2} \ldots l_{i_k\lambda_k} \sin\theta \; d\phi \, d\theta \, d\psi \quad \text{(A2.6)}$$

where each $l_{i_k\lambda_k}$ is the cosine of the angle between the space-fixed axis and the molecule-fixed axis. The rotational average tensor $I^{(k)}$ can be expressed as a linear combination of isotropic tensors, *i.e.* tensors that are invariant under rotation. It is important to notice that the way that the tensor $I^{(k)}$ is defined in equation (A2.6) implies that the rank of the tensor is $2k$. Each member of the linear combination is a product of two isotropic tensors, one referred to the space-fixed frame and the other to the molecule-fixed frame. An important property of these products is that the Latin and Greek indices do not mix; explicitly;

$$I^{(k)} = \sum_{st} m_{st}^{(k)} \mathbf{f}_s^{(k)} \mathbf{g}_t^{(k)} \quad \text{(A2.7)}$$

Here $\mathbf{f}_s^{(k)}$ ($\mathbf{g}_t^{(k)}$) is an k-rank tensor in the space-fixed (molecule-fixed) frame, and the tensor indices $i_1 \ldots i_k$ $(\lambda_1 \ldots \lambda_k)$ are suppressed for convenience. Then the rotational average of the quantum amplitude product emerges as;

$$\begin{aligned}\left\langle \left| M_{fi} \right|^2 \right\rangle = {} & N \sum_{e,m,e',m'=0}^{n} \sum_{ts} m_{ts}^{(r+r')} \left[\mathbf{f}_t^{(r+r')} \mathbf{S}^{(r)}_{e;\,m;\,n-e-m} \overline{\mathbf{S}}^{(r')}_{e';\,m';\,n-e'-m'} \right] \left[\mathbf{g}_s^{(r+r')} \mathbf{T}^{(r)}_{e;\,m;\,n-e-m} \overline{\mathbf{T}}^{(r')}_{e';\,m';\,n-e'-m'} \right] \\ & + N(N-1) \sum_{e,m,e',m'=0}^{n} \sum_{ts} m_{ts}^{(r+r')} \left[\mathbf{f}_t^{(r)} \mathbf{S}^{(r);\xi}_{e;\,m;\,n-e-m} \right] \left[\mathbf{g}_s^{(r)} \mathbf{T}^{(r);\xi}_{e;\,m;\,n-e-m} \right] \sum_{t's'} m_{t's'}^{(r')} \left[\mathbf{f}_{t'}^{(r')} \overline{\mathbf{S}}^{(r');\xi'}_{e';\,m';\,n-e'-m'} \right] \left[\mathbf{g}_{s'}^{(r')} \overline{\mathbf{T}}^{(r');\xi'}_{e';\,m';\,n-e'-m'} \right]\end{aligned} \quad \text{(A2.8)}$$

Finally, for calculational implementation a simpler notation can be adopted, where the rotational average of the molecular and radiation tensors is implied;

$$\begin{aligned}\left\langle \left| M_{fi} \right|^2 \right\rangle = {} & N \sum_{e,m,e',m'=0}^{n} \sum_{ts} m_{ts}^{(r+r')} \left\langle S^{(r)}_{e;\,m;\,n-e-m} \overline{S}^{(r')}_{e';\,m';\,n-e'-m'} \right\rangle \left\langle T^{(r)}_{e;\,m;\,n-e-m} \overline{T}^{(r')}_{e';\,m';\,n-e'-m'} \right\rangle + \\ & N(N-1) \sum_{e,m,e',m'=0}^{n} \sum_{ts} m_{ts}^{(r+r')} \left\langle S^{(r)}_{e;\,m;\,n-e-m} \right\rangle \left\langle T^{(r)}_{e;\,m;\,n-e-m} \right\rangle \times \\ & \times \sum_{t's'} m_{t's'}^{(r')} \left\langle \overline{S}^{(r')}_{e';\,m';\,n-e'-m'} \right\rangle \left\langle \overline{T}^{(r')}_{e';\,m';\,n-e'-m'} \right\rangle\end{aligned} \quad \text{(A2.9)}$$

Acknowledgements

We are pleased to acknowledge support from the Engineering and Physical Sciences Research Council for this work.

References

[1] P. W. Atkins, *Quanta: A Handbook of Concepts*, Clarendon Press, Oxford (1974)

[2] L. Allen, M. W. Beijersbergen, R. J. C. Spreeuw and J. P. Woerdman, *Phys. Rev. A* **45**, 8185-8189 (1992)

[3] L. Allen, M. J. Padgett and M. Babiker, *Prog. Opt.* **39**, 291-372 (1999)

[4] L. Allen, S. M. Barnett and M. J. Padgett, *Optical Angular Momentum*, IOP, Bristol, (2003)

[5] M. Padgett, J. Courtial and L. Allen, *Physics Today* **57** (5), 35-40 (2004)

[6] D. L. Andrews, L. C. Dávila Romero and M. Babiker, *Opt. Commun.* **237**, 133-139 (2004)

[7] L. D. Barron, *Molecular Light Scattering and Optical Activity*, University Press, Cambridge (1982)

[8] L. D. Barron, *Nature*, **405**, 895-896 (2000)

[9] B. Piccirillo, A. Vella and E. Santamato, *J. Opt. B-Quantum Semicl. Opt.*, **4**, S20-S24 (2002)

[10] C. Cohen-Tannoudji, J. Dupont-Roc and G. Grynberg, *Photons and Atoms: Introduction to Quantum Electrodynamics*, Wiley, New York (1989)

[11] D. L. Andrews and P. Allcock, *Optical Harmonics in Molecular Systems*, Wiley-VCH, Weinheim, Germany (2002)

[12] J. G. Kirkwood, *J. Chem. Phys.*, **5**, 479 (1937)

[13] D. P. Craig and T. Thirunamachandran, *Molecular Quantum Electrodynamics: an Introduction to Radiation-Molecule Interactions*, Dover Publications, Mineola, New York (1998)

[14] E. A. Power and T. Thirunamachandran, *Am. J. Phys.*, **50**, 757-757 (1982)

[15] E. A. Power and T. Thirunamachandran, *Phys. Rev. A*, **28**, 2671-2675 (1983)

[16] H. M. Nussenzveig, *Introduction to Quantum Optics*, Gordon and Breach, London (1973)

[17] L. C. Dávila Romero, D. L. Andrews and M. Babiker, *J. Opt. B: Q. and Semicl. Opt.*, **4**, S66 (2002)

[18] S. Naguleswaran and G. E. Stedman, *J. Phys. B-At. Mol. Opt.*, **29**, 4027-4040 (1996)

[19] D. L. Andrews, P. Allcock and A. A. Demidov, *Chem. Phys.*, **190**, 1-9 (1995)

[20] M. Babiker, C. R. Bennett, D. L. Andrews and L. C. Dávila Romero, *Phys. Rev. Lett.*, **89**, art. no.-143601 (2002)

[21] H. He, M. E. J. Friese, N. R. Heckenberg and H. Rubinsztein-Dunlop, *Phys. Rev. Lett.* **75**, 826-829 (1995)

[22] F. Araoka, T. Verbiest, K. Clays and A. Persoons, *Phys. Rev. A*, **71**, art. no.-055401 (2005)

[23] L. C. Dávila Romero, A. R. Carter, M. Babiker and D. L. Andrews, *Proceedings of SPIE, Nanomanipulation with light*, **5736**, 150-157 (2005)

[24] I. S. Gradshteæin and I. M. Ryzhik, *Table of Integrals, Series, and Products*, 4th ed., Academic Press, New York, pp. 1037-1039 (1965)

[25] W. Magnus and F. Oberhetlinger, ed., *Formulas and Theorems for the Special Functions of Mathematical Physics*, Chelsea Publishing, New York, (1949)

In: Trends in Chemical Physics Research
Editor: A.N. Linke, pp. 177-203
ISBN 1-59454-483-2

Chapter 9

"CHAOTIC" KINETICS, MACROSCOPIC FLUCTUATIONS AND LONG-TERM STABILITY

Maria K. Koleva* and L.A. Petrov
Insititute of Catalysis, Bulgarian Academy of Science
1113 Sofia Bulgaria, Fax: +359 2 9712967

Abstract

Our recent interest is focused on establishing the necessary conditions that guarantee a long-term stable evolution of both natural and artificial systems. Two necessary conditions, called global and local boundedness, are that a system stays stable if and only if the amount and the rate of exchange of energy and/or matter currently involved in any transition do not exceed the thresholds of stability of the system. The relationship between the local and global boundedness and the stability of the system introduces two new general properties of the state space and the motion in it, namely: the state space is always bounded, the successive steps of motion are always finite and involve only nearest neighbors. An immediate consequence of the boundedness is that the invariant measure of the state space is the normal distribution. The necessary condition for the asymptotic stability of the invariant measure is derived. It is found out that the state space exhibits strong chaotic properties regardless to the particularities of the system considered.

An example of kinetics, called chaotic one, that is compatible with both global and local boundedness is considered.

Introduction

A fundamental conjecture of the statistical mechanics is that for the systems not far from the equilibrium there always exists a single state at which the system arrives asymptotically starting at any initial state. After that the system can undergo fluctuations so that the motion in its state space has an invariant measure. So far the only way to construct such general frame is based on the assumption that the state space of the system is connected by a finite irreducible Markov chain that always converges to a single stationary distribution over

---

* E-mail address: mkoleva@bas.bg

configurations. The motion on a phase trajectory involves transitions between any two states, far distant included [1].

Our goal now is to study how this conjecture is modified under considering the development of the fluctuations in the course of time.

The involvement of the time in the state space motion is immediately interrelated with the issue of the stability of the system, namely: the system stays stable if and only if the amount and the rate of exchange of energy and/or matter currently involved in any transition does not exceed the thresholds of stability of the system. To elucidate this limitation let us consider an example that comes from the study of anomalous fluctuations developed by the dynamics of the local elementary processes. The study of this type processes involves numerical "shooting" of the sequence of spatial configurations that develops in the course of the time under the dynamics induced by given local rules and geometrical constraints. Under the supposition that the state space is connected by a Markovian chain, the aim is to find out the stationary distribution to which the state space converges. Yet, our question now is whether the transitions between any two states are admissible, i.e. which transitions do not violate the stability of the systems?! Next we present an example that certain transitions can cause system breakdown. Among the great variety of anomalous fluctuations we consider the bistability first established by [2]. Further it has been found out that it occurs both for finite [3] and infinite systems [4]. In the latter case a stable reactive state (with high oxygen and low CO coverage) coexists with a stable inactive near CO-poisoned state (high CO and low oxygen coverage). Under the supposition that only short-range interactions are involved in the elementary processes of adsorption and reaction, the global transition goes trough local transitions. Though Liu and Evans [3] have proved that the geometrical constraints can couple spatial fluctuations, the problem about coupling the local fluctuations at the infinite systems remains open. Indeed, the lack of boundary constraints and long-range interactions renders the local transitions non-correlated both in space and in time. This however, immediately introduces local strain, local overheating, sintering etc. At the next instant the spatial configuration changes and the local defects moves. Due time course their interaction can produce local reconstruction of the surface, can create mechanical defects etc. Eventually the process can yield either the system breakdown or the reaction termination.

There is a wide spectrum of works aimed to explore an effect of correlations of fluctuations in extended systems as an interplay among noise correlations, non-linearitity and spatial coupling. However, all the developed so far approaches to the task model the stochastic variables and noise sources as a Wiener process whose increments are independent and unbounded. Thus, eventhough a cooperation of the fluctuations is available, the sequence of spatio-temporal configurations through which the system arrives to the global coupling varies in uncontrolded way that it is incompatible with the idea of the boundedness.

Hence, a transition between two states does not violate the stability of the system if and only if the rates of energy and/or matter exchange permanently do not exceed their local thresholds of stability. This constraint, called hereafter local boundedness, selects a set of accessible states to any given one. Obviously, the set of accessible states and the rates of energy and/or matter exchange is specific to the system and even to the state. Yet, the local boundedness is matched in the following common constraints of the state space: both the "distance" between any accessible states and the "velocity" of the motion anywhere in the attractor and at any instant should be finite. The latter requirement is in a sharp contrast with the infinite velocity obtained in the case when the transitions between any two states are allowed [5] and [6].

Another important property of the state space is its global boundedness introduced by another necessary condition for keeping the evolution of any natural and artificial system stable, namely: the amount of energy and/or matter involved in the system is limited. So, the size of fluctuations should never exceed the macroscopic thresholds of stability of the system. The global boundedness renders the state space always confined in a finite volume. This allows to reenumerate the state space so that all accessible states to a given one are the nearest neighbors.

Thus, the relationship between the time development of fluctuations and the stability of the system introduces two new general properties of the state space and of its motion, namely: the state space is always bounded, the successive steps of motion are always finite and involve only nearest neighbours. It`s most important feature, proven in sec.1.2, is that it ensures a finite velocity of the motion in the state space.

Our task now is to account for the structure of a bounded state space subject to local boundedness and to answer the general question whether it has a stationary state and invariant measure.

We start with the question whether the state space is connected by a Markovian chain under the imposed constraints of the local and global boundedness. If so, we can apply the developed so far approaches such as Focker-Planck equation, master equation etc. to our problem. Otherwise, we should look for new approach(es) and to anticipate new properties.

Let us suppose that the transition from state j to neighbor state i depends only on whether the transition to j has happened. However, the transition to j depends on whether it comes from its nearest neighbourhood k. Hence, the transition to i is set on the chain of the previous transitions $...l...kj$. So, is our process non-Markovian though the Chapman-Kolmogorov relation holds?! It seems Markovian because the transition from j to i depends only on j. However, it is non-Markovian, because any admissible transition depends on the succession of the previous ones. (Examples of non-Markovian chains that fulfill Chapman-Kolmogorov relation are presented in [7]). The question now becomes whether the "non-Markovian" succession covers the whole history of the process. The clue is in the Lindedeberg theorem [7] that states: any bounded irregular sequence (BIS) has finite mean and finite variance. Then, our trajectory "loses" its memory any time it crosses the mean. Though any "walk" that starts at the mean comes back to it in a finite time, the size of the "walks" varies up to the size of the state space. It means that a non-Markovian "walk" can comprise most of the states.

As a result, we should look for a new approach to establish whether our state space has an invariant measure.

At first, let us specify some features of the state space trajectories. The existence of more than one accessible state makes each transition rate a multi-valued function. Indeed, at any instant only one transition takes place. However, it is arbitrarily chosen among the transitions to all accessible states. This introduces certain stochasticity of the transition rate even when all its selections can be carried out explicitly. The outcome is that the trajectory appears as a kind of a fractal Brownian "walk". The stochastisity induced by the choice of one selection among all available breaks any possible long-range periodicity (i.e. the incremental memory has finite size). So, the trajectories in our state space are fractal Brownian walks subjects to an incremental boundedness, namely: finite size of the increments (i.e. transition rates), finite size of the incremental memory and arbitrary selection rules.

In turn, the boundedness of the incremental memory size distinguishes a scale above which all the scales up to the thresholds of stability contribute uniformly to the creation of the fluctuations. Then, it is to be expected that the coarse-grained structure of the attractor exhibits universal properties insensitive to the details of the incremental statistics. In particular, it is to be expected that these properties are the same in any direction of the attractor. So, we can study the one-dimensional sequence produced by the projection of a trajectory onto any direction. We should recall that every such sequence belongs to the class of bounded irregular sequences (BIS) subject to incremental boundedness defined above.

Indeed, in sec.1 we prove that any one-dimensional BIS subject to incremental boundedness has certain universal properties. It is found out that coarse-grained structure of each BIS is a succession of well separated one from another successive excursions. An excursion is a trajectory of a walk originating at the mean value of a given sequence at the moment t and returning to it for the first time at the moment $t + \Delta$. The characteristics of each excursion are amplitude, duration and embedding interval. The latter implies that each excursion is loaded in a larger interval whose duration is interrelated with the duration of the excursion itself. The embedding does not allow any overlapping of the successive excursions and thus prevents the grow of the excursion amplitude to an arbitrary size. Thus, it preserves both the global and local boundedness arbitrarily long-term.

The incremental boundedness renders that each fluctuation has certain duration interrelated with its amplitude. By the use of this relation we prove in sec. 1.2 that the velocity of the motion on the state space trajectories is always finite. It is in accordance with our general requirement about the boundedness of the amount and rates of the energy and/or matter exchange involved in each transition.

An immediate consequence of the Lindeberg theorem is that every BIS is a subject to the Central Limit Theorem. So, the deviations from the mean are normally distributed. It should be stressed that this distribution is robust to the details of the "incremental" statistics, i.e. to the particularities of the transition rates and to the specific structure and properties of the sets of accessible states. Hence, the normal distribution appears as the target invariant measure of a bounded state space with bounded increments.

However, the Lindeberg theorem does not guarantee the asymptotic stability of the normal distribution as an invariant measure. The problem arises from the way in which the trajectories approach the boundaries of the state space, i.e. whether they "bump" into or "slide" past them. In the former case a motion termination is possible while in the latter case it proceeds. In other words, the question is whether there is mechanism that sustains the state space motion confined arbitrary long term.

The presence of the boundedness as the basic characteristic of our type state space suggests a parallel with the low-dimensional deterministic chaos - a phenomenon that occurs at the dynamics of simple deterministic systems. It is associated with unpredictability and great sensitivity to the initial conditions introduced by stretching and folding [8-9]. However, the deterministic chaos also exhibits boundedness: the folding is provided by the fact that the dynamics of the discussed systems is confined to a finite volume of the state space. The stretching happens along the unstable directories and gives rise to the unpredictability.

In particular, our attention is focused on the folding because its presence sustains the evolution of a chaotic system permanently bounded to a finite phase volume. Intuitively it seems that the folding in our state space is ensured by the presence of the thresholds of the stability. However, it is to be expected that the particularities of the boundary conditions imposed by the thresholds of stability make the folding sensitive to them and thus not universal. The question now becomes whether there is a folding insensitive to the details of the boundary conditions. We consider this problem together with the issue about the folding viewed as a necessary condition for keeping the evolution confined in a bounded state space arbitrary long time. From this point of view, the folding is to be associated with the largest fluctuations (i.e. deviations from the mean), namely those whose amplitude is of the order of the thresholds of stability.

In the sec. 2 we find out that an universal folding do exists whenever certain relation among the thresholds of stability, a parameter set on the incremental statistics and the variance of the sequence holds. The derivation of the relation involves characteristics of the coarse-grained large-scaled fluctuations established in sec.1.

On the way of deriving the necessary condition for existing of a universal folding we prove that our attractor is a transitive dense set of periodic orbits and the value of the Lyapunov exponent is insensitive to the initial point of a trajectory. According to [10-11] these properties are representative for the deterministic chaos. However, along with these our attractor exhibits other chaotic properties as well. It has been proven [12] that each BIS subject to an incremental boundedness has the following chaotic properties insensitive to the particularities of the incremental statistics:

(i). the power spectrum uniformly fits the shape $1/f^{\alpha(f)}$ where $\alpha(f)\to 1$ at $f\to 1/T$ (T is the length of the sequence) and $\alpha(f)$ monotonically increases to $p>2$ at $f\to\infty$.

(ii). the phase space attractor has non-integer correlation dimension $\nu(X)$ that monotonically decreases from $\nu(X)=d$ at the mean value to $\nu(X)=0$ at the boundaries of the attractor; d is the embedding (topological) dimension.

(iii). the Kolmogorov entropy is finite.

Thus, when the relation established in sec.2 holds, the coarse-grained structure of a bounded state space subject to incremental boundedness has an asymptotically stable invariant measure (the normal distribution) and the motion in the state space exhibits chaotic properties.

It is worth noting that the boundedness appears naturally in the above considerations through the obvious physical constraint that each system has its finite local and global thresholds of stability associated with the amount and the rate of the energy and/or matter exchange.

However, when the evolution of an extended many-body system is developed under local rules, the lack of correlation among the local spatio-temporal fluctuations renders a self-sustained boundedness impossible. Indeed, when the local fluctuations are unbounded and develop independently one from another, they would produce permanently non-correlated local defects and eventually would cause the system breakdown. This gives rise to the question about a physical mechanism that gives rise to both local and global boundedness at each point of the state space.

Recently one of us has proposed a general mechanism for coupling spatio-temporal fluctuations at surface reactions that proceed at infinite interfaces [13]. The mechanism is insensitive to the particularities of the reaction mechanism, surface properties etc. As a result of the coupling, at any value of the control parameters the system exerts macroscopic fluctuations that are compatible with the local and global boundedness. The important property of this coupling mechanism is its persistence, i.e. for each state of the system it determines a non-zero set of states accessible from that one. In turn, this provides the "continuity" of the state space trajectories. Moreover, the distance between any given and each of the accessible states is always finite.

We give a brief presentation of that mechanism in sec.3. The aim is to elucidate those its general properties that make the macroscopic evolution of the surface reactions to be the subject of the incremental boundedness. It turns out that the coupling mechanism ensures the boundedness of the rates of all the elementary processes. In turn, the evolutionary equations ensure the boundedness of the increments of the state variables. Since the incremental boundedness of the state variables gives rise to the chaotic properties of the corresponding state space, the evolutionary equations are called chaotic kinetics.

It should be stressed that the chaotic kinetics radically differs from the mean-field approach because it is brought about by a long-range coupling process while the mean field approach is founded on the premise that any system can be divided into subsystems whose fluctuations are independent one from another. However, this idea runs into controversy: it gives rise to a permanent induced non-homogeneity of the adlayer and a ubiquitous divergence of the rates of all the elementary processes. The coupling of fluctuations eliminates the induced non-homogeneity and ensures the boundedness of the rates. On the other hand, these properties are strongly related to the boundedness in its relation to the long-term stability. The spatial coherence is neccesary not only for providing the boundeness of the rates but it is also necessary for sustaining long-term stability of any extended system. Indeed, if the local fluctuations are unbounded and develop independently one from another, they would produce permanently local defects and eventually would cause the system breakdown. The permanent boundedness of the global rates renders the boundedness of the increments of the state variables at the motion in the state space. In turn, the boundedness of the increments meets the above stated general requirement that the distance between the successive states through which the system evolves is limited so that the system stays permanently within its thresholds of stability.

In sec. 4 we present several experimental evidences that for the chaotic kinetics.

1 Properties of the Large Scale Fluctuations

The task of the present section is to work out those properties of the BIS`es by the use of which we can prove the existence of a universal folding. In turn, it ensures the asymptotic stability of the invariant measure (normal distribution) of the attractor.

Let us recall that the boundedness of the incremental memory size distinguishes a scale above which all the scales up to the thresholds of stability contribute uniformly to the creation of the fluctuations. Thus, it is to be expected that the coarse-grained properties of the large-scaled fluctuations are insensitive to the particularities of the incremental statistics. It gives rise to the universality not only of the folding but to the universality of the coarse-grained structure of the attractor.

In the next subsection we prove that the coarse-grained large scale fluctuations of each BIS appear as a sequence of separated by non-zero intervals successive excursions. It should be stressed that this structure is a result of merely the global and the incremental boundedness.

Each excursion is characterised by its amplitude, duration and embedding interval. The separation of the successive excursions means embedding of each of them in a larger interval so that no other excursions can be found in that interval. The duration of the "embedding" interval is a multi-valued function whose properties are strongly related to the duration of the embedded excursion Δ itself: the range and the values of of the selection are set on Δ; the realisation of any excursion is always associated with the realisation of its embedding interval. Since the duration of the each embedding interval is a multi-valued function, its successive performances permanently introduce stochasticity through the random choice of one selection among all available. Thus, this induced stochastisity breaks any possible long-range periodicity (i.e. large-size memory) and helps the excursion sequence to preserve the chaotic properties listed in the Introduction.

The relation $\Delta \leftrightarrow A$ is set on a "blob" structure of the incremental walk that produces a BIS, namely: because of the finite size of the incremental memory, on coarse-grained scale any fractal Brownian walk can be considered as a symmetric random walk of "blobs" created by subwalks whose size counterparts the size of the incremental memory.

Our first task is to work out explicitly the relations $T \leftrightarrow \Delta \leftrightarrow A$, i.e. the relations between the duration of the embedding time interval T, respectively the duration Δ and the amplitude of the corresponding excursion A. .

1.1 Embedding Time Intervals. Relation $T \leftrightarrow \Delta$

The major role of the "embedding" is that it does not allow the overlapping of the successive excursions and thus prevents the grow of the excursion amplitude to an arbitrary size. So, the "embedding" preserves the global boundedness arbitrarily long-time.

The present task is to work out explicitly the relation between the duration of the embedding intervals and the duration of the corresponding excursions. That relation is based on the notion of an excursion: a trajectory of a walk originating at the mean value of a given sequence at time t and returning to it for the first time at time $t + \Delta$. Therefore, the probability for an excursion of duration Δ is determined by the degree of correlation between any two points of the sequence. On the other hand, the probability that any two points of a sequence, separated by interval η, have the same value is given by the autocorrelation function $G(\eta)$. A generic property of the BIS`es [12] is that the coarse-grained autocorrelation function of any of them can be defined for sequences of arbitrary but finite length T. Yet, its shape is universal, namely:

$$G(\eta,T) \propto 1-\left(\frac{\eta}{T}\right)^{\nu(\eta/T)}. \tag{1}$$

where $\nu\left(\frac{\eta}{T}\right)$ is an continuous everywhere monotonically decreasing between the following limits function:

$$\nu\left(\frac{\eta}{T}\right) \to p-1 \text{ as } \frac{\eta}{T} \to 0 \tag{2a}$$

$$\nu\left(\frac{\eta}{T}\right) \to 0 \text{ as } \frac{\eta}{T} \to 1. \tag{2}$$

Then, the probability that an excursion of duration Δ happens in an interval T reads:

$$J = \int_a^b x^{\pm\alpha(x)} dx = \frac{b^{\alpha(b)+1}}{1 \pm \alpha(b)} - \frac{a^{\alpha(a)+1}}{1 \pm \alpha(a)} \tag{3}$$

Thus,

$$P(\Delta,T) = \frac{1}{T}\int_0^{\Delta} G(\eta,T)d\eta = P(\Delta,T) = \int_0^{\Delta/T} \left(1-\varepsilon^{\nu(\varepsilon)}\right)d\varepsilon \tag{4}$$

By the use of the standard calculus [13], the integration of any power function of non-constant exponent $\alpha(x)$ ($\alpha(x)$ is an everywhere continuous function) reads

$$P(\Delta,T) = \frac{\Delta}{T}\left(1-\left(\frac{\Delta}{T}\right)^{\nu\left(\frac{\Delta}{T}\right)}\right). \tag{5}$$

The $P(\Delta,T)$ dependence only on the ratio Δ/T verifies the assumption that every excursion of duration Δ is "embedded" in an interval of duration T so that no other excursion happens in that interval.

The next step is to work out the shape of $P(\Delta,T)$. Its role is crucial for the behavior of the excursion sequences. To elucidate this point let us consider the two extreme cases:

(i). $P(\Delta, T)$ is a sharp single-peaked function. It ensures a single value of the most probable ratio $\frac{\Delta}{T}$. So, when the sequence involves identical excursions, their sequence matches a periodic behavior. The latter, however, introduces long-range correlations in the incremental statistics.

(ii). $P(\Delta, T)$ has a gently sloping maximum. Then, the relation between Δ and T behaves as a multi-value function: a range of nearly equiprobable values of T corresponds to each most probable Δ. Thus, this induced stochastisity breaks any long-range periodicity (i.e. large-size memory among increments).

The establishing of the particular shape of $P(\Delta, T)$ requires the knowledge about the explicit shape of $\nu\left(\frac{\Delta}{T}\right)$. Next it is figured out on the grounds of the proof that neither BIS sustained to an arbitrary length comprises any long-ranged increment correlations. The general restriction on the increment correlation size requires an uniform contribution to the power spectrum of all scales, i.e. there are no "special" frequencies at the power spectrum.

The only factor that can modify the power spectrum is the non-constant exponent $\alpha(f)$ of its shape $1/f^{\alpha(f)}$. The boundedness requires the monotonic decay of $\alpha(f)$ in the limits $[1, p]$ not specifying its shape [12]. Our task now is to establish the shape(s) of $\alpha(f)$ that fits the lack of long-range increment correlations. In virtue of the strict monotony of the power spectrum the required criterion is that neither any its component nor any its derivative of arbitrary order has a specific contribution. Simple calculations yield that it is achieved if and only if the shape of $\alpha(f)$ is the linear decay, namely:

$$\alpha(f) = (1 + \gamma f) \tag{6}$$

Eq. (6) provides that the $1/f^{\alpha(f)}$ derivative of an arbitrary order has the same sign throughout the entire frequency interval of the power spectrum. On the contrary, for any non-linear decay of $\alpha(f)$ each derivative of $1/f^{\alpha(f)}$ changes its sign at certain frequencies. Thus, only the linear decay does not introduce any additional scale to those inherent for the increment statistics.

Because of the diffeomorfism between $\alpha(f)$ and $\nu\left(\frac{\Delta}{T}\right)$ it is obvious that the shape of $\nu\left(\frac{\Delta}{T}\right)$ reads:

$$\nu\left(\frac{\Delta}{T}\right) = (p-1)\left(1 - \frac{\Delta}{T}\right). \tag{7}$$

The plot of $P(\Delta, T)$ with the above shape of $\nu\left(\frac{\Delta}{T}\right)$ shows that it has a gently sloping maximum: indeed, the values of $P(\Delta, T)$ in the range $\frac{\Delta}{T} \in [0.25, 0.4]$ vary by less than 7%. Outside this range $P(\Delta, T)$ decays sharply. Thus, though $P(\Delta, T)$ is a single-valued function, it provides a multi-valued relation between the most probable values of Δ and T, namely: a certain range of nearly equiprobable values of T is associated with each Δ. In the course of the time the multi-valued relation is exerted as a random choice of the duration of the "embedding" intervals even when the sequence comprises identical excursions.

1.2 Relation $A \leftrightarrow \Delta$. Symmetric Random Walk as the Global Attractor for the Fractal Brownian Motion. Finite Velocity

The general constraint of the incremental boundedness gives rise to the expectation that there is certain uniform relation between the amplitude of each excursion and its duration. The relation amplitude$\leftrightarrow$duration determines not only a property of an excursion but the "velocity" of the "motion" on that excursion. It has already been established in the Introduction that each state space trajectory can be considered as a fractal Brownian walk. So, being a one-dimensional projection of a state space trajectory, any presently considered BIS is a fractal Brownian walk as well. The latter provides certain relation between the amplitude A and the duration Δ of an excursion, namely: $\sqrt{\langle A^2 \rangle} \propto \Delta^{\beta(\Delta)}$, β is set by the particularity of the increment statistics; the averaging is over the sample realizations. The dependence of β on Δ comes from the interplay between the finite radius of the increment correlations a and the amplitude of the excursion itself that is limited only by the thresholds of stability.

Because of the finite size of the incremental memory, on coarse-grained scale any fractal Brownian walk can be considered as a symmetric random walk of "blobs" created by subwalks whose size is a counterpart of the incremental memory size. The implication that the incremental memory has a finite size renders that the incremental walks that create blobs has a finite length m; the particularities of the incremental statistics determines an exponent β, such that the $\sqrt{m.s.d.}$ of the blob creating walks equals m^{β}. Then, the large excursions can be approximated by a symmetric random walk with a constant step equal to the size of the blobs. Thus, the dependence of any large scale excursions on its duration reads:

$$\sqrt{\langle A^2 \rangle} \propto N^{0.5} m^{\beta} \tag{8}$$

where N is the number of the blobs.

It is obvious that when $N >> m$ the dependence tends to:

$$\sqrt{\langle A^2 \rangle} \propto N^{0.5} a \tag{9}$$

where a is considered constant independent of N. So, the symmetric random walk with constant step appears as the global attractor for any fractal Brownian motion.

An important outcome of eq.(8) is that it sets the velocity of the motion on the excursion. It is obvious that this relation provides a finite velocity whenever there is diffeomorfism between A and Δ.Evidently, it is provided by any non-zero but finite value of β. The finite incremental boundedness is manifested in a non-zero value of β. In addition, the fractal Brownian walk derivation of β implies that the latter is bounded in the range $(0,1]$. Any non-zero but finite value of β makes the size, duration and the structure of the blobs to depend strongly on the incremental statistics and thus to be specific to the BIS. To compare, suppose $\beta = 0$. In this case, the size and the duration of the blobs are independent from one another and appear as parameters.

Another outcome of our considerations is that the finite size of the "blobs" ensures the uniform convergence of the average to the mean of the original BIS. Indeed, the distinctive property of any fractal Brownian motion is that any exponent $\beta \neq 0.5$ arises from an arbitrary correlation between the current increment μ_i and the corresponding step τ_i. Then the average $\overline{A}$ reads:

$$\overline{A} = \sum_{i=1}^{N} \mu_i(\tau_i)\tau_i = \sum_{i=1}^{N} (-1)^{\gamma_i i} \tau_i^{\beta_i} \tag{10}$$

and correspondingly the m.s.d. :

$$\langle A^2 \rangle \propto \left\langle \sum_{i=1}^{N} (\mu_i(\tau_i)\tau_i)^2 \right\rangle = \left\langle \sum_{i=1}^{N} \tau_i^{2\beta_i} \right\rangle \tag{11}$$

where the averaging is over the different samples of the trajectory; The property of the above relations is that whenever the probabilities for γ_i odd and even are not permanently equal there is a correlation between the increment and the corresponding step. So $\overline{A}$ is certainly non-zero that immediately makes that the deviation from the mean non-zero. Moreover, eq.(10) yields that $\overline{A}$ can become arbitrarily large on increasing N.On the contrary, a permanent equal probability for γ_i odd and even means independence from one another of the increments and the steps. It yields $\overline{A} = 0$ which guarantees the uniform convergence of the average to the mean.

A major property of any BIS is the existence of mean and variance that is guaranteed by the Lindeberg theorem [7]. This brings about two very important consequences. The first one is that the amplitudes of the excursions are normally distributed. The second one is that the excursion sequence is a stationary process. Indeed, the boundedness and the finite-size memory render a uniform convergence of the average to the mean of every BIS. In turn, it provides the stationarity of the excursion. Then, the frequency of occurrence of an excursion of a size A is time-independent and reads:

$$P(A) = cA^{1/\beta(A)} \frac{\exp(-A^2/\sigma^2)}{\sigma} \tag{12}$$

The required probability $P(A)$ is given the duration $\Delta = A^{1/\beta(A)}$ of an excursion of amplitude A weighted by the probability for appearance of excursion of that size (normal distribution). σ is the variance of the BIS and in the present consideration is a parameter; $c = \frac{1}{\sigma^{1/\beta(\sigma)}}$ is the normalizing term. The stationarity of the excursion appearance ensures that $P(A)$ has the same value at every point of the sequence.

The behavior of any BIS is inherently related to the incremental statistics trough the explicit dependence of $P(A)$ on $\beta(A)$ in eq.(12). However, when the size of the required excursion is much larger than the size of the symmetric random walk, $\beta(A)$ turns constant equal to 0.5. Then, $P(A)$ gradually gets insensitive to the details of the incremental statistics. Thus, whenever $A >> a$ the behavior of the excursions becomes totally insensitive to the incremental statistics.

2 Universal Folding Mechanism of the BIS`ES

The goal of this section is to establish the necessary conditions for the asymptotic stability of the invariant measure (the normal distribution) of an attractor with global and incremental boundedness. The global boundedness suggests that the motion on such attractor creates a kind of a stretching and folding. Then, the asymptotic stability of the invariant measure is to be associated with folding that is insensitive to the way how a trajectory approaches the thresholds of stability. Thus, our aim is to find out the conditions under which such insensitive to the "boundaries" folding exists.

From the viewpoint of the deterministic chaos the folding is associated with a negative value of the Lyapunov exponent. As a measure of unpredictability the latter is the average measure how fast a trajectory deviates under arbitrary small perturbation of the initial conditions. On the other hand, from the point of view of a BIS`es, it is to be associated with the largest fluctuations, namely those whose amplitude is of the order of the thresholds of stability. Thus, our first task is to define the Lyapunov exponent in terms of the large scaled fluctuations and to show explicitly the dependence of its value and sign on their characteristics.

A necessary step for the universality of the folding is the insensitivity of the large-scaled fluctuations to the details of the incremental statistics. It has been already established that every coarse-grained BIS has 3 specific parameters: values of the threshold of stability A_{tr}, the variance σ and the power $\beta(A)$ in the relation amplitude$\leftrightarrow$duration of the excursions. Next we prove that certain relation among these parameters provides the existence of the target folding.

From the viewpoint of the deterministic chaos the folding is associated with a negative value of the Lyapunov exponent whose rigorous definition reads:

$$\xi = \lim_{t\to\infty} \frac{1}{t} \ln|U(t)| \tag{13a}$$

where

$$U(t) = X(t) - X^*(t) \tag{13b}$$

$X^*(t)$ is unperturbed trajectory and $U(t)$ is the average deviation from it. So $|U(t)|$ is the measure of all the available deviations from a given point X^*.

On the other hand, from the point of view of BIS, the Lyapunov exponent is to be associated with the large-scaled fluctuations since they give rise to the essential deviations from the mean.

Now we are ready to write down the asymptotic explicit expression for the average deviation from a trajectory that starts at X^* . The corresponding $|U(t)|$ set on the terms of the excursions reads:

$$|U(t)| = \int_{A^*}^{A_{tr}} AP(A)dA + \int_{A_{cgr}}^{A^*} AP(A)dA \tag{14}$$

A_{cgr} is the level of coarse-graining, i.e. averaging over all scales smaller than A_{cgr} This "smoothes out" all the excursions whose sizes are smaller than A_{cgr} and renders their contribution to the Layapunov exponent zero.

The separation into two terms each of which represents the deviations from A^* to larger and smaller amplitudes is formal. It is made only to elucidate the idea that starting at any point of the attractor one can reach any other through a sequence of excursions. Thus, the Lyapunov exponent ξ reads:

$$\xi = \ln \int_{A_{cgr}}^{A_{tr}} AP(A)dA \tag{15}$$

Here we come to the same result as the Oseledec theorem states [10], namely: the Lyapunov exponent for the chaotic systems does not depend on the initial point of the trajectory.

On the other hand, the stationarity of the excursion process makes the set of excursion sequence a dense set of periodic orbits. Moreover, it renders its transitivity as well: starting anywhere in the attractor a sequence of excursions can reach any other point in it. Some authors [11] list these properties as a definition of the chaos. Here they appear as a result of the global and the incremental boundedness of the BIS`es. It is worth noting that the chaotic properties listed in the Introduction are also a result of the global and the incremental boundedness. It supports the suggestion about the paramount role of the boundedness at defining the chaoticity.

It should be stresses that the chaotic properties listed in the Introduction are derived under the condition that all scales larger than the blob size a contribute uniformly to the stochastic properties of any BIS. The above chaotic properties also do not involve any specific scale larger than the blob size. However, it seems that it brings a contradiction: how the scale-free process of excursion sequence interferes with the boundary conditions imposed by the presence of the thresholds of stability. The contradiction is solved by the presence of folding since the latter makes the approach to a boundary a U-turn. Thus, the folding being a necessary condition for keeping the evolution of a BIS permanently confined in a finite attractor ensures that the chaoticity produced by the stretching and folding is a scale-free process.

It is to be expected that the size of an excursion determines its contribution to the stretching or folding of a trajectory. Indeed, the small size excursions are random walks whose frequency is essentially high (eq.(12)). Figuratively speaking, they "hold" any trajectory permanently deviated from the mean. So, the small size excursions most probably contribute to the stretching of the trajectories. On the contrary, the largest excursions are rather occasional and any trajectory subjected to them spends most of its time closest to the mean. So, they would provide the folding. The explicit revealing of the role of small and large excursions is made by the use of the coarse-graining: the role of the excursion size is carried out by scanning the ratio A_{cgr}/σ.

The ratio A_{cgr}/σ has two extreme cases:

(I) $\dfrac{A_{cgr}}{\sigma} << 1$, i.e. the contribution of the small excursions prevails. By the use of the steepest descent method, eq.(15) yields:

$$\xi \approx \ln \sigma \tag{16}$$

Eq.(16) tells that asymptotically any trajectory visits any point of the attractor so that the mean deviation from any initial point is the same for every trajectory and is bounded by the thresholds of the attractor itself. This makes the value of ξ positive. The latter justifies our speculation that the small size excursions contribute to the stretching. Further, visiting of any point of the attractor starting anywhere in it makes the motion on our attractor ergodic.

(II) (ii) $A_{cgr} >> \sigma$, i.e. large scale excursions contribution prevails. Eq.(15) yields:

$$\xi \approx \left(\frac{1}{\beta(A_{cgr})} + 1 \right) \ln \frac{A_{cgr}}{\sigma} + \ln \sigma - \frac{A_{cgr}^2}{\sigma^2} \tag{17}$$

While ξ from eq.(16) is always positive which provides stretching, eq.(17) opens the alternative for ξ being both positive or negative setting on the relation among $\beta(A_{cgr}),\sigma$ and A_{cgr}. The permanent presence of folding is necessary for keeping the evolution bounded in a finite size attractor arbitrary long time. The natural measure of the folding is the negative value of the Lyapunov exponent. Thus we come to the condition: the largest size fluctuations provide folding whenever A_{tr}, $\beta(A_{tr})$ and σ are such that $\xi<0$. So our target relation reads:

$$\xi \approx \left(\frac{1}{\beta(A_{tr})}+1\right)\ln\frac{A_{tr}}{\sigma}+\ln\sigma-\frac{A_{tr}^2}{\sigma^2}<0 \tag{18}$$

It is worth noting that the eq.(18) provides that the excursions "U-turn" without "feeling" the boundaries of the attractor. This makes the folding insensitive to the details of the thresholds of stability and the way they are approached.

It is worth noting that the folding is a broader notion than a tangent approach to a boundary. Both the folding and the tangent approach produce the same effect: they contribute to the convergence of a trajectory making it to depart from the threshold. Yet, the tangent approach itself is a property of the random walk that creates the excursions along with appropriate boundary conditions imposed, while the folding is provided by eq.(18) not involving any boundary conditions.

3 Diffusion-Induced Noise, Coupling of Fluctuations and "Chaotic" Kinetics

The goal of the present section is to give an example that meets the criteria for both local and global boundedness, namely: the accessible transitions are at the neighborhood of any given state, the incremental memory has finite size, the state space is bounded.

Yet, the focus is put on the physical mechanism that ensure the boundedness of the rates of the elementary processes. It is presented in the first subsection.

The second subsection elucidates some general properties of the evolutionary equations that are explicitly introduced by the boundedness of the rates. Among them is the property that the boundedness of the rates ensures the boundedness of the increments of the state variables. Since the incremental boundedness gives rise to the chaotic properties of the corresponding state space established in the previous section, the evolutionary equations are called chaotic kinetics.

3.1 Diffusion-Induced Noise and Coupling of Fluctuations

Next we consider the surface reactions that proceed at infinite interfaces gas/solid exposed to steady external constraints. Recently one of us put forth [12,14] a driving mechanism of a local quantum phenomenon, called diffusion-induced noise, that gives rise to rather unusual macroscopic effects. Its extraordinarity is founded on an apparent association with the stability of the system. Thus, the modelling of this phenomenon requires a new approach to the motion in the state space that explicitly involves the issue of stability. The importance of the diffusion-induced noise is that it is inevitable and generic for all the surface reactions and its presence and certain properties are insensitive to the particularities of the gas-solid system.

To elucidate the ubiquity of the considered phenomenon we present the driving mechanism of the diffusion-induced noise for the adsorption, since it is a step prerequisite of any surface reaction at any control parameter choice. The mechanism is based on the interplay between: (I) the lack of correlation between moments and points of the gas phase species trapping at the surface; (ii) the generic property of the most adlayers that no more than one species can be adsorbed at a single active site. That interplay causes fundamental changes of the properties of the overall probability for adsorption (correspondingly the adsorption rate). Given is a species trapped in a vacant site. Its further relaxation to the ground state can be interrupted by an adspecies that arrives at the same site and most probably occupies it. Thus the adspecies violates the further trapped species relaxation at that site, since no more than one species can be adsorbed at a single site. The trapped species can complete the adsorption if and only if after migration it finds another vacant site. The impact of the adspecies intervention to the trapped species probability for adsorption is twofold: first, it cannot be considered as a perturbation, since it changes the adsorption potential qualitatively, namely from attractive it becomes repulsive. That is why, that type of interaction has been called diffusion-induced non-perturbative interaction. Second, the lack of coherence between the trapping moment and the moment of adspecies arrival makes the probability for adsorption multi-valued function: each selection corresponds to a certain level of relaxation at which a diffusion-induced non-perturbative interaction happens. Therefore, the adspecies mobility brings about a fundamental duality of the probability for adsorption (and of the adsorption rate correspondingly): though each selection can be computed at microlevel, the establishing of a given selection is a stochastic process since it is a random choice of a single selection among all available.

It is obvious that the presence of the diffusion-induced noise is insensitive to the details of the adsorption Hamiltonian and thus to the particularities of the adspecies and the interface. Further, it is presented at each value of the control parameters. So, the diffusion-induced noise persists at each point of the state space and is generic for any surface reaction.

Further, since the diffusion-induced non-perturbative interactions are local events, the non-correlated mobility of the adspecies produces a lack of correlation between the established selections at any distance and at any instant. As a result, the produced adlayer would be always spatially non-homogeneous even in the academic case of identical adsorption and mobility properties of all types of adspecies. Outlining, the non-correlated diffusion-induced non-perturbative interactions always make the adsorption rates that come from different adsorption sites non-identical that immediately yields spatial non-homogeneity. Furthermore, the latter would be permanently sustained by the lack of coherence between the trapping moments and the adspecies mobility. In turn, the adlayer configuration would vary in an uncontrolled way which in a short time would cause either the reaction termination or the system breakdown. Thus, a stable long-term evolution is available if and only if there is a mechanism that suppresses the induced non-homogeneity. The importance of the problem is rendered by the fact that the latter should be ubiquitous for all the surface reactions at each parameter choice.

A successful mechanism that suppresses the induced non-homogeneity should couple the fluctuations so that eventually all the species even their rates. Coupling of the fluctuations is an essentially non-local event and thus requires the involvement of spatially extended excitations. Since the interactions among reacting species are short-ranged, the only available non-local excitations are the cooperative excitations of the lattice (surface, interface). A successful coupling needs a feedback that acts toward evening of the initially non-identical rates making the coupled species "response" to further perturbations coherent. It should be grounded on a strong coupling adlayer-lattice, namely: the energy of colliding species dissipates to local cooperative excitations of the lattice. In turn, the impact of these local modes on the colliding species is supposed large enough to induce a new transition that dissipates through the excitation of another local cooperative modes and so on. The feedback ceases its action whenever the colliding species response becomes coherent.

However, the used so far weak coupling approach to the interaction adspecies-lattice renders any feedback local and non-correlated both in space and time. It provides only local evening of the current states of the colliding species. Indeed, suppose that the collision energy dissipates through the excitement of local cooperative modes. The weak-coupling approach considers the impact of these excitations on the Hamiltonian a perturbation. Consequently, the latter cannot produce a new transition. As a result, the interaction colliding species-lattice stops.

Along with the above argument, the necessity of a new approach to the coupling adlayer-lattice is supported by the following consideration. Next it is elucidated how the enduring non-correlated mobility of the adspecies makes the scattering length divergent in the low-energy limit. In turn, it yields divergence of the cross-section and correspondingly the rates of all the elementary processes. To unravel this point let us start with recalling that at each elementary process a species changes it state from a bounded one to a free (reaction, desorption) one or vice versa (adsorption). So, each process involves crossing of the scattering threshold. On the other hand, near the scattering threshold, the impact of the adspecies on the reaction (adsorption) potential is large enough to cause fine-tuning of a highly excited state to the scattering threshold. Since the adspecies mobility is enduring and non-correlated, the adlayer configuration permanently varies. Consequently, it produces everlasting fluctuations of the adsorption (reaction) potential that make the fine-tuning a most probable phenomenon. This immediately renders the scattering length permanently divergent. Furthermore, the divergence is rather ubiquitous than exotic phenomenon since the low-energy limit exhibits universal properties in the following sense: at low enough energies the de Broglie wavelength becomes larger than the range of interaction. This makes the scattering insensitive to the inherent structure of the scatters [15]. Thus, the universality of the low-energy limit renders the divergence of the scattering length insensitive to the chemical identity of the species.

Outlining, the above considerations call for a new approach to the coupling adlayer-lattice particularly in its low-energy limit. It imposes the following requirements: insensitivity to the chemical identity of the species and the particularities of the lattice; ubiquity - insensitivity to the details of the reaction, surface and the values of the external constraints. Next, it should ensure permanent boundedness of the rates of all the elementary processes. And last but not least, it should give rise to a successful coupling mechanism that yields suppressing of the induced non-homogeneity.

To meet the above requirements we have introduced two major assumptions [13]. The first one is that any bounded Hamiltonian separates into a "rigid" and a "flexible" part under the disturbances induced by the non-correlated mobility of the adspecies. The "rigid" part is specific to the system - it comprises those low-excited states where the impact of the mobile adspecies can be treated as perturbation. On the contrary, the "flexible" parts are associated with those highly excited states where the impact of the mobile disturbances are large enough to be treated as perturbations. Further in [13], it has been proven that in the low-energy limit the "flexible" part of the Hamiltonian exhibits strong chaotic properties. In turn, the chaoticity supports the universality of the low-energy limit since the spectrum of a chaotic Hamiltonian is characterized by a single parameter regardless to the chemical identity of the species.

The chaotic part is supposed sensitive to the environment: the dissipation of any transition energy excites cooperative modes that in turn participate to the chaotic part. This causes a change of the chaotic part so that to induce a new transition. The next assumption is that the induced transitions are non-radiative and always dissipate through excitation of appropriate cooperative modes. Thus, our two assumptions give rise to a non-perturbative feedback between the species in the chaotic states and the lattice whose most prominent property is that it "works" permanently until the global "response" to further perturbations becomes coherent.

In turn, the coherent response sustains the boundedness of the scattering length since it is a process that prevents the fine-tuning. In addition, the transitions involved in the feedback do not contribute to the cross-sections. In turn, this keeps the cross-sections finite because the transitions involved in the feedback happens in the chaotic part of the spectrum where the energies are very small. It should be stressed, that a transition that involves a vanishingly small energy makes the cross section anomalous large. Thus, the coherent response and the lack of contribution of the far infrared transitions are the warrants of the cross-sections boundedness.

Further in [13] it is shown that the collisions between the species in chaotic states, i.e. the weakest perturbations that drive the feedback, dissipate through the excitation of acoustic phonons. The particular property of the feedback created on the dissipation through acoustic phonons (gapless modes) is that the coupling area continuously enlarges on decreasing of the energy that drives the feedback.

The universality of the feedback chaotic states$\Leftrightarrow$acoustic phonons is set on the insensitivity of the chaotic spectrum to the chemical identity of the reacting species (a chaotic spectrum is characterized by a single parameter) and the insensitivity of the acoustic phonons to the details of the adspecies configuration and the lattice itself (its dispersion relation involves a single parameter - sound velocity).

The feedback chaotic states$\Leftrightarrow$acoustic phonons "couples" the species so that all of them share the same rate. This rate, called global rate, always equals the individual rate that comes from certain local configuration. It has been proven that the feedback always selects that individual rate which initially is in the most favorable local configuration: such that the difference in the state of that species and its immediate neighbors is the smallest. It turns out that the established throughout the system global rate does not depend on the details of the spatio-temporal configuration of the "chaotic" species and the adspecies configuration. It is proven as well that the coupling is a scale-free process that does not blur the individual properties of the "most favourable" rate.

After the evening of the local rates is completed, the relaxation continues through the "rigid" part of the spectrum where the specific properties of the system emerge. It is worth noting that the weak-coupling approach is available at the "rigid" part of the spectrum.

The most pronounced differences among the local rates come from the undergoing of a diffusion-induced non-perturbative interaction. Thus, the particularity of the "most favourable" rate is determined by the established selection. Furthermore, the lack of correlation between the trapping moments and the adspecies mobility makes all the selections equiprobable. Therefore, at next coupling session the "most favorable" rate involves another selection of the local rate. Consequently, the random choice of a single selection among all available brings about everlasting variations of the global rate in the course of time.

The hallmarks of the global adsorption and reaction rates are: the boundedness, persistence at any point of the state space; everlasting variations.

However, the property of the greatest impact is that the variations of the global rates come from a long-range coupling process. Indeed, any long-range coupling of the fluctuations interferes with the fundamental idea of the statistical mechanics that every system that exerts short-ranged interactions can be divided into subsystems whose fluctuations are independent one from another. It should be stressed that this premise makes the lack of correlation in the motion of the adspecies plausible. In turn, the latter gives rise to the induced non-homogeneity and the divergence of the rates of the elementary processes.

Yet, the spatial coherence is necessary for sustaining long-term stability of any extended system and provides the boundedness of the rates. They are two aspects of the basic relation between the boundedeness of fluctuations and the long-term stability of the system that says: a system stays stable if the fluctuations that it exerts do not exceed its local and global thresholds of stability. Indeed, if the local fluctuations are unbounded and develop independently one from another, they would produce permanently local defects and eventually would cause the system breakdown. The permanent boundedness of the global rates renders the boundedness of the increments of the state variables. In turn, the boundedness of the increments meets the general requirement that the distance between the successive states through which the system evolves is limited so that the system stays permanently within its thresholds of stability. Some general properties of the equations that couples the global rates and the state variables are presented in the next section.

3.2 "Chaotic" Kinetics

The permanent bounded variations of the global adsorption and reaction rates give rise to rather unusual properties of the temporal behavior of the surface reactions that proceed at infinite interfaces gas/solids. Yet, some of these properties are generic for the macroscopic evolution of the latters. At steady external constraints the evolutionary equations read:

$$\frac{d\vec{X}}{dt} = \vec{\alpha}\hat{A}\left(\vec{X}\right) - \vec{\beta}\hat{R}\left(\vec{X}\right) \tag{19}$$

where $\vec{\alpha}$ and $\vec{\beta}$ are these parts of the rates that explicitly depend on the external constraints and thus, for steady conditions, appear as control parameters ; $\vec{X}$ is the vector of the reaction species concentrations; $\hat{A}(\vec{X})$ and $\hat{R}(\vec{X})$ are the matrices of the global adsorption and reaction rates of the reactants and the intermediates. The r.h.s. of eq.(19) matches the reaction mechanism. It should be stressed that at any instant each component of $\hat{A}(\vec{X})$ and $\hat{R}(\vec{X})$ is a selection randomly chosen among all available.

It should be stressed that though eqs.(19) ressemble the mean field equations they are fundamentally different because they involve global rates that come from a coupling process. On the contrary, the mean-field approach involves averaged local rates. One of the major properties of the kinetics described by the eqs.(19) is that the boundedness of the rates maintains the incremental boundedness of the state variables (the concentrations of the reactants and intermediates). In turn, as it was established in sec.2, the incremental boundedness gives rise to the strong chaotic properties of the state space. That is why, the evolution described by eqs.(19) is called "chaotic" kinetics.

To elucidate the role of the permanent variations of the global adsorption and reaction rates let us rewrite eq.(19) in the form:

$$\frac{d\vec{X}}{dt} = \vec{\alpha}\hat{A}_{av}(\vec{X}) - \vec{\beta}\hat{R}_{av}(\vec{X}) + \vec{\alpha}\hat{\mu}_{ai}(\vec{X}) - \vec{\beta}\hat{\mu}_{ri}(\vec{X}), \tag{20}$$

where $\hat{A}_{av}(\vec{X})$ and $\hat{R}_{av}(\vec{X})$ are the mean values of the adsorption and reaction rates at given parameter choice $\vec{\alpha}$ and $\vec{\beta}$; $\hat{\mu}_{ai}(\vec{X}) = \hat{A}(\vec{X}) - \hat{A}_{av}(\vec{X})$ and $\hat{\mu}_{ri}(\vec{X}) = \hat{R}(\vec{X}) - \hat{R}_{av}(\vec{X})$. Then, in the course of the time $\hat{\mu}_{ai}(\vec{X})$ and $\hat{\mu}_{ri}(\vec{X})$ appear as successive terms of corresponding zero-mean bounded irregular sequences. The subscript i serves to stress that only one selection, randomly chosen among all available, takes place at a given instant. $\hat{\mu}_{ai}(\vec{X})$ and $\hat{\mu}_{ri}(\vec{X})$ has Markovian property in the sense that the probability for occurrence of a given selection does not depend on the probability for the appearance of any selection at the previous instant.

A representative property of the eqs.(20) has been worked out in [12]. It has been established that at any parameter choice the power spectrum of the solution of eq.(20) comprises additively two parts. The first one is the power spectrum of $\vec{X}_{av}$ settled by the following equations:

$$\frac{d\vec{X}_{av}}{dt} = \vec{\alpha}\hat{A}_{av}(\vec{X}_{av}) - \vec{\beta}\hat{R}_{av}(\vec{X}_{av}) \tag{21}$$

Eqs.(21) is a system of autonomous ordinary differential equations, non-linear in general, and thus its solution can vary with the parameter choice, i.e. it is not limited to a single steady state only but e.g. it can be a limit cycle.

The second part of the power spectrum of $\vec{X}(t)$ is the power spectrum of the zero-mean BIS constituted by the variations of $\left(\vec{X}(t)-\vec{X}_{av}\right)$. It has been proven that the power spectrum of any zero-mean BIS subject to a bounded incremental statistics is a continuous band that fits the universal shape $1/f^{\alpha(f)}$, where $\alpha(f)\to 1$ at $f\to 1/T$ (T is the length of the sequence) and $\alpha(f)$ monotonically increases to $p>2$ at $f\to\infty$. This shape is insensitive to the details of the incremental statistics. In our case, this implies that the shape $1/f^{\alpha(f)}$ is robust to the particularities of the adsorption and reaction mechanism involved in $\hat{\mu}_{ai}\left(\vec{X}\right)$ and $\hat{\mu}_{ri}\left(\vec{X}\right)$.

To outline, the separation of the power spectrum into two parts is robust to the particularities of r.h.s. of eqs.(20). Thus, in particular, it is robust to the details of the reaction mechanism and is ubiquitous for the surface reactions.

It should be stressed that the power spectrum of the solution of neither system of ordinary differential equations (eqs.(21)) can comprise simultaneously a discrete and a continuous band in its power spectrum. So, the coexistence of a discrete band, coming e.g. from a limit cycle, and a continuous band of the above shape serves as an evidence that the macroscopic evolution of a system is a described by the equations of the type given by eqs.(19)-(20).

Another effect caused by the stochastic terms in eqs.(20) are the permanent deviations from the dynamical regime prescribed by eqs.(21) even when the control parameters are kept fixed. This is so because any difference $\left(\vec{X}(t)-\vec{X}_{av}\right)$ can effectively be presented as a solution of eqs.(21) at "shifted" control parameters. The latter immediately causes a change either of the characteristics of the original dynamical regime or it even induces a bifurcation. It is obvious that at a given parameter choice an induced bifurcation needs a fluctuation of certain size. On the other hand, the sequence constituted by $\left(\vec{X}(t)-\vec{X}_{av}\right)$ is a zero-mean BIS subject to bounded incremental statistics. So, the properties of the large-scaled fluctuations match those already established in sec.1. Two of these properties, the limited duration of any large scaled fluctuation and the embedding make each induced bifurcation to have a temporal effect - it appears at some instant and disappears at another one. Moreover, the succesive bifurcations are well-separated one from another.

Other general properties of the induced bifurcations are: (I) at each value of the control parameters an induced bifurcation needs development of a fluctuation of certain size; (ii) the corresponding power spectrum always comprises a continuous band of the shape $1/f^{\alpha(f)}$. As established in sec.1.2, on increasing of their size, the development of the large fluctuations becomes less and less sensitive to the details of the reaction mechanism.

The above properties make the induced bifurcations drastically different from those considered so far. Indeed, so far it has been supposed that the fluctuations give rise to macroscopic effects only near bifurcation points [16] . Here we come to the result that the diffusion-induced noise can generate temporal bifurcations at each value of the control parameters.

4 Comparison to the Experiment

The goal of the present section is to show an example that manifests the typical properties of a macroscopic evolution established in the previous subsection.

We have studied [17] the reaction system $HCOOH + O_2 + N_2$ supported over Pd catalyst (2 different charges of Heraeus gas purification catalyst containing $0.5\ wt\%\ Pd$ on $Al_2O_3 - SiO_2$). Its temporal behavior has been studied in the temperature interval $110 - 130\ ^\circ C$ by means of Kral type continuous flow calorimeter. The O_2 feed concentration was varied from $0.5 - 12\%$ while that of $HCOOH$ - from 0.5% to 8%. The contact time τ was varied from 0.3 to 13.6 $\sec$. The reactor was charged with 10 to $20\ mg$ of ground catalyst ($63 - 100\ \mu m$ sieve fraction).

The total BET surface area of the first charge ($No.68323/1977$) amounts to $380\ m^2/g$, the Pd metal surface area being $0.5\ m^2/g$. Approximating the Pd particles on the catalyst surface as cubes, the edge length has been calculated to be $a = 42$ Å while the $X - ray$ measurements gave $a < 50$ Å for the particle size. The size has also been estimated by means of transmission electron microscope - it varies from 12 to 50 Å. The respective characteristics of the second charge ($No.83247 / 1983$) are: $272\ m^2/g$ total surface area, $a = 22$ Å calculated, $a = 28$ Å (XRD) and $a = 8 - 35$ Å (TEM). The particle sizes of both charges are not significantly changed after the reaction.

The difference ΔT between the catalyst bed temperature and the feedstock at the reactor inlet was measured by a $NiCr - Ni$ thermocouple, digitized and continuously recorded. The sampling rate is 2 points per second. 80 time series of that difference have been recorded scanning the values of the feed concentrations and temperature of the feedstock at the two charges of the catalyst.

At all 80 time series the difference ΔT exhibits irregular variations whose amplitude does not exceed 10% of the average "shift" of the catalyst bed temperature from the feedstock one. However, there are occasional large variations whose amplitude is about 50% of the average "shift".

In order to find out whether the features of these variations match the characteristics of a macroscopic evolution described by the type of eqs.(20) we have worked out the power spectrum of each recorded time series. Our study has shown that:

(i) all 80 power spectra comprise a continuous band of the shape $1/f^{\alpha(f)}$, where $\alpha(f) \to 1$ at $f \to 1/T$ (T is the length of the sequence) and $\alpha(f)$ monotonically increases to $p > 2$ at $f \to \infty$. The shape is robust to the catalyst charge, feedstock temperature and the feed concentrations.

(ii) at certain values of the control parameters the power spectrum comprises a continuous band of the shape $1/f^{\alpha(f)}$ and a discrete band. Such coexistence is one of the hallmarks of a macroscopic evolution described by eqs.(20). Indeed, it justifies the additivity of the power spectrum - the discrete band comes from a system of ordinary differential equations (eqs.(21)) whereas the continuous band comes from the bounded stochasticity involved in eqs.(20). It should be stressed that this conclusion does not require any information about the particularities of the reaction mechanism and of the surface properties.

(iii) A large variation is developed at one of the time series. Its amplitude is about 50% of the average shift of the catalyst bed temperature from the feedstock one. The predominant amplitude of the variations is about 7%. In order to prove that the large variation induces a bifurcation we cut the time series into 3 successive parts: the first one involves the large variation, the second and the third one involve only "average" variations. The 3 parts are shown in Fig. 1a, 2a and 3a correspondingly. Their power spectra are presented in Fig.1b, 2b and 3b.

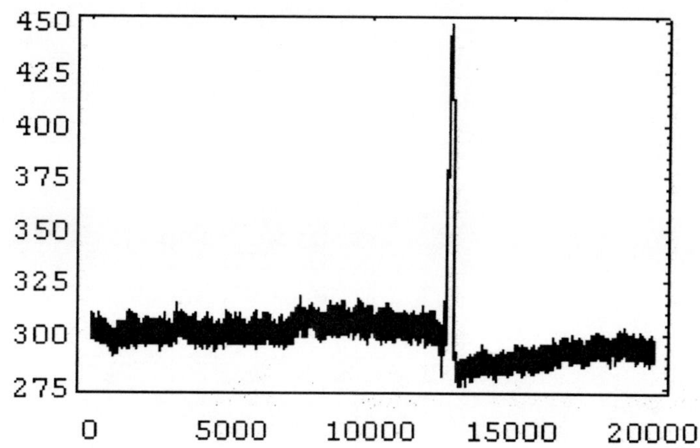

Fig.1(a) A time series in relative units of the catalyst temperature variation (°K) in the course of time (sec) at the oxidation of HCOOH at a limit cycle: first part

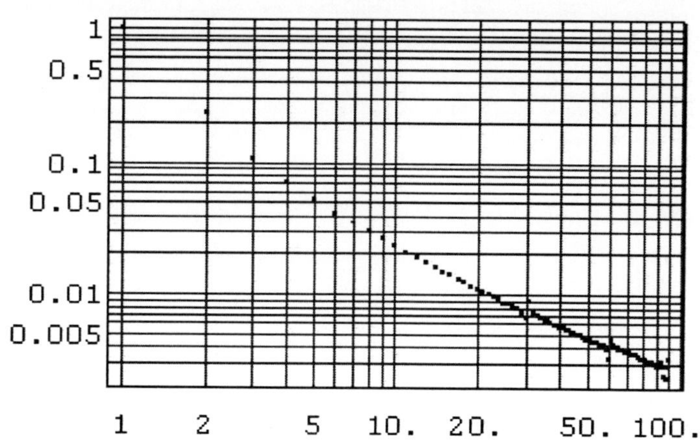

Fig.1(b) A power spectrum in relative units of the catalyst temperature variations (°K) in the course of time (sec) at the oxidation of HCOOH at a limit cycle : first part

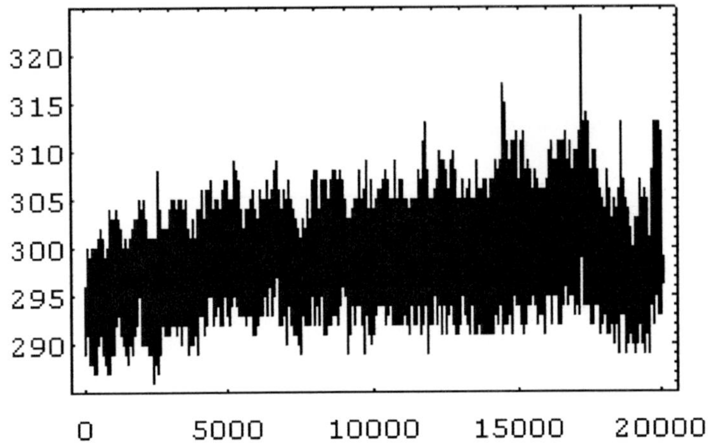

Fig.2(a) A time series in relative units of the catalyst temperature variation (°K) in the course of time (sec) at the oxidation of HCOOH at a limit cycle: second part

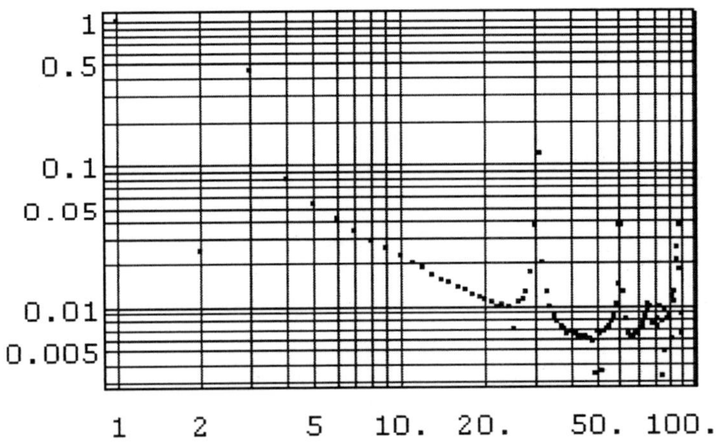

Fig.2(b) A power spectrum in relative units of the catalyst temperature variations (°K) in the course of time (sec) at the oxidation of HCOOH at a limit cycle : second part

It is obvious that the presence of the large variation strongly affects the amplitude of the discrete band in Fig1a - it is about 10 times smaller than that of the discrete band in Fig.2b and 3b. On the other hand, the period remains the same at all 3 power spectra. The great sensitivity of the amplitude of oscillations to the distance from the bifurcation point along with robustness of the period to that distance is the typical property of a limit cycle. Thus, the large variation in Fig.1a causes effectively a strong "shift" of the control parameters though physically they have been sustained fixed. The impact of the large variation is temporary - it lasts as long as the variation is essentially large. Indeed, the amplitude of the limit cycle at Fig.2b and 3b is significantly different from that of the Fig. 1b.

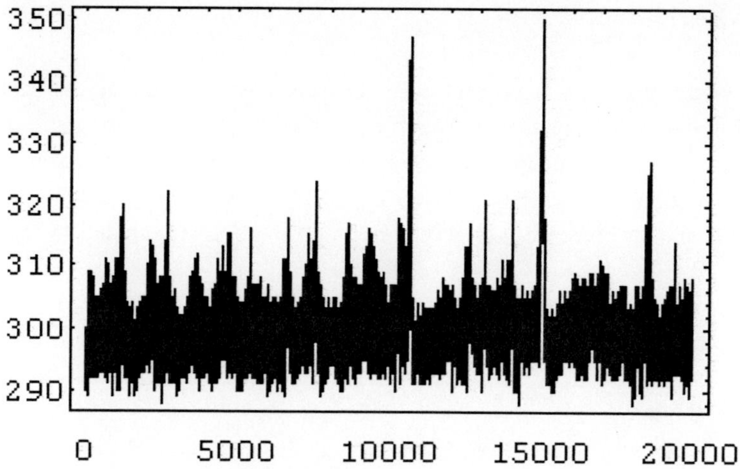

Fig.3(a) A time series in relative units of the catalyst temperature variation (°K) in the course of time (sec) at the oxidation of HCOOH at a limit cycle: third part

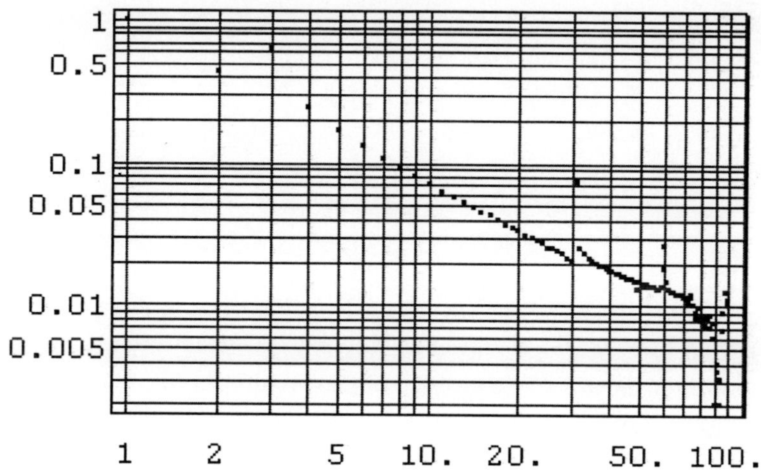

Fig.3(b) A power spectrum in relative units of the catalyst temperature variations (°K) in the course of time (sec) at the oxidation of HCOOH at a limit cycle : third part

The above results support the considerations of our previous subsection about the properties of the induced bifurcations - at each value of the control parameters an induced bifurcation needs a development of a fluctuation of certain size. The corresponding power spectrum always comprises a continuous band of the shape $1/f^{\alpha(f)}$. An induced bifurcation has a temporary effect - it lasts until the fluctuation is significant. The larger the fluctuation is the less sensitive to the details of the reaction mechanism is. That is why here we do not discuss the particularities of the reaction mechanism and the catalyst properties. They can answer questions such as: what is the particular bifurcation diagram, what are the thresholds of stability, does the variance of the fluctuations depend on the control parameters etc.

Conclusion

The major goal of the present chapter is to elucidate the impact of the global and local boundedness on the structure and the properties of the state space of the natural systems. The boundedness is a necessary condition for keeping the evolution of the natural and artificial systems stable, namely: a long-term stability is possible if and only if the amount and the rate of energy and/or matter currently exchanged in any transition is bounded by the thresholds of stability of the system.

The relationship between the local and global boundedness and the stability of the system introduces two new general properties of the state space and the motion in it, namely: the state space is always bounded, the successive steps of motion are always finite and involve only nearest neighbors.

An immediate consequence of the global boundedness is that the invariant measure of the state space is the normal distribution. Thus, the global boundedness makes the normal distribution ubiquitous for the natural systems.

The existence of more than one accessible state to any given state makes each transition rate a multi-valued function. Indeed, at any instant only one transition takes place. However, it is arbitrarily chosen among the transitions to all accessible states. This introduces certain stochasticity of the transition rate even when all its selections can be carried out explicitly. The outcome is that the trajectory appears as a kind of a fractal Brownian "walk". The stochastisity induced by the choice of one selection among all available breaks any possible long-range periodicity (i.e. the incremental memory has finite size). So, the trajectories in our state space are fractal Brownian walks subjects to an incremental boundedness, namely: finite size of the increments (i.e. transition rates), finite size of the incremental memory and arbitrary selection rules.

In turn, the boundedness of the incremental memory size distinguishes a scale above which all the scales up to the thresholds of stability contribute uniformly to the creation of the fluctuations. Then, the coarse-grained structure of the attractor exhibits universal properties insensitive to the details of the incremental statistics. It has been established that the properties of each coarse-grained BIS, respectively a trajectory, are set on 3 specific parameters: the threshold of stability, the variance of the sequence and the exponent of the relation amplitude$\leftrightarrow$duration of a fluctuation. The non-zero but finite value of the latter provides finite velocity of the motion on any state space trajectory.

The condition that guarantees the asymptotic stability of the invariant measure (i.e. the normal distribution) reveals the interrelation between the boundedness as a necessary condition for a BIS (trajectory) to demonstrate chaotic properties and the presence of stretching and folding for keeping the evolution of a trajectory confined in a finite phase volume arbitrarily long time. From this viewpoint, an unlimited in the time stable evolution of a trajectory is possible if and only if the folding does not "feel" the boundary conditions imposed by the presence of the thresholds of stability. Otherwise, the evolution strongly depends on the way the excursion approaches the threshold.

We have proved that when certain relation (eq.(18)) among the major characteristics of a BIS holds, a folding with the desired properties exists. As a result, the presence of the thresholds is unperceptible that ensures all scales, thresholds included, uniformly to participate in setting the chaoticity of a BIS. In turn, the uniform participation of all scales ensures an unlimited confinement of the BIS evolution in a finite attractor.

Consequently, on meeting eq.(18) a bounded state space subject to incremental boundedness appear as a global attractor whose invariant measure is the normal distribution. The universality along with the stationarity makes the attractor a transitive dense set of periodic orbits regardless to the particularities of the system. Thus, the structure of the attractor has steady properties insensitive to the development of any specific trajectory and/or the way of approaching the boundaries. The same properties render the motion on the attractor to exhibit the chaotic properties listed in the Introduction and established in [12]

A physical mechanism that gives rise to both local and global boundedness at each point of the state space is presented.

References

[1] van Kampen, N.G.; *Stochastic Processes in Physics and Chemistry*; North-Holland Publishing Company: Amsterdam, 1981

[2] Fichtorn, K.; Gulari E. & Ziff; *R. Phys. Rev. Lett.* 1989, vol.63(14), 1527-1530

[3] Liu, D.; Evans, J.W.; *J. Chem. Phys.,* 2002, Vol. 117, 7319-7328

[4] James, E.W.; Song C.;Evans, J.W.; *J. Chem. Phys.* 1999, Vol. 111, 6579-6589

[5] Zaslavsky, G. M.; *Phys. Rep.* 2002, Vol.371, 461-580

[6] Horshemke, W.; *Phys. Rev. E,* 1999, Vol.60(3), 2651-2663

[7] Feller, W.; In *An Introduction to the Theory of Probabilities*, J. Willey & Sons: New York; 1970

[8] Lichtenberg, A. J.; Lieberman M. A.; Regular and Stochastic Motion; *Appl. Math. Series*, Springer: Berlin, 1983 Vol. 38

[9] Zaslavsky, G. M.; *Phys. Rep.,* 1981, Vol.80, 157-250

[10] Oseledec V. I.; *Trans. Mosc. Math. Soc.,* 1968, Vol.19, 197-231

[11] Ruelle D.; *Chaotic Evolution of Strange Attractors*; Cambridge University Press: Cambridge, 1989

[12] Koleva, M. K.; Covachev V. C.; *Fluct. Noise Lett.* 2001 Vol.1, R131-R149

[13] Koleva, M. K.; Size-Independent Non-Equilibrium Fluctuations, http://arXiv.org/cond-mat/0212357

[14] Koleva, M.K.; *Bulg. Chem. Ind.* 1998, Vol. 69, 119-128

[15] Braater, E.; Hammer, H.-W.; http://arXiv.org/cond-mat/0410417

[16] Nicolis, G.; Prigogine, I.; *Self-Organisation in Non-Equilibrium Systems*; Wiley :New York, 1977

[17] Koleva, M.K.; Elyias, A.E.; Petrov, L.A.; *Metal-Ligand Interactions in Chemistry, Physics and Biology*; N. Russo, N. & Salahub, D.R.; NATO ASI Series C; Kluwer Academic Publishers: Dordrecht, 2000, Vol.546 pp.353-369

In: Trends in Chemical Physics Research
Editor: A.N. Linke, pp. 205-269
ISBN 1-59454-483-2

Chapter 10

ORBITAL-FREE MOLECULAR DYNAMICS: A LINEAR-SCALING DENSITY FUNCTIONAL THEORY METHOD FOR THE LARGE-SCALE SIMULATION OF METALLIC SYSTEMS

Andrés Aguado, David J. González, Luis E. González and José M. López
Departamento de F´sica Teórica, Atómica y Óptica
Universidad de Valladolid, Valladolid 47011, Spain

Abstract

Large scale computer simulations (those involving explicit consideration of a large number of atoms and/or long simulation times) are needed in order to get a proper understanding of many materials properties. Traditional *ab initio* electronic structure methods, which provide a fairly accurate description of atomic interactions, can not yet be routinely employed in large scale simulations, due to their prohibitive computational expense. The advent of so-called linear-scaling *ab initio* methods, which take advantage of the exponential decay of the one-body density matrix in real space, has much improved this situation for insulators and semiconductors. However, these methods can not be applied to metallic systems, where the absence of a finite band gap imposes a power-like decay of one-body density matrix and inhibits electron localization. In this contribution, we describe a natural scheme to obtain *ab initio* linear scaling for metallic systems. It involves the use of the density functional theory (DFT) within its Hohenberg-Kohn (HK) representation where the valence electron density stands as the basic variable, thus avoiding employment of auxiliary one-particle orbitals as in its Kohn-Sham (KS) representation. The electronic kinetic energy is expressed by an explicit functional of the density, and the interaction of valence electrons with ionic cores is represented by local pseudopotentials; the rest of the energy functional is the same as in traditional KS approaches. In order to perform molecular dynamics simulations, forces on atoms are obtained by means of Hellmann-Feynman theorem. We demonstrate applicability of the present methodology to the efficient simulation of problems which otherwise would be very difficult to simulate with the usual KS representation. Specifically, we provide new results for the melting-like transition in medium-size heterogeneous alkali clusters, as well as for the static and dynamic properties of bulk simple (*sp bonded*) liquid metals and the liquid-vapor interfaces.

1 Introduction

There is considerable interest nowadays in devising *ab initio* electronic structure methods whose computational efficiency scales linearly with system size. This is a very important issue as traditional orbital-based methods can not afford the efficient simulation of several properties of materials because of their inherent poor scaling. The determination of many materials properties (such as phase coexistence curves, for example) in molecular dynamics simulations can only be reliably performed by explicitly considering a large number of atoms along with sufficiently long simulation times, which heretofore has precluded the applicability of the usual *ab initio* schemes by the quantum chemistry community.

In this chapter, we describe a computational implementation of the so-called orbital-free molecular dynamics technique, which provides a natural way of performing linear scaling electronic structure calculations for metallic systems. The chapter is organized in the following way: section II includes a detailed description of the orbital-free molecular dynamics technique. Section III describes some applications of this technique to the efficient simulation of problems which are outside the present capabilities of orbital-based *ab initio* methods, namely: thermal properties and the melting-like transition of medium-size metal clusters, dynamical properties of bulk simple ($sp-$bonded) metals and their alloys, and the simulation of metal liquid surfaces. Section IV presents the main conclusions from this work, summarizes the main advantages and disadvantages of the orbital-free formulation, and gives some prospects for future work. Each separate section includes a brief introduction to the specific problem dealt with in that section, being this the reason for the brevity of this general introduction.

2 Theory

2.1 Orbital-Free Molecular Dynamics

The theorem derived by Hohenberg and Kohn (HK) in 1964 [1] establishes that knowledge of the electron density alone suffices to obtain all ground state properties of an electronic system. Combined with the Born-Oppenheimer approximation, it allows all the atomic properties of a system of electrons and nuclei to be obtained.

Within this formalism, the ground state energy of a system with N nuclei and N_e electrons is a functional of the electron density $n(\vec{r})$, and a function of the ion positions $\vec{R}_i (i = 1, 2, ..., N)$, with the following form (Hartree atomic units will be used throughout this work):

$$E[n,\vec{R}] = T_s[n] + \frac{1}{2}\int\int \frac{n(\vec{r})n(\vec{r}')}{|\vec{r}-\vec{r}'|} d\vec{r}d\vec{r}' + \int n(\vec{r})V_{ext}(\vec{r})d\vec{r} + E_{XC}[n] + E_{ion-ion}[\vec{R}], \quad (1)$$

where $\vec{R}$ is employed to denote the set of all atomic positions and the different terms are: the kinetic energy functional for a noninteracting electron gas with the same density as that of the interacting system of interest, the classical electron-electron repulsion energy (Hartree term), the interaction energy between the electrons and any external potential, the exchange-correlation energy functional, and the classical Coulomb repulsion between positive ions. In practice, we take $n(\vec{r})$ to be the electron density due just to valence electrons, and thus $V_{ext}(\vec{r})$ always contains the external potential provided by the instantaneous configuration of ions (here, each ion contains a nucleus plus the corresponding set of *core* electrons).

Three key approximations in the energy functional involve $T_s[n]$, $E_{XC}[n]$, and the electron-ion interaction. Kohn and Sham (KS) [2] introduced a set of N_e orthonormal auxiliary orbitals verifying

$$n(\vec{r}) = {}_{i=1}^{N_e}\ \psi_i(\vec{r})^*\psi_i(\vec{r}) \quad (2)$$

and

$$T_s[n] = {}_{i=1}^{N_e} \int \psi_i(\vec{r})^*(-\frac{1}{2}\ {}^2)\psi_i(\vec{r})d\vec{r}. \quad (3)$$

In terms of these auxiliary orbitals (which are themselves obtained by selfconsistent solution of KS equations [2]), the expression for $T_s[n]$ is exact, and thus the only approximations in KS-pseudopotential theory are those due to the exchange-correlation functional and to the transferability of ionic pseudopotentials. However, use of the KS scheme involves the iterative solution of N_e functional differential equations for each configuration of the ions. In essence, these equations contain two kinds of terms: the effect of the Hamiltonian on the orbitals, and a set of constraining forces which preserve orthonormality of the orbitals. The computational cost of evaluating those two terms can be shown to increase as $\Omega^2 \ln\Omega$ and Ω^3, respectively, where Ω is the volume of the system. This precludes the routine employment of traditional KS approaches in large-scale computer simulations.

However, if a sufficiently good approximation for $T_s[n]$ is available, we can proceed with HK density functional theory and find the ground state electron density for a given set of ion positions through the variational principle

$$\frac{\delta}{\delta n(\vec{r})}(E[n,\vec{R}] - \mu \int n(\vec{r})d\vec{r}) = 0, \quad (4)$$

where μ is the electron chemical potential chosen to give the desired number of electrons N_e. Writing

$$V(\vec{r}) = \frac{\delta E}{\delta n(\vec{r})} = \frac{\delta T_s}{\delta n(\vec{r})} + \int \frac{n(\vec{r}')}{|\vec{r}-\vec{r}'|}d\vec{r}' + V_{ext}(\vec{r}) + V_{XC}(\vec{r}), \quad (5)$$

with $V_{XC}(\vec{r}) = \frac{\delta E_{XC}[n]}{\delta n(\vec{r})}$, the electronic ground state is found by solving self-consistently the equations

$$V(\vec{r}) = \mu, \tag{6}$$

$$\int n(\vec{r})d\vec{r} = N_e. \tag{7}$$

It is convenient to work in terms of a single effective orbital $\psi(\vec{r})$ rather than $n(\vec{r})$, where

$$n(\vec{r}) = \psi(\vec{r})^2, \tag{8}$$

and to vary ψ rather than n. This has the advantage of maintaining n nonnegative if ψ is real. Assuming now that our computer experiments are performed under periodic boundary conditions, we expand $\psi(\vec{r})$ in a basis set formed by the plane waves periodic in the superlattice

$$\psi(\vec{r}) = \sum_{\vec{G}} C_{\vec{G}} e^{-i\vec{G}\vec{r}}, \tag{9}$$

$$C_{\vec{G}} = \frac{1}{\Omega}\int_{\Omega} d\vec{r}\psi(\vec{r})e^{i\vec{G}\vec{r}}. \tag{10}$$

where $\vec{G}$ is a reciprocal lattice vector of the superlattice. Following Car and Parrinello [3] we regard the coefficients $C_{\vec{G}}$ as generalized coordinates of a set of fictitious classical particles each of mass m, and write down the Lagrangian for the whole system of electrons and ions as

$$L(C,\dot{C},\vec{R}) = \sum_{\vec{G}} \frac{1}{2}m \mid \dot{C}_{\vec{G}} \mid^2 + \sum_{i=1}^{N} \frac{1}{2} M_i \dot{\vec{R}}_i^2 - E[n,\vec{R}], \tag{11}$$

where M_i is the mass of the ith ion, and the constraint to be considered is

$$\int_{\Omega} d\vec{r}n(\vec{r}) = \Omega \sum_{\vec{G}} C^*_{\vec{G}} C_{\vec{G}} = N_e. \tag{12}$$

The Lagrange equations of motion are

$$\frac{d^2 C_{\vec{G}}}{dt^2} = \frac{\Omega}{m}(\mu C_{\vec{G}} - \sum_{\vec{G}'} V_{\vec{G}-\vec{G}'} C_{\vec{G}'}), \tag{13}$$

$$\frac{d^2 \vec{R}_i}{dt^2} = -\frac{1}{M_i}\frac{\delta E}{\delta \vec{R}_i}, \tag{14}$$

where $\sum_{\vec{G}'} V_{\vec{G}-\vec{G}'} C_{\vec{G}'} = \frac{1}{\Omega}\int_{\Omega} d\vec{r} V(\vec{r})\psi(\vec{r})e^{i\vec{G}\vec{r}}$. The "molecular dynamics" of the electrons alone (eq. 13), holding the ions fixed, can be used to determine the electronic ground state for each ionic configuration. Specifically, the electron MD run is started from some arbitrary initial state (i.e. a set of coefficients $C_{\vec{G}}$), and the electron velocities are scaled down repeatedly, or equivalently the electron temperature is gradually reduced, until the system cools down or anneals into its ground state.

The force on an ion at $\vec{R}_i$ is given by $\vec{F}_i = -\delta E/\delta\vec{R}_i$, but if the electrons are in their ground state for the particular configuration of ions $\{\vec{R}\}$, so that $n(\vec{r})$ is the ground state electron density, then

$$\vec{F}_i = -\frac{E[n,\vec{R}]}{\vec{R}_i} - \int \frac{\delta E[n,\vec{R}]}{\delta n(\vec{r})}\frac{\delta n(\vec{r})}{\delta \vec{R}_i} = -\frac{E[n,\vec{R}]}{\vec{R}_i}, \tag{15}$$

because E is stationary with respect to n at the ground state (this is the Hellman-Feynman theorem). Only the explicit dependence of E on the ion positions enters the force so that only the electron-ion and Madelung energies contribute. If (15) is used to calculate the forces care must be taken to keep the electrons in the ground state for the current ion configuration, because if $\delta n(\vec{r})$ is the deviation of the electron density from the current ground state, the error in the "Hellman-Feynman" force is first order in δn. Thus, we first locate the electronic ground state for a particular configuration of the ions through dynamical annealing, and then the ions are allowed to move according to equation (15). Thus, our scheme for *ab initio* MD differs from the Car-Parrinello approach [3], where the electron and ion subsystems are allowed to evolve conjointly by employing the same time step in equations (13–14). By finding the correct ground state density for each atomic configuration (Born-Oppenheimer MD), we can adopt the natural time step of the ionic system for integration of eq. (14). This is determined by the highest vibrational frequency of the atomic system, which is usually much lower than typical frequencies of oscillation for the electron coefficients $C_{\vec{G}}$, due to the difference between electron and ion masses. Integration of both electronic and atomic equations of motion is performed by employing Verlet algorithm [4]. Technical details, such as the choice of appropriate time steps or electronic fictitious masses, are different in general for each specific problem, and will be separately described in Section III.

The explicit functional of the density we employ for the electronic energy is much superior in computational speed and memory requirements to the conventional KS orbital approach, allowing the treatment of larger systems for longer simulation times. First, we need to solve M equations of type (13), where M is the number of plane waves, instead of $N_e \times M$ orbital coefficient equations found with the KS functional; secondly, orthogonality constraints are avoided. Thus, the scaling of computer time with system size is $\Omega\ln\Omega$, that is, approximately linear. Memory requirements also scale linearly, as opposed to quadratic scaling in KS approaches. Linear scaling is thus obtained in a natural way, without invoking the existence of a band gap in the spectrum of quasiparticles, as it does in the KS scheme [5]. Therefore, the orbital-free (OF) technique is particularly appealing for metallic systems, which do not have a finite band gap. A related advantage for extended metal systems is that Brillouin-zone sampling, which is necessary to obtain accurate KS results for metals, is not needed in OF calculations as it only affects the wavefunction, not the density itself.

At a practical level, the fundamental difference between KS and HK realizations of DFT is that an explicit approximate expression for the functional dependence of the electronic kinetic energy on the electron density is needed in HK-DFT. This accounts for the computational superiority of HK-DFT, discussed in the previous paragraph, but at the same time results in comparatively less accurate results. Nevertheless, finding of a very accurate expression for $T_s[n]$ would not immediately imply that OF calculations with fully predictive power are possible for metals. The reason is that the exibility provided by the auxiliary orbitals is additionally exploited in KS-DFT by defining different pseudopotentials for atomic orbitals of each angular momentum. Thus, the interaction of *core*−electrons with $s-$ and $p-$like valence electrons, for example, is described by different *channels* of a non-local pseudopotential [6]. In HK-DFT, only a representation of the electron density is available, and thus employment of *local* pseudopotentials is mandatory. We will describe in the next two subsections the kinetic energy functionals and local pseudopotentials employed in the present study.

As concerns the exchange-correlation functional, we employ the local density approximation (LDA)

$$E_{XC}[n] = \int_{\Omega} n(\vec{r})\varepsilon_{xc}(n(\vec{r}))d\vec{r}, \tag{16}$$

$$V_{XC}[n] = \frac{\delta E_{XC}}{\delta n} \mid_{n(\vec{r})} = \frac{}{n}(n\varepsilon_{xc}) \mid_{n(\vec{r})}, \tag{17}$$

where $\varepsilon_{xc}(n)$ is the exchange-correlation energy per electron for a uniform gas of density n. Specifically, we use the Perdew and Zunger parameterization [7] of the electron gas results of Ceperley and Alder [8].

2.2 The Electronic Kinetic Energy Functional

There have been several attempts to derive approximate expressions for $T_s[n]$ [9, 10, 11, 12, 13, 14, 15, 16, 17, 18, 19, 20, 21, 22, 23, 24, 25, 26]. The most appealing approaches try to build known exact limits into a computationally tractable form for $T_s[n]$. For example, for a system with uniform density $n = n_0$, the local density approximation is correct:

$$T_s[n] = T_{TF} = \frac{3}{10}(3\pi^2)^{2/3} \int d\vec{r} n^{5/3}(\vec{r}), \tag{18}$$

which is the Thomas-Fermi (TF) expression for $T_s[n]$. This is also the correct expression in the high-density limit $r_s \to 0$, and that is why it is relatively simple to evaluate the electronic contribution to the pressure in the interior of stars, for example. It is expected to work reasonably well for those cases where $n(\vec{r})$ changes very smoothly in the scale of the inverse of the Fermi wavevector k_f, but this is not true for the electron densities of atoms, molecules or solids. It is not surprising then that TF formula has serious shortcomings: the electron density at an atomic nucleus is infinite, and its tail does not show an exponential but instead a power law decay; also, there is no chemical binding between atoms with the TF formula.

An exact $T_s[n]$ expression for a system with a long-wavelength (or slowly varying) density perturbation is provided by the gradient expansion (GE)

$$T_{GE} = T_{GE}^0 + T_{GE}^2 + T_{GE}^4 + \ldots \tag{19}$$

with

$$T_{GE}^0 = T_{TF} \tag{20}$$

$$T_{GE}^{2} = \frac{1}{72}\int d\vec{r}\frac{|\ n\ |^{2}}{n} \tag{21}$$

$$T_{GE}^{4} = \frac{1}{540}(3\pi^{2})^{-2/3}\int d\vec{r}n^{1/3}\left((\frac{{}^{2}n}{n})^{2} - \frac{9}{8}(\frac{{}^{2}n}{n})\,|\,\frac{n}{n}\,|^{2} + \frac{1}{3}\,|\,\frac{n}{n}\,|^{4} \right). \tag{22}$$

Higher order terms in the gradient expansion may be derived but produce divergences in the regions of vanishing density. Also, the functional derivative of T_{GE}^{4} diverges in the tail region, so its use in fully selfconsistent calculations is problematic. Nevertheless, GE2 already corrects the nuclear and tail atomic problems and leads to stable binding of atoms, thus significantly improving the description of inhomogeneous densities over TF approximation.

The exact kinetic energy functional for one-electron systems and two-electron systems in the Hartree-Fock approximation, and the leading term for general systems with a rapidly varying (short-wavelength) perturbation in the density, has the von Weizsäcker (vW) form

$$T_{vW} = \frac{1}{8}\int d\vec{r}\frac{|\ n\ |^{2}}{n} = 9T_{GE}^{2}. \tag{23}$$

For small-amplitude perturbations of any wavelength, about an average electron density n_0, linear response theory (LRT) gives the correct form for the kinetic energy

$$\frac{\delta^{2}T}{\delta n^{2}}\Big|_{n_0} = -\frac{1}{\chi_0}, \tag{24}$$

where χ_0 is the response function for a noninteracting homogeneous electron gas, given by Lindhard's expression [1]. It shows a derivative discontinuity at a wavevector $k = 2k_f$, which is an important ingredient for a correct description of screening in metals, and gives rise, in particular, to Friedel oscillations. We note here that TF approximation is equivalent to approximating χ_0 by its value at $k = 0$ (uniform gas limit). The expansion in gradients of the density is equivalent to a power-series expansion of χ_0 about $k = 0$. As a power expansion can not extrapolate properly through a derivative discontinuity, no Friedel oscillations will appear. vW expression amounts to approximating χ_0 by its asymptotic value at $k \to \bullet$. A gradient expansion, starting from this rapidly varying limit, can be devised, with the TF term as the lowest-order correction. This expansion will not reproduce Friedel oscillations either, for the same reasons.

Most of the reasonably accurate $T_s[n]$ functionals reported up to date are designed to reduce to these limiting forms under the appropriate circumstances. Wang *et al.* [26] have pointed out that, if the $T_s[n]$ functional is going to be employed in variational calculations, its functional derivative with respect to density should also be required to be accurate. In particular, once selfconsistency has been achieved, the equality $\frac{\delta T_s[n(\vec{r})]}{\delta n(\vec{r})} = \mu - V_{KS}(\vec{r})$, where $V_{KS}(\vec{r})$ is the Kohn-Sham potential, should be verified.

Next we describe the specific kinetic energy functional that we employ in the OF description of extended systems (those where an average density n_0 may be defined). A description of other functionals, obtained from similar physical requirements (reproduction of the exact limits (18–24)), may be found in the literature [20, 21, 22, 23, 24]. Our T_s expression is a simplified version of that proposed by Garc´a-González *et al.* [20], and has the form

$$T_s[n] = T_{vW}[n] + T_\beta[n], \tag{25}$$

$$T_\beta = \frac{3}{10}\int d\vec{r}\, n(\vec{r})^{5/3-2\beta}\tilde{k}(\vec{r})^2, \tag{26}$$

$$\tilde{k}(\vec{r}) = (2k_f^0)^3 \int d\vec{s}\, k(\vec{s})w_\beta(2k_F^0|\vec{r}-\vec{s}|), \tag{27}$$

where $k(\vec{r}) = (3\pi^2)^{1/3}n(\vec{r})^\beta$, k_f^0 is the Fermi wave vector corresponding to the mean electron density n_0, and $w_\beta(x)$ is a weight function, determined by requiring the correct recovery of both the LRT and the uniform density limits. Note that $\tilde{k}(\vec{r})$ is obtained as a convolution, which can be performed rapidly with usual Fast-Fourier-Transform (FFT) techniques. It is most important to mention that this functional is positive definite for any value of β. Moreover, in the $k_f^0 \to 0$ limit when the mean electron density vanishes (for instance in the tail region of a metal surface), the von Weizsäcker term is recovered if $\beta = 4/9$, whereas for other values of β the limit is $T_{vW} + CT_{TF}$. For a range of values of β, namely $0.5 < \beta < 0.566$, the functional fulfills other useful mathematical requirements related to the finite character of the functional derivative for small densities. In the applications we have used a value of $\beta = 0.51$, close to $4/9$ and for which the coefficient C in the $k_f^0 \to 0$ limit is 0.046.

For finite systems such as atomic clusters, which show spatial regions where the density varies greatly, there is not unique way of defining a uniform reference system. In these cases, a practical solution is to try to reproduce limits (18–23) by means of a semilocal functional, depending on both the density and its gradient

$$T_s^{GGA}[n] = c_0 \int d\vec{r} n^{5/3}(\vec{r})\Phi(s), \tag{28}$$

where $\Phi(s)$ is an unknown function of the dimensionless variable s, which is related to the density gradient through

$$s = \frac{c_2}{c_0} \mid \frac{n}{n^{4/3}} \mid^2, \tag{29}$$

with $c_0 = \frac{3}{10}(3\pi^2)^{2/3}$ and $c_2 = \frac{1}{72}$. This form is similar to the generalized gradient approximation (GGA) developed for the exchange-correlation functional [27], and can satisfy both TF and vW limits, as well as the gradient expansion through second order, GE2 limit. Therefore, it can also reproduce the LRT limit up to order k^2 in a series expansion of the response function about $k = 0$. The function $\Phi(s)$ may be constructed by a convergent resummation of a subseries of the gradient expansion, thus effectively incorporating part of the higher$-k$ effects. Proceeding this way, Perdew has shown [16] that GE2 approximation is indeed extremely close to the best possible GGA. Also, starting from the one-body Green's function, Yang has shown [11] that GE2 is the result of representing both diagonal and off-diagonal elements of the first-order reduced density matrix with the same accuracy. Further improvement over GGA may be attained by incorporating the laplacian of the density into the kinetic energy functional [26], which would provide a meta-generalized gradient approximation (or meta-GGA, employing again the same terminology adopted for the several approximations to the E_{XC} functional [28]). Also, the nonlocal kinetic energy functional based on LRT may be tried on clusters by employing some definition of the cluster volume V (for example, the volume of a sphere of radius equal to the mean gyration radius of the cluster) and defining somewhat arbitrarily the reference density as $n_0 = N_e/V$. This has been done in recent simulations of the melting process in Na clusters [29, 30] and shown to lead to accurate results. Improvement along these lines still represents work in progress and this is the reason why we have adopted the GE2 approximation in the cluster studies presented here. It will be demonstrated in section III that this approximation is quite accurate, as long as energetic and thermal properties of clusters are involved.

2.3 Local Pseudopotentials

There are two basic methods for pseudopotential construction. The first one is modeling, i.e. a model functional form is assumed with several parameters to be fitted so as to reproduce some property, previously determined by either experiment or from all electron *ab initio* calculations. The second route to pseudopotential construction is the *ab initio* method, whose basic idea is as follows. First, consider a "reference state" for the atom we are studying, which is simple enough so as to compute accurately, via KS-DFT calculations, its electronic properties. Second, since the interest is on the valence electrons, we eliminate all those effects on the valence properties which are due to the presence of core electrons; this step is called "pseudizing" the valence properties. Third, consider what is the pseudopotential going to be used for, and how is it going to be used. Usually it will be used for the calculation of the ground state energy of a complex system, probably composed of many atoms. This calculation can be carried out using several different theoretical approaches, for instance, KS-DFT, LRT or OF-DFT. So the third step selects the appropriate theoretical approach to be used in the study of the complex system. Finally, substitute our atom in the "reference state" by an effective potential, and apply the selected theoretical approach to calculate those valence properties considered in the second step. The pseudopotential is that effective potential which reproduces those valence properties.

Therefore, in the *ab initio* construction of the pseudopotential, three basic ingredients have to be specified: (i) the reference state, (ii) the valence properties on which we are going to focus, and (iii) the theoretical approach used in the study of the complex system. We employ this procedure in the generation of local pseudopotentials for OF-DFT molecular dynamics simulations of extended systems (bulk liquids and surfaces).

The environment of an atom in a metal is quite different from free space. Therefore it seems appropriate to use a reference state which closely resembles this environment, while being simple enough so that KS-DFT can still be applied to find the valence electronic properties. Then, we choose as reference state an atom embedded in a uniform background (jellium) of density equal to the mean electron density of the complex system (the metal at some specified thermodynamic state), in which a spherical cavity has been made around the atom. This system still has spherical symmetry, like the free atom, and the main complication is that now the valence states are unbound, i.e., they are scattering states forming a continuum occupied up to the Fermi energy. The considered valence property is the "displaced" valence electron density, $\Delta n(r)$, that is, the change in the electron density induced in the jellium due to the presence of the atom and the cavity. This is pseudized by eliminating the core-orthogonality oscillations leading to $\Delta n_{\rm ps}(r)$. The theoretical approach used for the complex system is OF-DFT. Therefore, we must find an effective potential that, when inserted into the jellium together with the cavity, will produce, within the OF-DFT theory, the same pseudized displaced density as that computed previously. Translated into formulae, it proceeds as follows. When the functional derivatives of the energy functional are performed, the Euler equation for our pseudopotential in the jellium-cavity system is

$$\mu_T(r) + V_{\rm ext}(r) + V_H(r) + V_{\rm xc}(r) - \mu = 0\,, \tag{30}$$

where each term is the derivative of the corresponding term in Eq. (1), namely,

$$\mu_T(r) = \mu_W(r) + \mu_\beta(r) = \frac{\delta T_{vW}}{\delta n(\vec{r})} + \frac{\delta T_\beta}{\delta n(\vec{r})}\,, \tag{31}$$

$$V_{\rm ext}(r) = v_{\rm ps}(r) + v_{\rm cav}(r) + v_{\rm jell}(r)\,, \tag{32}$$

$$V_H(r) = \int d\vec{s} n(s)/|\vec{r}-\vec{s}|\,, \tag{33}$$

with $n(r) = n_0 + \Delta n(r)$, and $V_{\rm xc}(r)$ is the exchange-correlation potential. All the magnitudes depend on r only, due to the spherical symmetry. Inserting Eq. (31–33) into Eq. (30) we obtain an algebraic equation for the unknown $v_{\rm ps}(r)$ in terms of the known $n(r)$, whereas the constant μ is just an energy origin, which is fixed so as $v_{\rm ps}(r)$ decays to zero for large distances. The pseudopotential generated this way is density-dependent, thus obtaining high accuracy in the electron-ion interactions at the cost of sacrificing transferability.

We do not have at present a sound way of generalizing this pseudopotential generation procedure to finite atomic systems such as clusters, because each atom in a cluster is under the action of a different environment, and thus there is no uniquely appropriate reference state. Therefore, it is sensible to adhere to the traditional scheme of generating pseudopotentials in a single reference state, in the hope they will be transferable to arbitrary environments. This is indeed accomplished with nonlocal pseudopotentials and, as long as *core*-electrons are really inert from the chemical point of view, there is not any compelling reason why there should not be so for the local pseudopotentials. An important step towards the realization of transferable local pseudopotentials has been recently reported by Carter *et al.* [31]

The model pseudopotential we employ in cluster calculations has the "evanescent-core" functional form given by Fiolhais *et al.* [32]:

$$v_{ps}(r) = -\frac{Z}{R}\left(\frac{1}{x}[1-(1+\beta x)e^{-\alpha x}] - Ae^{-x}\right), \tag{34}$$

where Z is the valence, R is a core decay length, $\alpha > 0$ and $x = r/R$. Only α and R are considered as adjustable parameters, as A and β are fixed by requiring that the first and third derivatives of the pseudopotential vanish at $r = 0$ (these conditions make the Fourier Transform of the pseudopotential to decay as fast as k^{-8} in the region of large wavenumbers, which is computationally convenient). The two parameters are adjusted by requiring the pseudopotential to reproduce the equilibrium volumes and valence interstitial electron numbers of bulk crystalline metals. The transferability of such a pseudopotential to cluster environments has been shown to be reasonable, at least for the case of the alkalis [33]. In any case, parameters in v_{ps} could be readjusted by fitting to some cluster properties (for example, interatomic forces) evaluated by means of an accurate *ab initio* method as, for example, KS-DFT. This approach has been followed in recent work on clusters [29, 30] and shown to lead to much better quantitative agreement with orbital-based *ab initio* results. In this work we will stick to the original parameter values given by Fiolhais *et al.* [32], as the semiquantitative accuracy provided by them is sufficient for our purposes, as shown in the next section.

3 Results

3.1 OFMD Study of the Melting-Like Transition in Alkali Clusters

3.1.1 Background

A series of recent experiments by Haberland and coworkers [34, 35, 36, 37, 38, 39, 40, 41] has renewed the interest in the melting transition of small atomic clusters. These experiments reveal that melting temperature T_m of sodium clusters Na_N^+ shows a non-monotonic dependence on cluster size N, contrary to expectations based on classical arguments stating that melting temperature of a finite liquid drop should monotonically decrease as its radius shrinks. The detailed evolution of T_m with N shows enhanced stability of the solid phase for selected ranges of cluster sizes, some of which are close to well known electronic and/or geometrical (icosahedral) shell closings, while some others are not in correspondence with any known feature. A first guess is that both electronic and structural effects play a role in explaining the observed results, but the latest experimental work [41] points to a much more important role of structure in determining the melting point. In this background section, we summarize the efforts of the computational physics community in finding an explanation for the experimental results on sodium and also other alkali homogeneous clusters.

Computer simulation may provide physical insights into the microscopic details of the melting-like transition in clusters and, hopefully, reproduce and explain the experimental $T_m(N)$ curve. Some recent attempts in this direction have been given by Manninen *et al.* [42] and Calvo and Spiegelman [43]. Manninen *et al.* perform molecular dynamics (MD) simulations of melting in Na_N ($N = 40 - 355$) and reproduce non-monotonic $T_m(N)$ behavior. However, due to computational limitations, interatomic forces in all clusters with $N \geq 93$ are calculated employing a phenomenological potential, and the $T_m(N)$ curve is not in good agreement with the experimental one for $N \geq 142$. The reasons for this discrepancy may be: a) purely electronic effects (not well described by any phenomenological potential) have a strong in uence; b) interatomic forces provided by that potential might be not accurate enough so that correct ground state (GS) isomers (for example, those predicted by an *ab initio* methodology) are not reproduced; c) even if the potential correctly reproduces the *ab initio* energetic ordering of isomers, it is very difficult to locate the GS isomer when so many atoms are involved; etc. Agreement with experiment is much better for smaller clusters. Specifically, Manninen *et al.* [42] reproduce the experimental result $T_m(Na_{55}^+) > T_m(Na_{93}^+)$. The work by Calvo and Spiegelman [43] finds interesting premelting effects, associated to a surface melting stage, for some cluster sizes around $N = 130$. They observe that premelting and melting temperatures are close to each other, thus providing a tentative explanation for the apparent absence of premelting effects in the experiments. Surface melting had also been previously observed in OFMD simulations by Aguado *et al.* [44, 45], which also succeeded in reproducing experimental melting temperatures of selected Na clusters in the range $N = 8 - 142$ to within a few percent (an exception was $N = 55$, for which the experimental melting temperature is substantially higher than the OFMD result). For the smallest-size clusters (Na_8 and Na_{20}), the premelting phenomena are associated to isomerization transitions involving just specific subsets of atoms, which occur at temperatures lower than that corresponding to homogeneous melting [44]. More recently, Aguado *et al.* [29, 30] have developed a more accurate realization of the OFMD technique, essentially by employing pseudopotentials adapted to the cluster environment. These pseudopotentials are obtained from a force-matching procedure, in which KS-DFT interatomic forces for a representative sample of cluster configurations are taken as the target to be reproduced by OF-DFT. These very recent calculations result in a much better (quantitative) agreement with the experimental calorimetric results than any previous simulation study. In particular, the surface premelting mechanisms have been unambiguously identified and shown to explain the experimental size variation of the entropy of melting [41]. The premelting stage consists of surface isomerizations which basically sample the same ground state isomer, with only permutation of atom positions, being this the reason for the absence of any premelting signature in the caloric curve, a result which also agrees with experiment. The oscillations in the $T_m(N)$ curve are shown to be the result of geometric packing effects, while the electronic shell closing effects are at most of secondary importance to the melting process. This observation, apart of being in good agreement with experimental observations [41], is somewhat expected as the number of electrons does not change upon melting.

Low-temperature structures of Na clusters are usually icosahedral, and thus present a well defined distribution of atoms in concentric shells, that is, the local density of atoms is very low or zero in the intershell regions. An interesting question then is to what degree the way the melting transition proceeds is determined by this radial order. Aguado *et al.* [46] have analyzed this problem by performing OFMD simulations of the melting-like transition in Na clusters showing no radial order (amorphous clusters). Results from this study show that the melting transition is not abrupt in amorphous Na clusters, but develops in an almost continuous way upon increasing of temperature, which means there are no appreciable free energy barriers against diffusion of atoms, just kinetic impediment at low temperatures. Therefore, the transition proceeds without any significant latent heat of fusion. The melting-like transition in icosahedral clusters of K, Rb, and Cs of several sizes has been studied by Aguado [47]. Qualitative features of the melting transition were found to be similar to those of Na clusters, which means a separate surface premelting stage was observed in all cases. The homogeneous melting temperature decreases with increasing atomic number as in the bulk limit, but the percentage value of this reduction in T_m is larger than for bulk materials. The height of the specific heat peaks decreases and their width increases with increasing atomic number, and the premelting effects are more important the heavier the alkali atom. These trends were rationalized in terms of physical features of the local pseudopotentials employed to describe electron-ion interactions. More recently, Aguado and López [48] have studied the melting of Cs_{55} employing cluster-adapted pseudopotentials obtained from a force-matching procedure, as described above. These more accurate results show that Cs_{55} melts homogeneously, without a separate surface premelting stage. Previous calculations with bulk-adapted pseudopotentials may therefore have magnified the importance of surface premelting, which is expected to occur only in incomplete icosahedra (those with atomic vacancies in their surface shell). Other studies on melting of alkali clusters may be found in the literature [49, 50, 51, 52, 53, 54, 55, 56, 57], as well as other examples of application of OFMD technique to cluster science [58, 59, 60, 61, 62].

Consideration of the melting-like transition in impurity-doped metal clusters (and in multicomponent clusters in general) is even more complicate from a computational point of view. First of all, compositional ordering increases significantly the number of different isomer configurations, and thus the search for the GS isomer is much more time consuming. Also, surface segregation of the component with lowest surface energy may result in specific features such as partial melting of one component or incomplete mixing. Finally, and most important for the melting problem, the presence of impurities in an otherwise homogeneous cluster might alter structural properties, in uencing thermal properties and, in particular, the location of the melting point. Although most of the work on multicomponent clusters has been devoted to structural properties, a few molecular dynamics simulations exist studying their freezing [63] and melting [64, 65, 66, 67] transitions, and the in u-ence of temperature on segregation [68]. Very recently, Aguado *et al.* [69, 70] have performed OFMD simulations of the melting transition in the impurity-doped clusters Li_1Na_{54}, K_1Na_{54} and Cs_1Na_{54}. These clusters adopt icosahedral structures at low temperatures, with the Li impurity substituting the most internal Na atom, and the K and Cs impurities occupying a surface vertex position, as expected from surface energy arguments. Doping of Na_{55} was found to significantly reduce melting temperatures, as measured from the main peak in the specific heat, but the melting transition proceeds in a temperature interval which is wider than that corresponding to homogeneous Na_{55}. New premelting effects specific to multicomponent clusters (apart from surface melting) were observed. These are related to different temperatures for activation of diffusive motion of atoms of each species.

More recently, Aguado *et al.* [48, 71, 72] have studied the structural and thermal properties of several binary and ternary heterogeneous alkali clusters, namely, $Li_{13}Na_{42}$, $Na_{13}Cs_{42}$, $Li_{13}Na_{30}Cs_{12}$, $Li_{13}Na_{32}Cs_{42}$ and $Li_{55}Cs_{42}$. These clusters provide a representative sample of the structural motifs expected in heterogeneous alkali clusters. In those cases with a large size mismatch and a strong preference of one of the components for surface seggregation (due to the differences in surface tensions), polyicosahedral structures emerge. In a polyicosahedral structure, each interior atom in the cluster is at the center of a 13-atom icosahedron, so this structure is formed by interpenetrating 13-atom icosahedra. The coordination number is thus large in these structures, which are nevertheless energetically disfavoured for homogeneous metal clusters already at quite small sizes due to a large associated bond strain. In multicomponent clusters, on the contrary, when the size mismatch is sufficiently important, the component with smaller atomic radius can release the core strain because of its shorter natural interatomic distance, explaining the high stability of polyicosahedral growing, which leads, for example, to a compact anti-Mackay icosahedron for the composition $A_{13}B_{32}$. This kind of structural motif is observed in $Na_{13}Cs_{42}$, $Li_{13}Na_{30}Cs_{12}$, and $Li_{13}Na_{32}Cs_{42}$. In the case of $Li_{13}Na_{42}$, some mixing of Li and Na atoms is observed to be energetically preferred, as the size and surface tension mismatches between these two atomic species is not sufficiently large. The resulting structure looks then quite amorphous due to the partial mixing, as the distribution of interatomic distances is much braoder than in a perfectly seggregated core-shell nanoparticle. Irrespective of the apparently amorphous structure, some tendency towards polyicosahedral packing has been identified [71]. For $Li_{55}Cs_{42}$, the size mismatch is so large that polyicosahedral ordering is locally frustrated, that is, some Li atoms can not achieve a coordination number of 13 due to atomic packing considerations, and a more complex structural behavior emerges [72]. Some of these clusters, namely $Li_{13}Na_{32}Cs_{42}$ and $Li_{55}Cs_{42}$ form compact core-shell nanoparticles, that is, the surface shell (in this case formed by 42 Cs atoms) is of monatomic thickness and does not contain any atomic vacancy. They show a substantial thermal stability, which has been analyzed in a very recent work [48]. The rest of them show either partial mixing of the atomic components ($Li_{13}Na_{42}$) or incomplete surface layers ($Na_{13}Cs_{42}$ and $Li_{13}Na_{30}Cs_{12}$). In these cases, premelting effects are preponderant and the thermal stability is decreased, that is, these clusters have lower melting points. The structural properties of binary metal and Lennard-Jones clusters have also been analyzed, by means of parametrized interaction potential models, in a number of recent publications [73, 74, 75, 76, 77, 78, 79, 80, 81, 82, 83, 84, 85]

In this chapter, we further demonstrate applicability of the OFMD technique to the efficient simulation of thermal processes and melting transitions in multicomponent atomic clusters. In particular, we simulate the melting transition in Rb_1Na_{54} and $Cs_{12}Na_{43}$ clusters.

3.1.2 Technical Details

The cluster is placed in a unit cell of a cubic superlattice, and the set of plane waves periodic in the superlattice is used as a basis set to expand the valence density. The size of the supercell is large enough so that periodic images do not interact with each other. Present calculations use a supercell of edge 35 A. and an energy cut-off in the plane wave expansion of the density of 20 Ryd. A 96×96×96 grid is used to perform Fast Fourier Transforms. The equations of motion are integrated using the velocity Verlet algorithm [4] with a time step equal to 4(3) fs for temperatures lower (higher) than 100 K. These choices resulted in a conservation of the total energy better than 0.1 %.

An important first step in the simulation of the melting-like transition in clusters is the location of the ground state isomer. As the number of different isomers is known to increase with size in an exponential way, it is a very hard task to find the minimum energy isomer of a cluster with 55 particles. It is prohibitive, for example, to employ well-behaved methods such as basin-hopping [86] or genetic algorithms [87] in conjunction with an OF evaluation of cluster energies. Instead, one has to adopt structures that are likely to have the main characteristics of the ground state. For Rb_1Na_{54}, we have taken complete icosahedral and cuboctahedral structures, as the most plausible candidates, and have considered all possible inequivalent positions for the substitutional impurity. For $Cs_{12}Na_{43}$, we have just tested icosahedral structures, with different allocations of the 12 Cs atoms on the surface of the icosahedron (from our previous work [69] we already know that interior sites will be energetically unfavorable for Cs atoms). In order to explicitly check the reliability of OF predictions, we optimize the same structures employing pseudopotential KS-DFT calculations as implemented in SIESTA. [5] This code employs standard norm-conserving pseudopotentials [88] in their fully non-local form [89] and a exible linear combination of atomic orbitals (LCAO) basis set for the description of valence states. [90] The local density approximation is adopted for exchange and correlation effects to be consistent with the reference OF calculations (this way, we test the effect of the approximate kinetic energy functional and local pseudopotential). Standard double-zeta plus polarization (DZP) basis sets were adopted for each atom. This basis size is found to converge the energies to within 1 meV/atom if the range of the basis functions is carefully optimized. The fineness of the real space grid employed to calculate Hartree and exchange-correlation terms is measured by the maximum kinetic energy of the plane waves that can be represented in the grid without aliasing. A value of 20 Rydbergs is employed, which is also consistent with the OF calculations. In both OF and SIESTA calculations, we employed a conjugate gradients technique to obtain optimal geometries for each different isomer.

Starting from an average temperature of 50 K, several molecular dynamics simulation runs at different constant energies were performed in order to obtain the caloric curves. Different values of the average temperature were obtained by isokinetic thermalization runs (approx. 5 ps long) previous to each constant energy run, where the atom velocities are scaled each time step in order to maintain a target temperature. The final atomic configuration of each run served as the starting geometry for the next isokinetic run at a higher temperature. A typical simulation time for the constant energy runs of solidlike clusters is 50 ps, but runs longer than 200 ps were performed for those energies where the melting transition sets in. For energies corresponding to liquidlike clusters, the length of the simulations was always in the range 70–100 ps. The total simulated time for each cluster was approximately 1.5 ns.

In order to characterize the thermal behavior of the clusters as a function of increasing internal energy, and the solid-like to liquid-like transitions, we monitor: (a) global quantities that are calculated from time averages over a whole trajectory at a given energy; (b) time dependent quantities that are calculated from averages over well separated time origins along a single trajectory.

First, we define the internal temperature T of the cluster as [91, 92]

$$T = \frac{2}{(3N-6)k_B} < E_{kin} >_t, \tag{35}$$

where k_B is Boltzmann constant, E_{kin} is the ionic kinetic energy, and $<>_t$ represents the time average over a whole trajectory. All the global quantities described below will be plotted as functions of this internal temperature.

We evaluate the following melting indicators: (1) the caloric curve, which gives the dependence of average temperature on total energy. A change of slope is observed in the caloric curve at the melting temperature, which connects the solid and liquid branches. From the height of the rounded discontinuity we get an estimation for the latent heat of fusion; (2) the specific heat per particle (in units of the Boltzmann constant) defined by [92]

$$C_v = [N - N(1 - \frac{2}{3N-6}) < E_{kin} >_t \ < E_{kin}^{-1} >_t]^{-1}. \tag{36}$$

This magnitude is related to uctuations in the ionic kinetic energy, and has peaks (corresponding to slope changes in the caloric curve) associated to some phase transitions; (3) the root-mean-square bond length uctuation,

$$\delta = \frac{2}{N(N-1)} {}_{i<j} \frac{\sqrt{< R_{ij}^2 >_t - < R_{ij} >_t^2}}{< R_{ij} >_t}, \tag{37}$$

where R_{ij} is the distance between atoms i and j. By restricting the sum to specific pairs of atoms, one can also evaluate separate uctuations for Na–Na, Cs–Cs, Rb–Na and Cs–Na bond lengths. This magnitude characterizes the degree of mobility of atoms (a sort of index of rigidity of the cluster), and thus increases abruptly at isomerization or melting transitions. For the bulk, a sharp increase in δ gives the Lindemann criterion for melting; (4) the diffusion coefficient of atoms of species A,

$$D_A = \frac{1}{6}\frac{d}{dt} < r_A^2(t) >, \tag{38}$$

which is obtained from the long time behavior of the corresponding mean square displacement $< r_A^2(t) > = \frac{1}{N_A n_t} {}_{j=1}^{n_t} {}_{i=1}^{N_A} [\vec{R}_i(t_{0_j} + t) - \vec{R}_i(t_{0_j})]^2$, where n_t is the number of time origins, t_{0_j}, considered along a trajectory, and N_A the number of atoms of species A. $< r_A^2(t) >$ is a time-dependent quantity that also serves as a measure of the rigidity of the cluster [91]. Its slope for large t is proportional to the diffusion coefficient. Thus, a at $< r_A^2(t) >$ indicates that atoms of species A are vibrating about their equilibrium positions. When diffusive motion of atoms A starts, the slope of $< r_A^2(t) >$ becomes positive; (5) short-time averages of the "atomic equivalence indexes",[53] defined for each atom by

$$\sigma_i(t) = {}_j \mid \vec{R}_i(t) - \vec{R}_j(t) \mid, \tag{39}$$

which contain very detailed structural information. The degeneracies in $\sigma_i(t)$ are due to the specific symmetry of the isomer under consideration so their temporal evolution allows for a detailed examination of the melting mechanism; (6) the average over a whole dynamical trajectory of the radial atomic distribution function $g(r)$, defined by

$$dN_{at} = g(r)dr \tag{40}$$

where $dN_{at}(r)$ is the number of atoms at distances from the center of mass between r and $r + dr$. At low temperatures, this quantity exemplifies the distribution of atoms in radial concentric shells. This radial order should not appear in a liquid cluster, in principle, and thus $g(r)$ is also a good melting indicator; (7) in order to have some measure also of the spatial ordering of atoms within each radial shell, we employ partial pair distribution functions, defined by

$$g_{AB}(r) = \frac{{}_{i\in A, j\in B}\,\delta(r - r_{ij})}{\frac{4\pi}{3}\rho[(r+\Delta r)^3 - r^3]}, \tag{41}$$

where A, B represent the atomic species, Δr is the width of the bins employed to discretize the r variable, and ρ is a reference density. As there is no way of defining a proper density in finite systems, we just use an approximate bulk value. The several pair distribution functions will tend to zero for large distances, but for the purpose of analyzing ordering of atoms this is not important. If required, summation in eq. (41) is restricted to specific subsets of atoms, which serves to analyze the spatial ordering of atoms belonging to different radial concentric shells of a cluster; (8) finally, in order to analyze variations in the global shape of the cluster, we have evaluated the root-mean-square radius of the cluster r, a shape deformation parameter β measuring the degree of asphericity, and another parameter γ quantifying its triaxiality. Following Bulgac and Kusnezov,[50] these parameters may be obtained from the three principal moments of inertia (with all the atoms assumed to have the same mass, $m = 1$ amu), $I_1 \geq I_2 \geq I_3$, as follows:

$$I_k = \frac{2}{3}r^2\left(1 + \beta sin(\gamma + \frac{(4k-3)\pi}{6})\right), \tag{42}$$

$$I_1 + I_2 + I_3 = 2r^2. \tag{43}$$

$\gamma = 0$ corresponds to an axially symmetric prolate aggregate (cigar-like), while for $\gamma = \pi/3$ we have an oblate axially symmetric shape (pancake). For intermediate γ values, cluster deformation is triaxial. If $\beta = 0$, the shape is spherical and the γ parameter has not any meaning. If $\gamma = \pi/3$, the maximum value of β is 0.5, corresponding to a disk of zero thickness. If $\gamma = 0$, the maximum value of β is 1, corresponding to a linear chain.

3.1.3 Melting Transition in Rb_1Na_{54}

There are four geometrically inequivalent sites in a perfect 55-atom icosahedron. Introducing a substitutional impurity in each of these sites generates four different Rb_1Na_{54} isomers, whose energies are compared in Table I. We do not include results for corresponding cuboctahedral isomers as these are located at higher energies. As expected from size mismatch and surface energy arguments, Rb prefers to be located at a vertex site of the cluster surface (see Fig. 1c in Ref.[69]). Moreover, OF calculations reproduce the energetic ordering of isomers predicted by SIESTA, which supports the employment of our OF procedure to study thermal properties. From a structural point of view, the only appreciable difference between OF and SIESTA calculations is that OF distances are systematically longer by slightly less than 1 %, an error which lies well within the expected accuracy of our methodology.

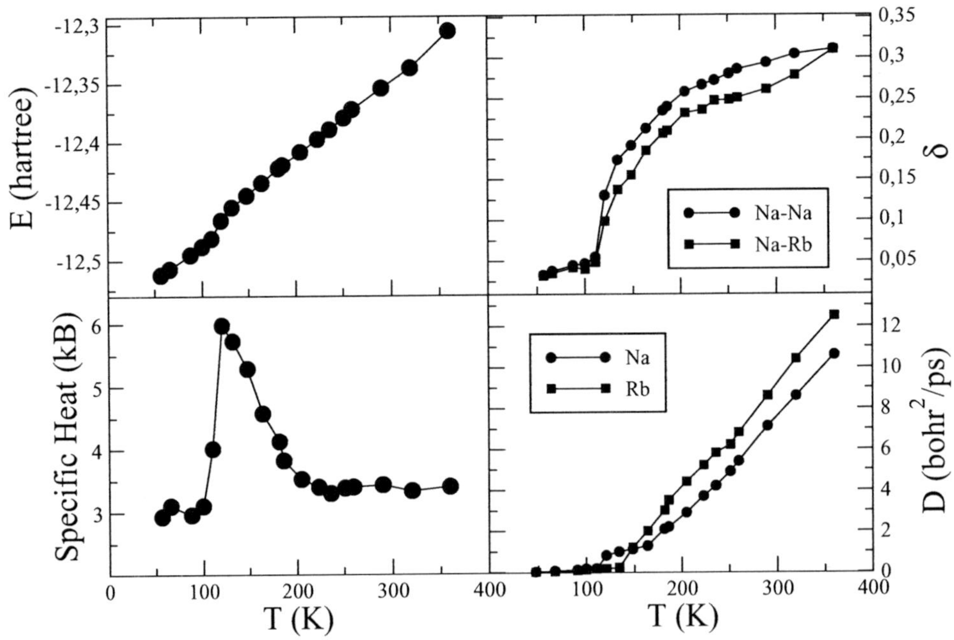

Figure 1: Caloric and specific heat curves (left side) and partial rms bond length uctuations and diffusion constants (right side) of Rb_1Na_{54}, taking the internal cluster temperature as the independent variable.

In Figure 1, we show the caloric curve, specific heat, partial rms bond-length uctuations and diffusion constants of Rb_1Na_{54}, as a function of average temperature. Both the caloric curve, specific heat and δ lead to the same value for the estimated melting temperature (approx. 120 K), which is a good check for the convergence of our results. The latent heat of fusion, estimated from the height of the caloric curve step, is 0.32 eV.

The temperature evolution of the specific heat displays just one main peak centered at T_m=120 K. At this temperature, there is an abrupt increase in the phase space volume available to the system. Analysis of the temperature evolution of diffusion constants, however, shows that the Rb atom does not participate in diffusion events until a temperature of 140 K is reached. This is significantly higher than T_m and shows that melting proceeds in two well defined steps: (1) at $T = T_m$, all sodium atoms (both surface and inner-shell atoms) can freely diffuse across the whole cluster volume, but Rb atom stays in its vertex position at the

Table 1: Energies (in meV/atom) of several isomers of Rb_1Na_{54}, taking the energy of the corresponding ground state (GS) isomer as a reference. Rb atom is located at (a) the center of the icosahedron; (b) any of the 12 equivalent positions within the first icosahedral shell; (c) a vertex position of the surface shell (again, the multiplicity of this site is 12); (d) any of the 30 non-vertex sites of the outer shell.

	a	b	c	d
SIESTA	11.7	9.4	GS	1.3
OF	12.1	10.2	GS	1.1

surface, without participating in diffusion processes; (2) at $T = 140$ K, free energy barriers against Rb diffusion may be overcome and the whole cluster is uid-like. This two-step melting process is validated by visual inspection of the movement of atoms on a computer screen. The second step is not re ected in the specific heat because the relative increase in the available phase space associated to it is very small. It is not re ected either in the partial bond length uctuations because diffusion of only Na atoms is enough to induce a step in the thermal evolution of Na-Rb bond length uctuations. It only can be appreciated in indicators which depend on the identity of each single atom, as is the case for the diffusion constants. Segregation of Rb to the surface liquid layer is observed for all temperatures considered in this work. The melting mechanism is thus analogous to that found in Cs_1Na_{54} [69].

Figure 2 shows the temperature evolution of averaged radial atomic densities. At T=70 K, atoms are grouped in shells. For the case at hand, we find the Na atom at the center of the icosahedron, then a group of 12 atoms forming the first icosahedral shell; next, a group of those 30 atoms of the outer shell which are capping the edges of the inner Na_{13} icosahedron; and finally, 12 atoms at vertex positions of the surface. Within this surface shell, Rb can be identified as a small bump at larger r due to its larger nearest neighbor distances. Although no mixing between different shells is observed, some intrashell disorder appears for temperatures approaching T_m (for $T = 90$ and $T = 110$, the minimum separating the last two peaks in $g(r)$ is not so well defined). This disorder, which appears only at the surface layers, is induced by temporary excursions of some atoms to positions of low average radial atomic density. After these temporary excursions, each atom comes back to its original site, that is, this disorder does not involve intrashell diffusion. At a higher temperature ($T = 150$ K), the amplitude of these motions is large enough to induce mixing between different shells and melting of the whole cluster. Note that there is not a separate surface-melting stage, as observed for example in Na_{20} or Na_{92} [44, 45], because both surface and inner Na atoms start to diffuse at the same temperature, $T = 120$ K. This is similar to the homogeneous melting process observed for Na_{55} [45]. The only difference is that, at $T = 120$ K, Rb atom is not able to diffuse yet. For all temperatures between 120 and 150 K, however, the $g(r)$ plots look quite similar, that is, this indicator is not useful to detect the second step in the melting transition. Another feature extracted from Fig. 2 is that, even at a relatively high tempera-

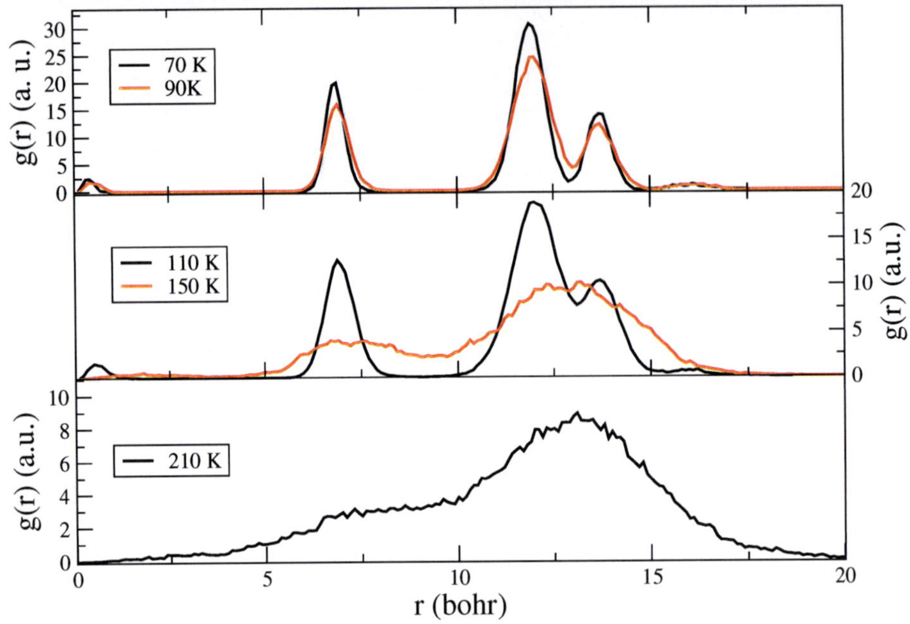

Figure 2: Time averaged radial atomic density distributions of Rb_1Na_{54}, at some representative temperatures.

ture of 210 K, some radial order remains, as the $g(r)$ function is not strictly parabolic. This means that there are some values of the radial distance which the atoms, on average, try to avoid, leading to oscillations in the density profile which are similar to those observed at the liquid surface of bulk metals (see next section on liquid metal surfaces). This has been discussed in our recent publication [69].

It is also interesting to analyze the temperature dependence of the global cluster shape, as this can be inferred from ion mobility experiments. [93] These results are shown in Fig. 3. The mean square radius increase in a pronounced way across the melting transition (by approx. 0.5 a.u.), in line with the expectation that a larger free volume is needed in order to "activate" the liquid state. It is also seen that thermal expansion is larger in the liquid phase, as expected. is close to zero in the solid phase, which means the cluster is almost spherical. Upon melting, the asphericity degree increases markedly, and continues to increase in the liquid phase. Similarly to the Cs_1Na_{54} case [69], the triaxiality parameter does not decrease with temperature in a pronounced way. As Rb is much heavier than Na, it tends to favor oblate shape distortions, as it prefers to be located along the axis with the largest moment of inertia. This tendency seems to be in competition with the general preference towards prolate distortions at high temperatures found in Li_1Na_{54} [69] and pure Na clusters [45]. We can not conclude from an analysis of Fig. 3 whether an oblate or prolate shape will prevail at higher temperatures.

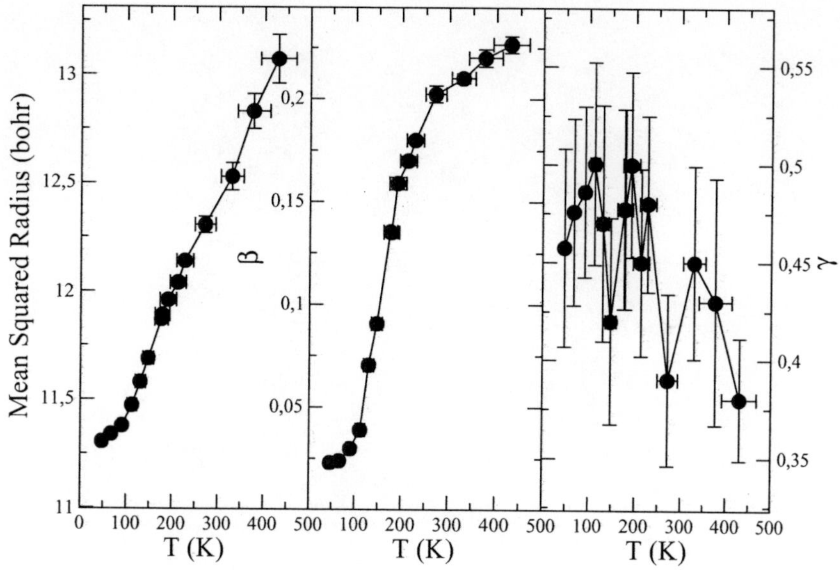

Figure 3: Temperature evolution of the mean square radius r (left), asphericity parameter β (middle), and triaxiality parameter γ (right) of Rb_1Na_{54}. Horizontal error bars represent temperature uctuations (not shown in previous plots); vertical error bars are covariances of cluster shape parameters.

3.1.4 Melting Transition in $Cs_{12}Na_{43}$

There are many possibilities for distribution of 12 Cs atoms on the 42 surface sites of a Na_{55} icosahedron, and we have not tested all of them for obvious reasons. We have just generated some representative structures with different values for the number of Cs-Cs bonds. This number is maximum when the two atomic species do not mix and clustering of Cs atoms at the surface appears; it is minimum (equal to zero) when the first coordination shell of each Cs atom is formed only by Na atoms (the number of Na-Cs bonds is then maximal, given the imposed restriction that Cs atoms must occupy only surface sites). In Figure 4 we show the minimum-energy isomer obtained in our search. Cs atoms occupy the 12 surface vertex positions of the icosahedral structure, which means that "surface solvation" of Cs is energetically preferred at zero temperature. Clustering of Cs atoms at the surface is found to be unfavorable as the volume of the underlying Na icosahedron is too small to accommodate 12 close Cs atoms without significant structural distortions. Our calculations, however, can not rule out the possible existence of a lower-energy isomer with a non-icosahedral structure. We have tried to perform some unbiased searches for the GS structure of $Cs_{12}Na_{43}$ by using a simulated annealing technique, but the rate of temperature reduction we can afford is always too fast and amorphous structures (with higher energies than the isomer shown in Fig. 4) emerge. We thus take the isomer shown in Fig. 4 as a reasonable approximation to the real ground state isomer. Figure 5 shows the caloric curve and the temperature evolution of specific heat, partial rms bond-length uctuations, and diffusion constants. Again,

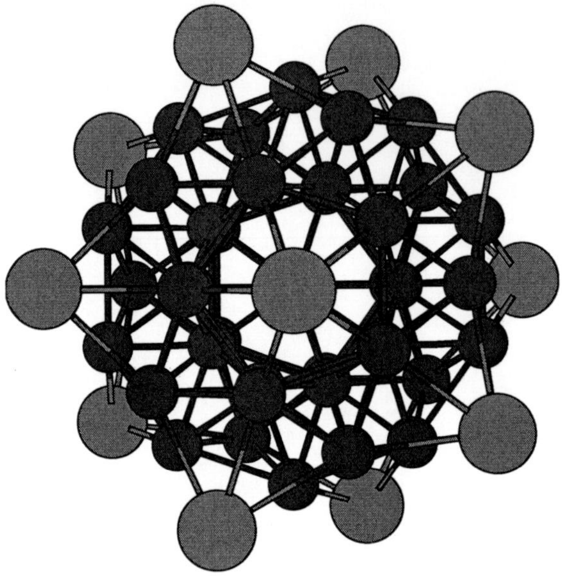

Figure 4: Ground state icosahedral isomer of $Cs_{12}Na_{43}$, as predicted from OF-DFT calculations. Red large and blue small balls represent Cs and Na atoms, respectively.

all the indicators lead to the same value for the estimated melting temperature of 130 K. The latent heat of fusion, estimated from the height of the caloric curve step, is 0.18 eV. Contrary to the situation found in Rb_1Na_{54}, activation of diffusion for both atomic species occurs at the same temperature. At temperatures close to 300 K, an increase in the specific heat is observed (also appreciated in the caloric curve as a slope change) which corresponds to incipient evaporation (see below).

The melting indicators shown in Fig. 5 are not helpful at all in elucidating atomistic details of the melting transition; this is a frequent situation indeed. In order to obtain some physical insight on melting mechanisms, we plot the radial atomic distribution functions $g(r)$ (Figure 6), atomic equivalence indexes $\sigma_i(t)$ (Figure 7) and partial pair distribution functions $g_{ij}(r)$ (Figure 8) for several representative temperatures. At 70 K, atoms are distributed in well-separated radial shells, and the three partial pair distribution functions for surface atoms show several peaks indicating that the tangential order is maintained at this temperature. We note here that, given the large size mismatch between Cs and Na atoms, the 12 Cs atoms form a well separate shell (in fact, the separation between Cs and Na surface radial shells is comparable to that between surface and inner Na radial shells–see fig. 6 and 7a).

At T=90 K, Fig. 6 demonstrates that the distribution of atoms in radial shells is still preserved, although some overlap between Cs and Na surface peaks appears. Fig. 7b tells us that there is already substantial thermal disorder, but atoms in different radial shells do not mix yet, supporting the conclusions from Fig. 6. Also, Fig. 5 shows that diffusion constants are still very small at this temperature, so thermal disorder does not involve sub-

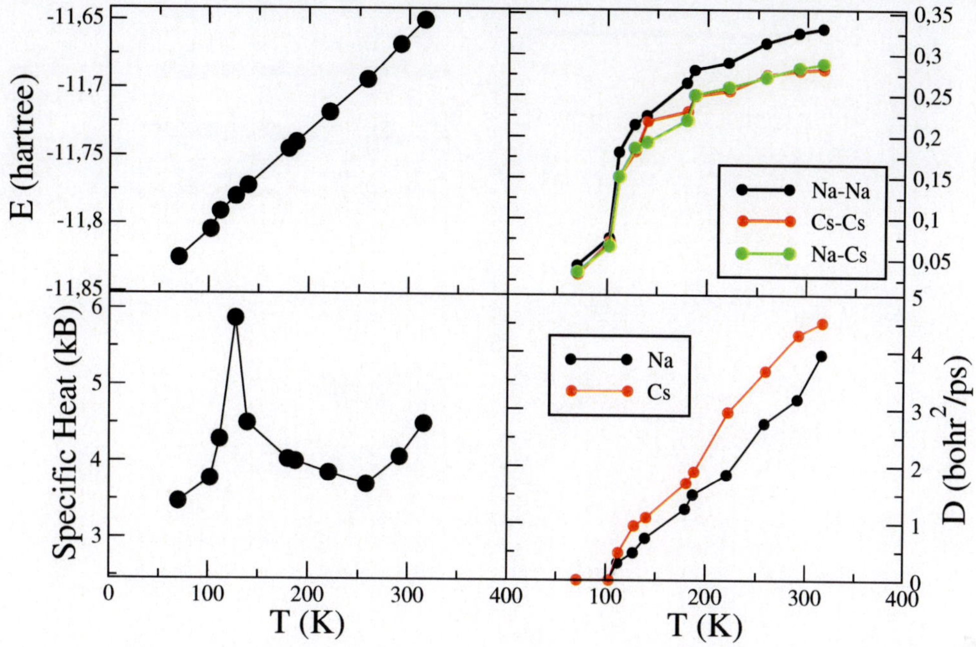

Figure 5: Caloric and specific heat curves (left side) and partial rms bond length uctuations and diffusion constants (right side) of $Cs_{12}Na_{43}$, taking the internal cluster temperature as the independent variable.

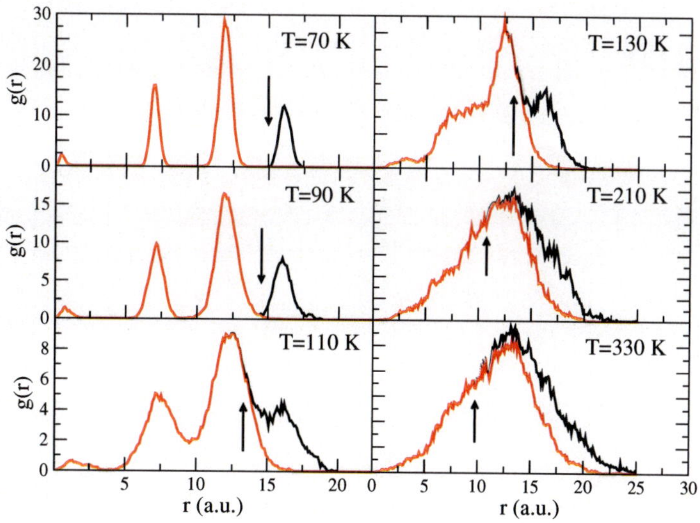

Figure 6: Time averaged radial atomic density distributions of $Cs_{12}Na_{43}$, at some representative temperatures. The red line represents the contribution of Na atoms to $g(r)$, while the black line is the total $g(r)$. Only the red line is seen for those distances where both functions have the same value. The arrow indicates the smallest radial distance for which $g_{Na}(r)$ and $g(r)$ do not coincide (maximum penetration depth of Cs atoms into the cluster).

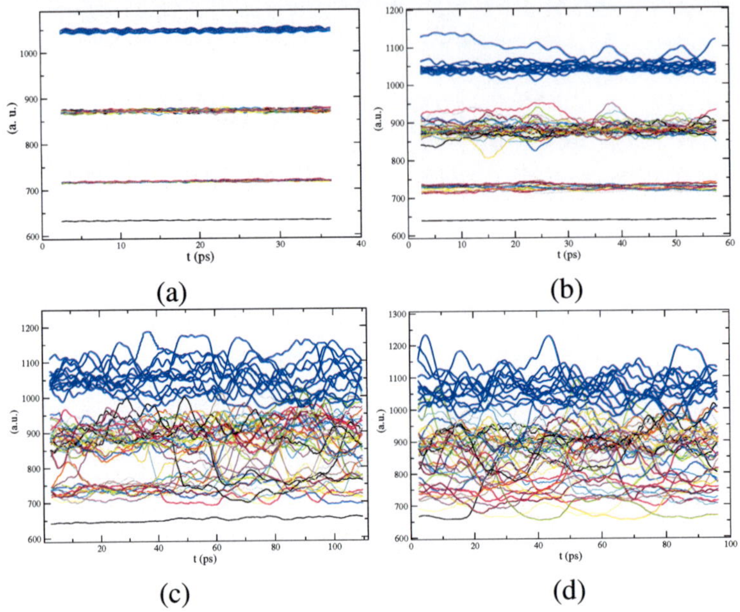

Figure 7: Time evolution of atomic equivalence indexes of $Cs_{12}Na_{43}$, averaged over 1000 time steps, for representative temperature values: 70 K (7a), 90 K (7b), 110 K (7c) and 130 K (7d). The curves corresponding to the Cs atoms are represented by bold blue lines.

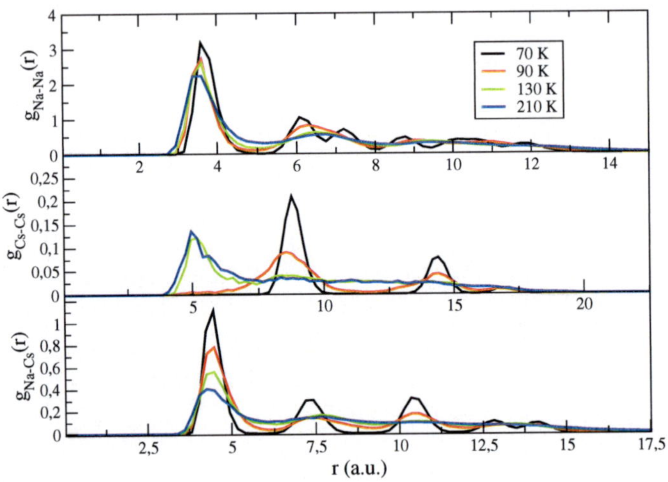

Figure 8: Partial pair distribution functions of $Cs_{12}Na_{43}$ at four representative temperatures. These functions were obtained by considering only atoms in the surface shell, in order to show more clearly the angular order in the outer cluster region (in order to define the "surface shell" we employ the minimum in $g(r)$ separating surface and inner Na radial shells. This is only possible at sufficiently low temperatures. For higher temperatures, a more or less arbitrary value of 10 a.u. was used to define the boundary between surface and inner shells. The curves are rather insensitive to the specific value employed).

stantial intrashell diffusion either. From Fig. 8, we see that long-range angular order at the surface shells is somewhat washed out in the pair distribution functions. Also the first peak in g_{Na-Na} shifts to a shorter distance, and g_{Cs-Cs}, although preserving the peaks positions on average, shows that temporary close approaches between Cs atoms exist. By visual inspection of the MD movie in a computer screen, we obtain the following picture of thermal disorder: In Fig. 4, each Cs atom is surrounded by a pentagon of Na surface atoms, and each Na surface atom is surrounded by an hexagon containing 2 Cs and 4 Na surface atoms, with the 2 Cs atoms at opposing vertices of the hexagon (let us denote this hexagon as CsNaNaCsNaNa). At 90 K, some new isomers begin to be visited, involving Na atoms surrounded by pentagons and Cs atoms surrounded by hexagons. Na and Cs atoms remain, however, at approximately the same radial distance (induced by the size mismatch effect) as in the GS isomer. As the distance between Cs and Na surface layers is large (see previous paragraph), it is relatively easy for the atoms to swap "angular sites", without swapping "radial sites", thus allowing the temporary creation of "defects": Na atoms surrounded by CsNaCsNaNaNa hexagons, for example. During these visits, the distance between some Cs atoms is substantially shorter than in the GS isomer, providing thus an explanation for the results in Fig. 8. The new isomers visited through this process explain that the specific heat value at 90 K is equal to 3.8, substantially higher than its value (3.0) for a cold solid cluster.

At T=110–130 K, substantial overlap between Cs and Na surface peaks, and between surface and inner Na peaks, are visible in $g(r)$. Atomic equivalence indexes in Fig. 7c and 7d show that intershell diffusion of atoms is present. The only difference between the $\sigma_i(t)$ curves at 110 and 130 K is that the innermost Na atom is not able to diffuse at 110 K but interchanges its position with atoms in external shells at 130 K, being this the reason why the melting is complete only at 130 K. The partial pair distribution functions at 130 K (Fig. 8) are typical of a liquid material. A drastic change in g_{Cs-Cs} is observed upon melting. Specifically, the main peak in g_{Cs-Cs} shifts to a substantially shorter distance, implying partial clustering of Cs atoms at high temperatures. We have tried to start some simulated annealing runs from this kind of high-temperature isomers and have not found any other isomer which has a lower energy (at zero temperature) than that shown in Fig. 4. Our calculations then predict that clustering of Cs atoms is only preferable at high temperatures, so it must be driven by entropic considerations. The entropy of mixing decreases upon clustering of Cs atoms, but the larger volume of the liquid cluster amounts to an important increase in configurational entropy. Also, with a larger volume, and with the structural freedom of a liquid cluster, clustering of Cs atoms does not involve as large structural distortions as at zero temperature conditions.

With increasing temperature, the pair distribution functions (Fig. 8) do not change much. The only appreciable change is a continuous increase in volume (thermal expansion) which is accompanied by an increasingly higher mixing between Cs and Na shells (see arrows in Fig. 6). However, segregation of Cs atoms to the surface layer is maintained up to the highest temperature considered. At temperatures close to 300 K, thermal uctuations are large enough that some Cs atoms are temporarily quite far from the rest of the cluster, meaning that evaporation of Cs atoms is close. Evaporation of Cs atoms was first observed at ≈ 340 K (results not shown in the figures). Examination of these evaporation processes and of the boiling transition is however beyond the scope of the present work.

The results for the temperature evolution of the global cluster shape are shown in Figure 9. The same qualitative features in the mean square radius and asphericity parameter β as observed for Rb_1Na_{54} and other alkali clusters [69] are detected, suggesting that they are general for all alkali clusters. Also, the presence of heavy ions (Rb or Cs) in a Na cluster seems to suppress the appearance of cigar-like shape at high temperatures (these cigar-like shapes are observed with lighter impurity atoms such as lithium). The reasons for this behavior have been explained in the previous subsection.

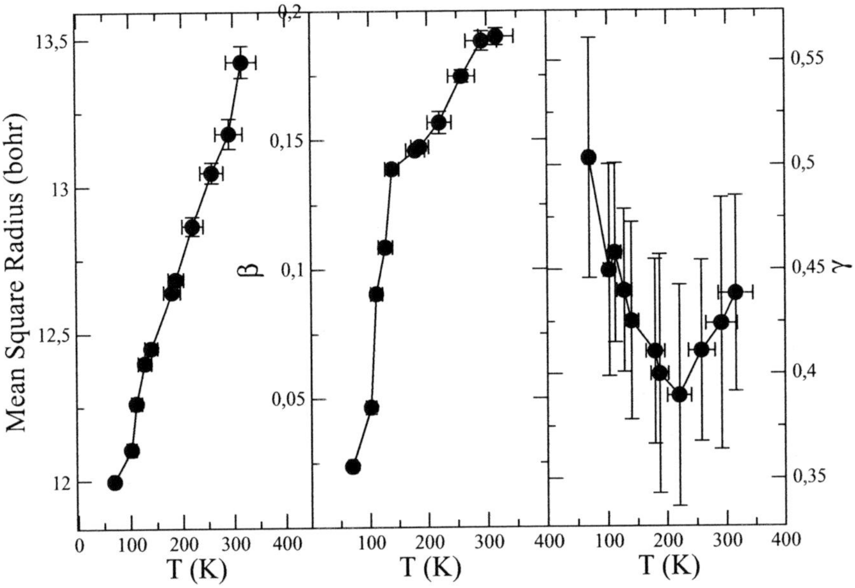

Figure 9: Temperature evolution of the mean square radius r (left), asphericity parameter β (middle), and triaxiality parameter γ (right) of $Cs_{12}Na_{43}$. Horizontal error bars represent temperature uctuations (not shown in previous plots); vertical error bars are covariances of cluster shape parameters.

3.2 OFMD Study of Bulk Simple Liquid Metals and Alloys

The study of the dynamical properties of liquid metals has already produced much experimental and theoretical work [94]. Most experimental investigations have been performed by means of inelastic neutron scattering (INS) which has been in the last 30 years the only available tool for studying the dynamics of condensed matter in the kinetic region. Recently, the development of new synchrotron radiation facilities has opened the possibility of using X rays too. This process has also been stimulated by MD simulations because of its ability to determine several dynamical magnitudes which are not accessible to experiment and, in this way, they have supplemented the experimental information. On the theoretical side, the development of microscopic theories such as the memory function formalism or the kinetic theory along with the realization that the decay of several time-dependent properties can be explained by the interplay of two different dynamical processes has created a theoretical framework whose application to simple liquids has lead to good qualitative results for several dynamical magnitudes [94, 95, 96].

Less attention has been devoted to the dynamics of liquid binary alloys, although the last twenty years have witnessed an increasing effort in this direction. Following a classical molecular dynamics (CMD) study of liquid Na-K [97], the field was stimulated by the results of a CMD simulation of liquid Li_4Pb by Jacucci *et al.* [98], where a new, high-frequency mode, supported by the Li atoms only (the so-called "fast sound") was found. Subsequently, several theoretical [99, 100, 101], computer simulations [102, 103, 104] and experimental [103, 105, 106, 107] studies have investigated the existence and properties of the collective excitations in liquid binary systems. On the theoretical side, the application of the kinetic theory for a model binary mixture of hard spheres (the revised Enskog theory–RET) [99, 100] and the recent generalized collective model (GCM) approach [101] have given useful insights into the nature of the collective excitations. For binary systems with disparate masses, two main branches (low- and high-frequency ones) of collective excitations have been found to contribute to the longitudinal dynamics, although its origin is still unclear. However, the transverse dynamics has been much less investigated, mainly because its is not visible in scattering experiments and only the MD simulations can provide information about the transverse excitations. At this stage we mention that the recent application of the GCM approach, which combines CMD simulations with the memory function formalism, to binary Lennard-Jones uids and some binary liquid alloys has shown the existence of transverse optic modes, which arise in connection with the concentration uctuations [101].

This section provides further insight into the capability of the OFMD simulation method to describe the structural and dynamical properties of simple (*sp* bonded) liquid metals and alloys. The following sections present results for liquid Li, Mg, Al and the Li-Na, and Li-Ba liquid binary alloys.

3.2.1 Dynamic Properties of Liquids

The basic magnitudes for describing the dynamic properties of a uid system are the time correlation functions of several dynamic variables of the system. We will first define some dynamical variables and their correlation functions will be considered later.

The self number density of particle j is defined as

$$\rho_{s;j}(\vec{r},t) = \delta(\vec{r} - \vec{R}_j(t)) , \tag{44}$$

where $\vec{R}_j(t)$ denotes the position of particle j at time t. Its associated space Fourier transform (FT) will be denoted as $\rho_{s;j}(\vec{q},t)$.

The (collective) number density of particles of a given type, is defined as the sum of all the self number densities of particles of that type. We will assume throughout that we are dealing with one-component systems although the generalization to multicomponent mixtures is straightforward. Thus, the number density for a one-component system is defined as

$$\rho(\vec{r},t) = \sum_j \delta(\vec{r} - \vec{R}_j(t)) \,, \tag{45}$$

and its space FT will be denoted as $\rho(\vec{q},t)$

The current density of particles of a given type is also an important dynamic variable, defined as

$$\vec{j}(\vec{r},t) = \sum_j \vec{V}_j(t)\, \delta(\vec{r} - \vec{R}_j(t)) \,, \tag{46}$$

where $\vec{V}_j(t)$ is the velocity of particle j at time t. In fact the particle's velocity itself is also an important dynamic variable which will be considered later on. The corresponding space FT, $\vec{j}(\vec{q},t)$, is usually decomposed into a longitudinal (parallel to $\vec{q}$) component $\vec{j}^L(\vec{q},t)$ and a transverse (perpendicular to $\vec{q}$) component $\vec{j}^T(\vec{q},t)$, with $\vec{j}(\vec{q},t) = \vec{j}^L(q,t) + \vec{j}^T(q,t)$.

There are also other dynamic variables related to the dynamic properties of a system, like the stress tensor, the energy current density, etc., that will not be discussed here. Instead, we refer the reader to the many books and reviews on the subject, see, for instance References [94, 108].

The time correlation functions between dynamic variables are the basic magnitudes to describe the dynamic properties. Particularly important are the time autocorrelation functions of the dynamic variables considered in the previous subsection.

The time autocorrelation function associated with $\rho_{s;j}(\vec{q},t)$ defines the self intermediate scattering function

$$F_s(q,t) = \left\langle \rho_{s;j}(\vec{q},t+t_0)\, \rho_{s;j}(-\vec{q},t_0) \right\rangle \,, \tag{47}$$

where, assuming that we are dealing with homogeneous and isotropic bulk systems, the average $\langle \cdots \rangle$ is carried out over time origins t_0, particles of the same type (i.e. all the particles) j, and directions of wave vector $\vec{q}$. The FT to real space of $F_s(q,t)$ is the van Hove self function $G_s(r,t)$, which is related to the probability that a given particle has moved a distance r in a time interval t. On the other hand, the FT of $F_s(q,t)$ into frequency domain is the self dynamic structure factor $S_s(q,\omega)$, which is directly related to the incoherent intensity obtained in an INS experiment, i.e. it is an experimentally accessible property.

The intermediate scattering function is defined as

$$F(q,t) = \frac{1}{N} \langle \rho(\vec{q},t+t_0)\, \rho(-\vec{q},t_0) \rangle \,, \tag{48}$$

where N is the number of particles, and the average is now over time origins t_0 and over directions of $\vec{q}$. For multicomponent mixtures the factor $1/N$ is replaced by $1/\sqrt{N_\alpha N_\beta}$ in the definition of $F_{\alpha\beta}(q,t)$, where α and β run from 1 to the number of different species and N_α is the number of particles of species α. The FT to real space of $F(q,t)$ is the van Hove function $G(r,t)$, which is related to the probability that if one particle is at the origin at a certain time, then any particle (the same or a different one) will be at a distance r from the origin after a time interval t. The FT into the frequency domain of $F(q,t)$ is the dynamic structure factor $S(q,\omega)$, which is also experimentally accessible, since it is directly related to the coherent intensity obtained in an INS or X-ray scattering experiments. Finally we note that $F(q,t=0) = S(q)$ with this latter one being the static structure factor.

The longitudinal and transverse current correlation functions are defined as

$$C^L(q,t) = \frac{1}{N}\left\langle \vec{j}_l(\vec{q},t+t_0)\cdot\vec{j}_l(-\vec{q},t_0)\right\rangle \tag{49}$$

$$C^T(q,t) = \frac{1}{2N}\left\langle \vec{j}_t(\vec{q},t+t_0)\cdot\vec{j}_t(-\vec{q},t_0)\right\rangle . \tag{50}$$

Due to its definition, the longitudinal current correlation function is directly related to the intermediate scattering function, or, in the frequency domain, to the dynamic structure factor

$$C^L(q,t) = -\frac{1}{q^2}\frac{d^2F(q,t)}{dt^2}, \qquad C^L(q,\omega) = \frac{\omega^2}{q^2}S(q,\omega)\,, \tag{51}$$

and can therefore be measured experimentally. The situation is however different for the transverse current correlation function, which is not accessible by experiment.

Another correlation function, not directly accessible to experiment is the velocity autocorrelation function, $Z(t)$, and the closely related mean square displacement, $\delta r^2(t)$, of a given particle in the system

$$Z(t) = \left\langle \vec{V}_j(t+t_0)\cdot\vec{V}_j(t_0)\right\rangle \tag{52}$$

$$\begin{aligned}\delta r^2(t) &= \left\langle \left|\vec{R}_j(t+t_0) - \vec{R}_j(t_0)\right|^2\right\rangle \\ &= 2\int_0^t d\tau\,(t-\tau)Z(\tau)\,,\end{aligned} \tag{53}$$

where, for a homogeneous bulk system the average is over time origins and over particles of the same type. Both functions are related to the diffusion coefficient, D, which is a measurable quantity, either as the time integral of $Z(t)$ or as the slope of $\delta r^2(t)$ (divided by 6) at long times.

As already mentioned, in the case of multicomponent mixtures, the previous definitions are easily generalized. This will lead to magnitudes such as the partial intermediate scattering functions, $F_{\alpha\beta}(q,t)$, whose FT into the frequency domain gives the partial dynamic structure factors, $S_{\alpha\beta}(q,\omega)$. Moreover, $F_{\alpha\beta}(q,t=0) = S_{\alpha\beta}(q)$ which are the partial static structure factors. The total static structure factor, $S_T(q)$, which is the magnitude usually measured by neutron scattering, is weighted combination of the $S_{\alpha\beta}(q)$. Similarly, there will also appear the partial longitudinal and transverse current correlation functions $C^L_{\alpha\beta}(q,t=0)$ and $C^T_{\alpha\beta}(q,t=0)$, respectively.

3.2.2 Liquid Li

During the last decade, liquid Li has undergone a quite thorough study of its static and dynamic properties, by means of experimental, theoretical and MD simulation methods, including CMD, KS-DFT and OF-DFT. Among the dynamical studies, we mention the INS experiments of Verkerk and coworkers,[109] the inelastic X-ray scattering (IXS) experiments of Sinn *et al.* [110], and those of Scopigno *et al.* [111] On the theoretical side, González *et al.* [112] have constructed effective interatomic pair potentials, based on local pseudopotentials, which they used to compute by both theoretical [112, 113] and CMD [114] methods, the static and dynamic properties of liquid Li; their results provided a good description of the static and dynamic properties [109, 110]. On the other hand, Kresse *et al.* [115] have performed the only KS-DFT simulations (to our knowledge) where the dynamic structure factor has been computed; they used a nonlocal pseudopotential of the so called ultrasoft type,[116] which are designed especially for the elements of the first row of the periodic table and for transition metals. OFMD simulations have been carried out by Anta *et al.* [117] who have studied the static and dynamic liquid structure; they used a kinetic energy functional which recovers the quadratic response function of the noninteracting homogeneous electron gas along with a local pseudopotential constructed by including in an approximate way the ionic structure of the liquid. The obtained $S(k,\omega)$ compared rather well with the IXS data [110] for all wave vectors considered. On the contrary, the KS-DFT results showed important disagreements in the small q region, as the obtained values of $S(q,\omega)$ were about three hundred times larger than the experimental and the OFMD ones.

We have also performed OFMD simulations by using the simpler T_s functional of Eq. (25) combined with a pseudopotential constructed as indicated in section 2.3. We considered 205 particles enclosed in a cubic box of appropriate side such that the number density is $\rho = 0.044512$ Å^{-3}, which is the experimental value at T=470 K. The cutoff energy in the planewave expansion was 15 Rydbergs, which amounts to 15900 coefficients. The timestep was 1 fs and the simulation lasted for 18 ps, after an initial equilibration of 5 ps.

Figure 10 shows the obtained static structure factor, $S(q)$, along with the corresponding X-ray [118] and neutron [119] diffraction data. The OFMD results compare rather well with the experimental ones, being closer to the neutron data. At small q values, our results are somewhat smaller than the neutron data, but still the agreement is rather good. The extrapolation of $S(q)$ to $q \to 0$, by using a quadratic expression $S(q) = S(0) + aq^2$, leads to an isothermal compressibility $\kappa_T = S(0)/(\rho k_B T) \approx 0.8 \times 10^{-10}$ Pa^{-1}, to be compared with the experimental value [120] of 1.02×10^{-10} Pa^{-1}.

The diffusion coefficient, derived from the slope of the mean square displacement, is $D = 0.66$ Å^2/ps (these units will be used throughout) which is close to the experimental value [121] of 0.64. The KS-DFT simulations gave $D = 0.66$ whereas Anta's OFMD simulations [117] produced a lower value, namely $D = 0.55$.

From the obtained $F(q,t)$, the corresponding $S(q,\omega)$ are readily computed and Figure 11 presents the results for several q-values ($q/q_p = 0.29, 0.5, 1$ and 1.41, with q_p being the position of the main peak in $S(q)$). The figure also includes the IXS data [110, 111] as

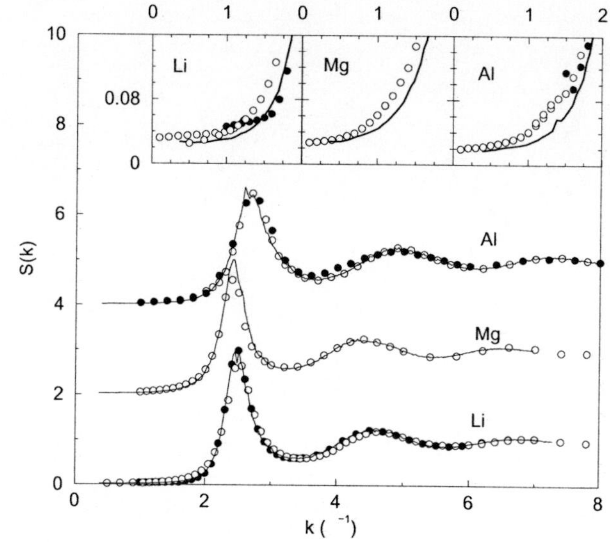

Figure 10: Static structure factor, $S(q)$, for the three metals studied. Note that the results for Mg have been shifted upward by 2 units and those for Al by 4 units. The insets show the low q behaviour. Full lines: present OFMD simulations. Open circles: X-ray diffraction data. Closed circles: neutron diffraction data.

well as the OFMD results of Anta *et al.* [117] For small q values, the experimental $S(q,\omega)$ exhibit well defined side peaks, which are typical of collective excitations and they persist up to $q \approx 2/3\,q_p$. For $q = q_p$ a typical increase in the height and narrowing in frequency is found (the deGennes narrowing) whereas for larger q a monotonic decay of $S(q,\omega)$ is obtained, leading to the free particle behavior. Both sets of IXS experimental data show good agreement, although the height of the peaks in Scopigno's data is somewhat higher. On the other hand, the OFMD results show an overall qualitative agreement with experiment although for small q-values they underestimate the heights of the peaks especially the central one. Comparison with the OFMD results of Anta *et al.* shows some discrepancies for small q values which we ascribe to our use of a simpler T_s functional; however, for larger q values both OFMD results almost coincide. From the positions of the peaks of the dynamic structure factor, denoted by $\omega_m(q)$, or from those of the spectra of the longitudinal current correlation function $C^L(q,\omega)$, denoted as $\omega_l(q)$, a corresponding dispersion relation is obtained, which is shown in Figure 12.

The isothermal sound velocity, obtained from the above calculated isothermal compressibility, is 4900 m/s. By using the experimental ratio of specific heats [120], namely $\gamma = 1.06$, this leads to an adiabatic sound velocity c_s=5050 m/s which is somewhat higher than the experimental one obtained from the adiabatic compressibility [120] c_s=4550 m/s. This is already apparent from Figure 11 where the OFMD side peak is somewhat shifted to the right of the experimental one (although there is also a certain q mismatch). The hydrodynamic dispersion $\omega = c_s q$ (with the theoretical c_s) is shown in Figure 12, where a positive

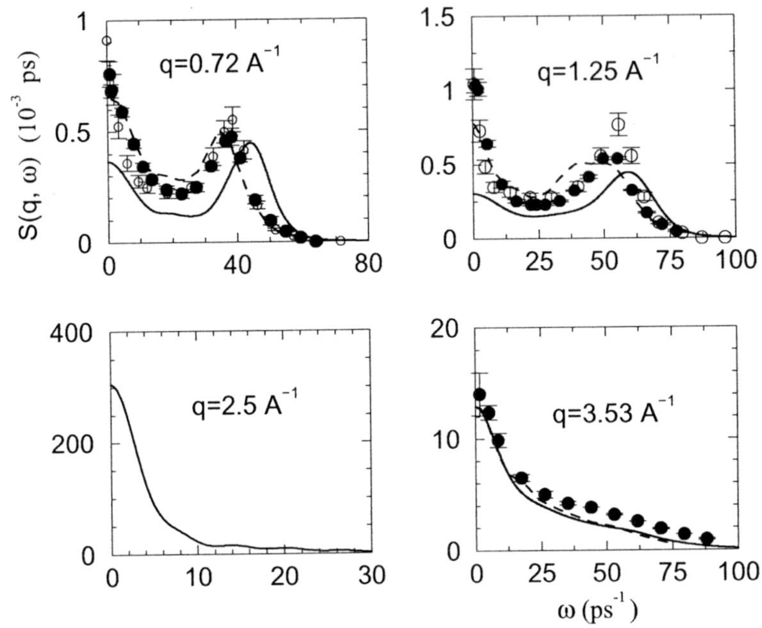

Figure 11: Dynamic structure factor, $S(q,\omega)$, for liquid Li. Full lines: present OFMD simulations ($q = 0.72, 1.25, 2.50$ and 3.53 A^{-1}). Dashed lines: Anta *et al*'s OFMD results ($q = 0.69, 1.26$ and 3.51 A^{-1}). Open symbols: inelastic X-ray scattering data; circles: Ref. [110] (q values shown on the graphs), triangles: Ref. [111] ($q = 0.70$ and 1.28 A^{-1}).

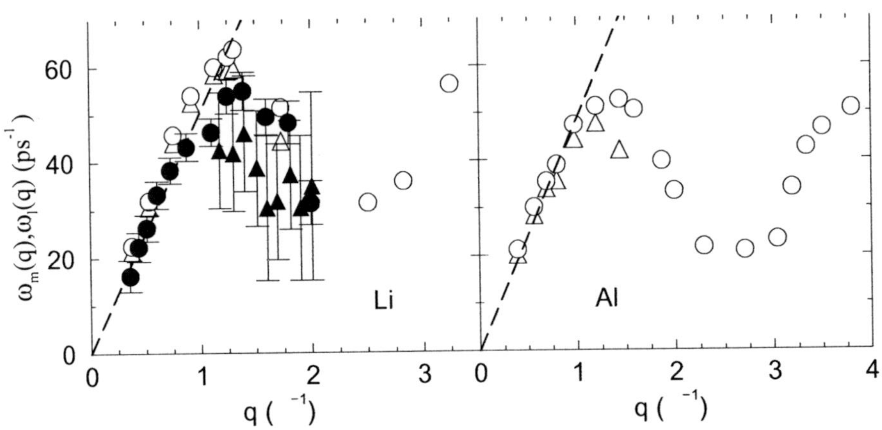

Figure 12: Dispersion relation of longitudinal collective modes for liquid Li and Al. Open triangles and circles: $\omega_m(q)$ and $\omega_l(q)$ respectively. (the error bars are within symbol size). Dashed lines: hydrodynamic dispersion (see text). Full circles: IXS data of Ref. [110]. Full triangles: INS data of Ref. [109].

dispersion is observed, i.e., the side peaks in $S(q,\omega)$ and $C^L(q,\omega)$ appear at frequencies higher than those corresponding to the hydrodynamic regime, which is only recovered for rather small values of q, not available in our simulations. Besides liquid Li where the experimental positive dispersion shows a maximum at $q \approx 0.55$ A^{-1},[110, 111] this effect has also been experimentally observed in other liquid metals near melting, such as Na,[122] Rb,[123] Cs,[124] and recently Al (see later).

3.2.3 Liquid Mg

In comparison with the alkali metals, not much work either theoretical or experimental has yet been performed on the liquid alkaline-earth metals. Concerning the dynamical properties, we are only aware of CMD results of Alemany *et al.* [125, 126], who report values of the diffusion coefficients (for Mg is $D = 0.665$ A^2/ps) and the intermediate and self intermediate scattering functions.

Restricting to the static structural properties, several studies have already been performed. Among them, we mention the KS-DFT simulations of deWijs *et al.* [127] and the OFMD simulations of Anta *et al*,[117] which report results for the static structure factor and for the electron-ion correlation functions. We have also performed OFMD simulations of liquid Mg, at T=953 K and $\rho = 0.03829$ A^{-3}, by using 205 particles, a timestep of 1.5 fs and a total averaging time of 25.5 ps after equilibration. The cutoff energy was 42.25 Rydbergs and the number of planewaves was around 85000. Comparison with the KS-DFT simulations (which used with a cutoff energy of 12 Rydbergs and 500000 coefficients) shows clearly the important computational advantages achieved in OFMD simulations, in spite of the larger cutoff required by local pseudopotentials.

The results obtained for $S(q)$, see Figure 10, show good agreement with the experimental X-ray data, even at the small q region. The diffusion coefficient obtained within the OFMD simulations is $D = 0.53$ A^2/ps, which is somewhat smaller than the value from the CMD simulations. The obtained dynamic structure factors are shown in Figure 13, for several q values (at similar q/q_p values as those shown for Li). The overall behavior of $S(q,\omega)$ is very similar to that of liquid Li and other liquid metals, showing longitudinal collective modes for small wave vectors, a deGennes narrowing at q_p and a gradual transition to the free particle regime for higher q. In Figure 13 we have also included the $S(q,\omega)$ obtained from the CMD $F(q,t)$ data of Alemany *et al.* It is observed that whereas the side peaks obtained in CMD and OFMD appear at the same frequency, however the central peak is much higher in CMD simulations; this is due to a marked diffusive behavior in the CMD $F(q,t)$ results, which is absent (or very weak) in the OFMD results. Nevertheless, both approaches lead to a very similar dispersion relation. Moreover, since there are no experimental data for γ, it is difficult to assert the possible existence of a positive dispersion in liquid Mg; however, the plot of $\omega_m(q)/q$ versus q exhibits a maximum (see Figure 14) indicative of some positive dispersion. Recent IXS experiments [128] also show a similar behavior.

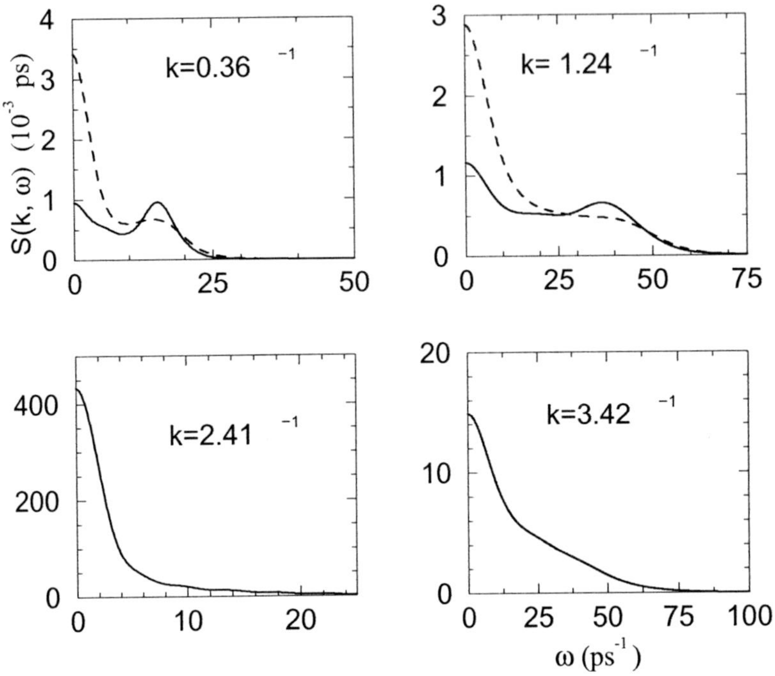

Figure 13: Dynamic structure factor of liquid Mg. Full lines: OFMD results. Dashed lines: CMD results.

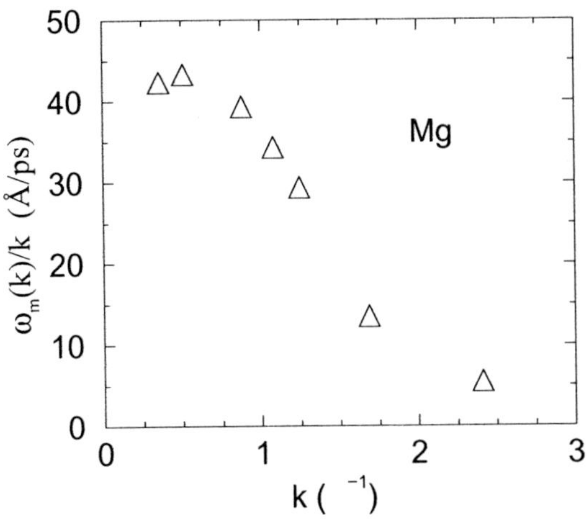

Figure 14: Phase velocity, $\omega_m(q)/q$, of longitudinal collective modes for liquid Mg.

3.2.4 Liquid Al

Most studies of liquid Al have been performed using CMD with pair potentials derived from either local [129] or nonlocal [130] pseudopotentials. There has also been one KS-DFT study [131] in which the static structure and the diffusion coefficient ($D = 0.60$ A^2/ps) were obtained, and an OFMD simulation [117] which only focused on static structural properties, showing good agreement with the experimental structure factor of the liquid. Here we report OFMD results (using a different T_s functional and a different local pseudopotential from those in reference [117]) for the static and dynamic structure of the system. These results refer to a thermodynamic state with T=1000 K and $\rho = 0.0524$ A^{-3}, which is close to the triple point. We have used 500 particles, the time step was 0.0015 ps, equilibration lasted for 3 ps and averaging was performed during another 30 ps (20000 configurations). The cutoff energy for the plane-wave expansion was 30 Ryd. which amounts to around 85000 planewave coefficients.

The obtained OFMD static structure factor is shown in Figure 10 together with the neutron [132] and X-ray [118] diffraction data. Comparison between both sets of experimental data shows small differences in the region between 2 and 5 A^{-1}, with the OFMD results between both sets, although closer to the X-ray ones.

The isothermal compressibility, κ_T, obtained from the $S(q \to 0)$ was found to be 2.374 x 10^{-11} Pa^{-1}, also close to the experimental value [133] of $2.4(\pm 0.1)$ x 10^{-11} Pa^{-1}. The calculated diffusion coefficient is $D = 0.49$ A^2/ps; unfortunately, no experimental data are available for this, but a recent KS-DFT calculation [134] for liquid Al at 1000 K, using 64 particles, gave values within the range $D = 0.58 - 0.62$.

The calculated $S(q,\omega)$ results are shown in Figure 15 for several q-values, along with the recent IXS data of Scopigno *et al.* [135] for liquid Al at T=1000 K. The $S(q,\omega)$ shows well defined sidepeaks up to $q \approx 1.6$ A^{-1}, which amounts to $\approx 3/5\ q_p$. The results qualitatively reproduce the general shape of the experimental IXS data, with some minor discrepancies for the height and positions of the sidepeaks. The dispersion relation, $\omega_m(q)$, is shown in Figure 12 along with the dispersion, $\omega_l(q)$, derived from the longitudinal currents. Extrapolating to $q \to 0$ the OFMD results for $S(q)$ and using the experimental value of γ ≈ 1.25 [136], gives $c_s \approx 4850$ m/s, which compares well with the experimental value [133] of ≈ 4700 m/s. Note also the positive dispersion in Figure 12; moreover a similar behavior is exhibited in the IXS data of Scopigno *et al.* [135] with a maximum at ≈ 0.45 A^{-1}.

The transverse current correlation function, $C^T(q,t)$, is not associated with any measurable magnitude and can only be determined by MD simulations. Its shape evolves from a Gaussian, in both q and t, in the free particle limit ($q \to \bullet$), towards a Gaussian in q for the hydrodynamic limit ($q \to 0$), (namely $C^T(q \to 0,t) = (1/\beta m)\ exp(-q^2\eta \mid t \mid /M\rho)$ where η is the shear viscosity, M is the ionic mass and β=$(k_B T)^{-1}$ is the inverse temperature times the Boltzmann constant) whereas for intermediate q-values, $C^T(q,t)$ exhibits a more complicated behavior. Its associated spectrum $C^T(q,\omega)$ is shown in Figure 16. where an inelastic peak is observed at $q = 0.29$ A^{-1}; as q increases the peak becomes better defined and persists up to q values around $3q_p$. The associated peak frequency increases with q towards a maximum value reached at $q \approx q_p$; it then attens and $C^T(q,\omega)$ evolves towards

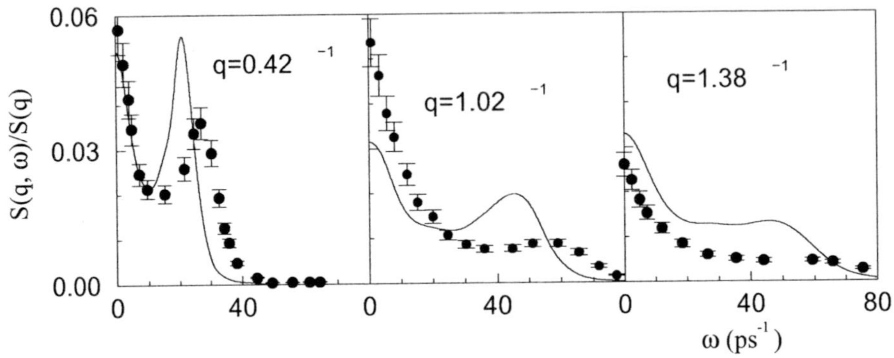

Figure 15: Dynamic structure factor of liquid Al. Symbols: IXS results (q values shown on the graphs). Full line: OFMD results ($q = 0.40, 0.98$ and 1.44 A^{-1}).

a Gaussian shape as q increases. By the way, we note that a similar behavior has also been observed for the alkali metals [94]. A shear viscosity coefficient, $\eta = 1.38$ GPa ps was obtained from the calculated $C^T(q,t)$, to be compared with the experimental data $\eta = 1.26$ GPa ps [137]. The KS-DFT simulation of Alfe and Gillan [134] gave values within the range 1.4 - 2.2 Gpa ps.

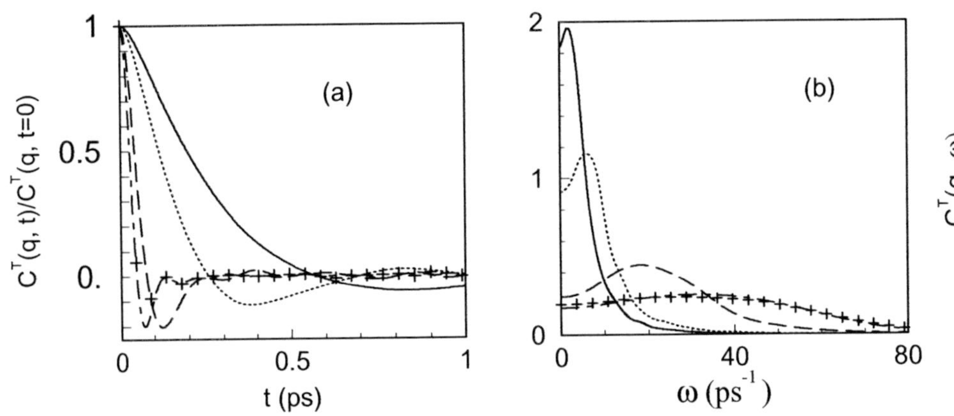

Figure 16: OFMD transverse current correlation, $C^T(q,t)$, and its spectrum, $C^T(q,\omega)$, at several q-values (in A^{-1}) for liquid Al at T =943 K. q=0.29 (full curve), q=0.42 (dotted curve), q=0.98 (dashed curve), q= 1.99 (dash-dotted curve) and q=2.70 (pluses).

3.2.5 Liquid LiNa

The Li-Na liquid alloy has attracted much attention being a typical phase separating system with a consolute point at $T \approx 577$ K and $x_{\rm Li} \approx 0.64$, close to the "zero alloy" composition ($x_{\rm Li} \approx 0.61$) for which the total static structure factor, $S_T(q)$, reduces to the concentration-

concentration partial structure factor $S_{CC}(q)$, namely, $S_T(q) = S_{CC}(q)/x_{\rm Li}x_{\rm Na}$. $S_T(q)$ has been measured at several temperatures and concentrations [138, 139], and at the zero alloy composition it shows a strong small-angle scattering, the subsequent oscillations are rather weak and there is a characteristic ripple around $q \approx 2.2$ A^{-1}. The zero alloy has recently been investigated through KS-DFT simulations. Hoshino and coworkers [140] used 100 atoms, an energy cutoff of 10 Ryd., and their $S_T(q)$ reproduces the main experimental trends, although the results are extremely noisy. Also, Costa Cabral et al. [141] have used 108 atoms, an energy cutoff of 9 Ryd. and similar nonlocal pseudopotentials, and the properties were obtained averaging over the 300-400 configurations. Their $S_T(q)$ qualitatively reproduces the experimental one, with some quantitative differences throughout the whole range of q.

OFMD simulations have been performed for liquid Li-Na alloy at concentrations $c_{\rm Li}$ = 0.40, 0.61 and 0.80 with respective experimental ionic number densities [142] ρ=0.02824, 0.03218 and 0.03682 A^{-3}. The simulation used 2000 particles, the cutoff energy was 8 Ryd., the equilibration took 5 ps and the calculation of properties was made averaging over 50 ps. [143]

Figure 17 shows the calculated total neutron weighted structure factor $S_T(q)$, along with the experimental one. The agreement experiment/simulation is very satisfactory, with the OFMD simulations reproducing rather accurately all the basic features of the experiment, namely, its quick increase when $q \to 0$ as well as the position and amplitude of the oscillations. Moreover, the two peaks at $q \approx 2$ and 2.5 A^{-1} for the $Li_{0.61}Na_{0.39}$ alloy are also accurately reproduced. The KS-DFT simulations of Costa Cabral *et al.* [141] for the $Li_{0.61}Na_{0.39}$ alloy have also provided a fair description of the experimental results, although the ripple is somewhat underestimated, possibly due to the small number of configurations and/or particles used in the simulation. Their results are shown in Figure 17 along with the corresponding KS-DFT simulations of Hoshino *et al.* [140] which show a similar agreement with experiment although are much noisier.

The partial dynamic structure factors provide information about the existence of propagating longitudinal modes. These are usually re ected, for a fixed q value, as side peaks in the $S_{\alpha\beta}(q,\omega)$ as a function of ω. Figure 18 shows the OFMD calculated partial dynamic structure factors $S_{\rm LiLi}(q,\omega)$ and $S_{\rm NaNa}(q,\omega)$ for the $Li_{0.61}Na_{0.39}$ alloy at three representative q-values. At the smallest q-value allowed by the simulations, q= 0.158 A^{-1}, both $S_{\rm LiLi}(q,\omega)$ and $S_{\rm NaNa}(q,\omega)$ show clear side peaks (also known as Brillouin peaks) at a common frequency, ω_B, which is the typical behavior in the hydrodynamic region and indicates a propagating sound mode with an adiabatic velocity of propagation $c_h = \omega_B/q \approx 3000$ m/s. $S_{\rm LiLi}(q,\omega)$ exhibits Brillouin peaks for the other two q-values of Figure 18 and continues to show side peaks or shoulders up to $q \approx 1.5$ A^{-1}, whereas $S_{\rm NaNa}(q,\omega)$ has only a shoulder for $q = 0.48$ A^{-1} and this feature disappears at $q \approx 0.9$ A^{-1}. Similar results are obtained for the other two concentrations, leading to calculated adiabatic velocities of propagation c_h = 3100 and 2530 m/s for x_{Li}= 0.80 and 0.40 respectively. We are not aware of any experimental measurement of the adiabatic sound velocity in this alloy; however, OFMD calculations (at T = 590 K) for the limits $x_{\rm Li}$ =0 (pure Na) and $x_{\rm Li}$ =1 (pure Li) give c_s = 2330 m/s and

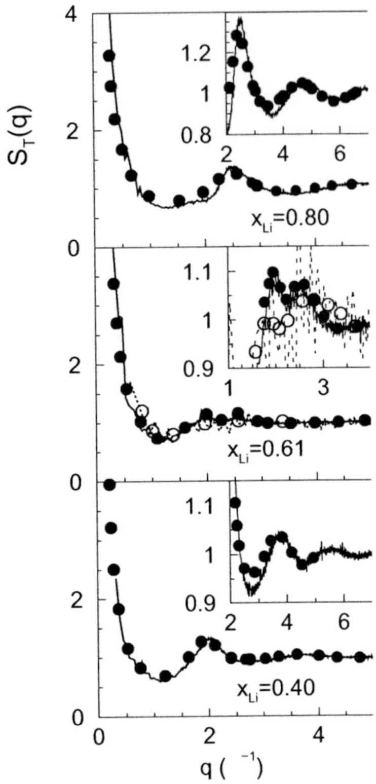

Figure 17: Total static structure factor, $S_T(q)$ for the liquid Li-Na alloy at T=590 K and three concentrations. Continuous lines are the OFMD results whereas the full circles are the experimental neutron diffraction data. In the figure for x_{Li}=0.61, the dotted line and the open circles are the KS-DFT results of Hoshino *et al.* [140] and Costa Cabral *et al.*[141] The inset shows an enlargement, by a factor of ten, of the main peak's region.

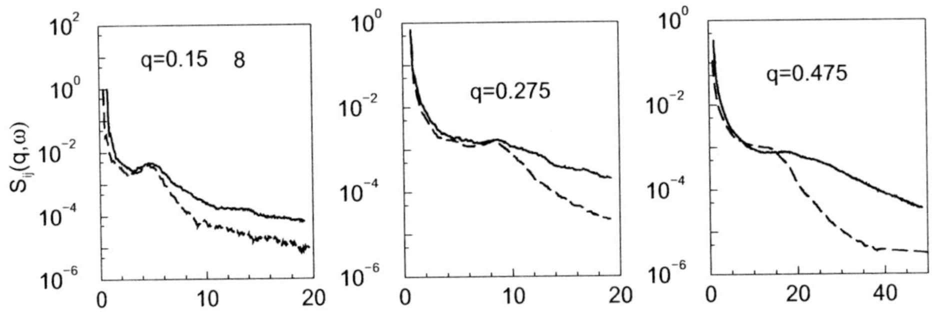

Figure 18: OFMD results for the partial dynamic structure factors (logarithmic scale) in liquid $Li_{0.61}Na_{0.39}$ at T=590 K. $S_{LiLi}(q,\omega)$ (full lines) and $S_{NaNa}(q,\omega)$ (dashed lines) as a function of ω (ps^{-1}) for three q-values.

$c_s = 4640$ m/s respectively, which are close to the corresponding experimental values [144] of 2410 and 4450 m/s.

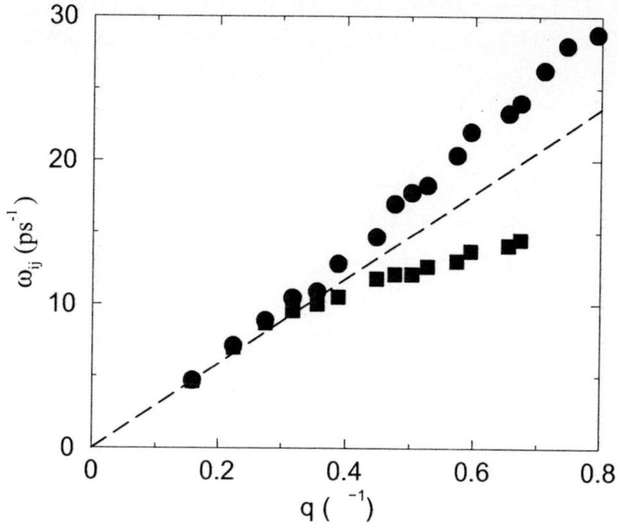

Figure 19: OFMD results for the dispersion curves of the collective modes in $S_{\mathrm{LiLi}}(q,\omega)$ and $S_{\mathrm{NaNa}}(q,\omega)$ (full circles and squares respectively) in the liquid $\mathrm{Li}_{0.61}\mathrm{Na}_{0.39}$ at T=590 K.

From the position of the side peaks in $S_{\mathrm{LiLi}}(q,\omega)$ and $S_{\mathrm{NaNa}}(q,\omega)$, the dispersion curves $\omega_{\mathrm{LiLi}}(q)$ and $\omega_{\mathrm{NaNa}}(q)$ are easily obtained. Those are plotted in Figure 19 for the $\mathrm{Li}_{0.61}\mathrm{Na}_{0.39}$ alloy. Up to $q \approx 0.3$ A^{-1}, $\omega_{\mathrm{LiLi}}(q)$ and $\omega_{\mathrm{NaNa}}(q)$ show the same linear behavior of hydrodynamic collective excitations propagating with speed $c_h \approx 3000$ m/s. Above this q-value, the dispersion curve splits into two branches signaling the onset of dynamic decoupling of the species, and giving rise to fast and slow sound modes in the alloy. The fast mode involves the Li particles only, with a phase velocity $c_{\mathrm{fast}} = \omega_{\mathrm{LiLi}}(q)\ /q \approx 3800$ m/s, whereas the slow mode phase velocity is $c_{\mathrm{slow}} = \omega_{\mathrm{NaNa}}(q)\ /q \approx 2100$ m/s, at q=0.65 A^{-1}. In fact, it has been theoretically predicted [99, 100], that the phase velocity associated with the fast mode should take a value close to the adiabatic sound velocity of the corresponding light particle uid, in this case a pure Li system obtained by substituting all the Na particles by Li particles. Consequently, OFMD simulations for pure Li at 590 K and a total ionic number density corresponding to the $\mathrm{Li}_{0.61}\mathrm{Na}_{0.39}$ alloy, produced an adiabatic sound velocity of 3900 m/s, which is very similar to the velocity of the fast mode. Denoting by q_h the upper limit of the hydrodynamic region, we stress that the OFMD results show that when q decreases towards q_h, the fast sound mode undergoes a continuous transition into the hydrodynamic sound mode and this process occurs at $0.2 \leq q \leq 0.4$ A^{-1}.

There has been some argument concerning the way in which the two kinetic modes behave when approaching q_h from above. Some theoretical and CMD results for He-Ar and He-Ne mixtures have suggested that either the fast and slow sound merge into the hydrodynamic sound [102, 103], or that the fast sound disappears at q_h, and the slow sound merges into the hydrodynamic one, with a predicted value [100, 102, 103] for $q_h \approx 0.07$ A^{-1}. A similar value for q_h has also been obtained for liquid Li_4Pb by both INS and by CMD calculations [104, 107]. However, recent CMD calculations [145] prompted by the INS measurements of Bafile *et al.* [146] for the dynamic structure of a $He_{0.77}Ne_{0.23}$ gas mixture at two densities, have shown a clear crossover from hydrodynamics to fast and slow modes at $0.2 \leq q \leq 0.5$ A^{-1} and gave $q_h \approx 0.2$, which is substantially greater than previous estimates, and quantitatively agrees with the OFMD results for the liquid Li-Na alloy.

The alloy shear viscosity can be also derived from a generalization of the ideas previously presented in liquid Al. Now a total transverse current correlation function $C_{tt}^T(q,t) = \langle j_t^T(q,t)\, j_t^{T*}(q,0)\rangle$ is defined in which $j_t^T(q,t) = x_1^{1/2} M_1 j_1^T(q,t) + x_2^{1/2} M_2 j_2^T(q,t)$ is a total transverse current, with x_i and M_i denoting the concentration and ionic mass of component i. Likewise the $j_i^T(q,t)$ refers to component i and is defined according to equation (46) . In the hydrodynamic limit [94]

$$C_{tt}^T(q \to 0,\, t) = (\bar{m}/\beta) \exp\{-q^2 \eta \mid t \mid / \bar{M} \rho\}\,, \tag{54}$$

where $\bar{m} = x_1 M_1 + x_2 M_2$ and η is the alloy shear viscosity. By extrapolating to $q = 0$ the calculated OFMD results for the transverse currents we have obtained the values η =3.3, 3.1 and 2.9 GPa ps for the concentrations x_{Li}= 0.80, 0.61 and 0.20 respectively. Although no experimental data are yet available for the shear viscosity in the alloy, we note that for the limiting cases x_{Li}=1 and 0 (which refers to pure Li and pure Na at T=590 K) the OFMD values are η = 4.0 and 3.2 (GPa ps units) whereas the experimental data are 4.3 and 3.3 respectively. [147]

For simple liquid alloys, η shows either a linear variation with concentration or a small negative deviation from linearity. In contrast, large positive deviations are exhibited by those alloys with heterocoordinating tendencies. For liquid alloys with homocoordinating tendencies, the available experimental data [137] show a mixed behavior although a majority of those systems exhibit a negative departure from linearity. Also, some semiempirical models [137] give a connection between positive (negative) deviations from the linear law and a negative (positive) enthalpy of mixing in the alloy. Although no experimental data are available for the enthalpy of mixing, several theoretical calculations have predicted positive values which, combined with the previous semiempirical models, would suggest a negative deviation of the alloy shear viscosity from a linear law. This is the trend followed by the OFMD results which, moreover, are in qualitative agreement with expectations for a homocoordinating tendency in the alloy.

3.2.6 Liquid LiBa

The experimental work performed on the $Li_{1-x}Ba_x$ alloy is rather scarce. Even its phase diagram is not yet fully settled as two rather different phase diagrams have been published [148, 149]. Whereas one finds an eutectic composition at x_{Ba}=0.10 and an intermetallic compound Li_4Ba, the other phase diagram locates the eutectic composition at x_{Ba}=0.20

and reports two intermetallic compounds $LiBa_2$ and $LiBa_6$. Ruppersberg and coworkers [150] have measured the density, compressibility, heat capacity and surface tension of the liquid Li-Ba alloy at several concentrations and temperatures; moreover, NS experiments were performed [151] for x_{Ba} = 0.12, 0.30 and T=575 K (the latter corresponding to the zero alloy composition) and x_{Ba} = 0.59 and T= 775 K. The obtained total static structure factors, S_T(q), showed a marked composition dependence and the data for $S_{cc}(q \rightarrow 0)$ suggested weak homocoordinating (heterocoordinating) tendencies for x_{Ba} = 0.12 and 0.30 (x_{Ba} =0.59).

A similar scarcity of results is found at the theoretical side, as we are aware of just two calculations. First, we mention the structural study [151] performed using the hard-sphere (HS) model within the Percus-Yevick analytical solution for HS binary mixtures. The obtained results for the $S_T(q)$ showed a qualitative agreement with experiment although the amplitude of the oscillations was grossly overestimated and significant discrepancies appeared when $q \rightarrow 0$. Subsequently, a more elaborate approach was taken by González *et al.* [152] who studied the static structure of liquid $Li_{0.88}Ba_{0.12}$ alloy by using effective interatomic pair potentials [112] and the modified hypernetted chain (MHNC) theory [153] to obtain the liquid static structure. Their calculated $S_T(q)$ substantially improved the previous HS-based calculations, but no other magnitudes were evaluated.

OFMD simulations have been performed for liquid Li-Ba alloy at concentrations $x_{\rm Li}$ = 0.88, 0.71 and 0.41 with respective experimental ionic number densities [150] ρ=0.03639, 0.02788 and 0.02052 A^{-3} and temperatures T= 575, 575 and 775 K. [154] The simulation used 1200 particles for x_{Li}=0.88 and 600 particles for the other two states. First, we note that a previous OFMD calculation for pure liquid Ba at thermodynamic conditions near its triple point provided a fair description of several static and dynamic properties, although the results were not so accurate as those already attained for liquid Li, Mg and Al. The calculated OFMD S(q) for liquid Ba qualitatively agrees with the experimental data [118], but it overestimates the height of the main peak and shows some dephasage in the oscillations.

Figure 20 shows the calculated total neutron weighted structure factor $S_T(q)$, along with the experimental one. The calculated OF-AIMD $S_T(q)$ qualitatively reproduce the main features of the experiment [151], although there is some underestimation for the amplitude of the oscillations and when $q \rightarrow 0$ some discrepancies appear for x_{Li}=0.88 and 0.70. Figure 21 shows the calculated partial dynamic structure factors, $S_{\rm LiLi}(q,\omega)$ and $S_{\rm BaBa}(q,\omega)$ at the lowest wavevector allowed by the simulation, i.e. q_{min} = 0.19, 0.22 and 0.20 A^{-1} for x_{Li} = 0.88, 0.70 and 0.41 respectively. Clear side peaks are observed in the $S_{\alpha\beta}(q,\omega)$ from which an estimation of the adiabatic sound velocity in the alloy can be obtained, i.e. c_s = 2500 m/s (for $x_{\rm Li}$ =0.88), c_s = 2000 m/s ($x_{\rm Li}$ =0.70) and c_s = 1700 m/s ($x_{\rm Li}$ =0.41) which compare well with the experimental [150] values of 2550, 1900 and 1500 m/s respectively.

The dispersion relations, as derived from the peak positions, are shown in figure 22. At the three concentrations, we observe a splitting into fast and slow kinetic modes. Note that the difference in slopes between both modes is substantially greater than in the Li-Na alloy, because of the greater mass difference (and therefore, of the oscillation frequencies) be-

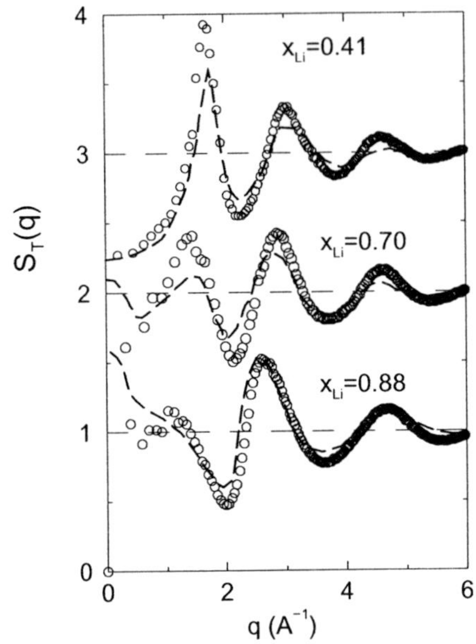

Figure 20: Total static structure factors, $S_T(q)$, for the liquid Li-Ba alloy at $x_{Li} = 0.88$, 0.70 and T=575 K and $x_{Ba} = 0.41$ and T= 775 K. Open circles are the OFMD results for $S_T(q)$ whereas the broken line are the experimental neutron diffraction data of Ruppersberg [151].

tween the Li and Ba ions. Moreover, the hydrodynamic regime is now attained at a smaller q as it becomes harder for the lighter Li ions to oscillate with the same frequency as the much heavier Ba ions because of the increased difference between their natural oscillation frequencies. It is also observed that the fast sound mode still appears at the smallest considered concentration of the heavy Ba particles, namely $x_{Li} = 0.88$, which is in compliance with the RET predictions which aver that the increase of the mass ratio and/or the decrease of the number density widens the concentration range for the existence of both modes.

By following the procedure outlined in the previous section, we have calculated the total transverse current correlation function and the alloy shear viscosity, η, whose values are 0.60, 1.40 and 1.65 GPa ps for $x_{Li} = 0.88$, 0.70 and 0.41 respectively. No experimental data are available for the alloy shear viscosities, but the calculated OFMD value for pure Ba near the triple point is $\eta_{\rm Ba} = 2.2 \pm 0.2$ whereas its reported experimental value is (in GPa ps units) $\eta_{\rm Ba}^{\rm exp} = 1.82$ [137]. By using the calculated OFMD value for pure Li at 575 K, namely $\eta_{\rm Li} = 0.42$, the present results for the alloy suggest some negative deviation from the linear law at x_{Li}=0.88, which conforms with the mild homocoordinating tendencies exhibited at this concentration [151].

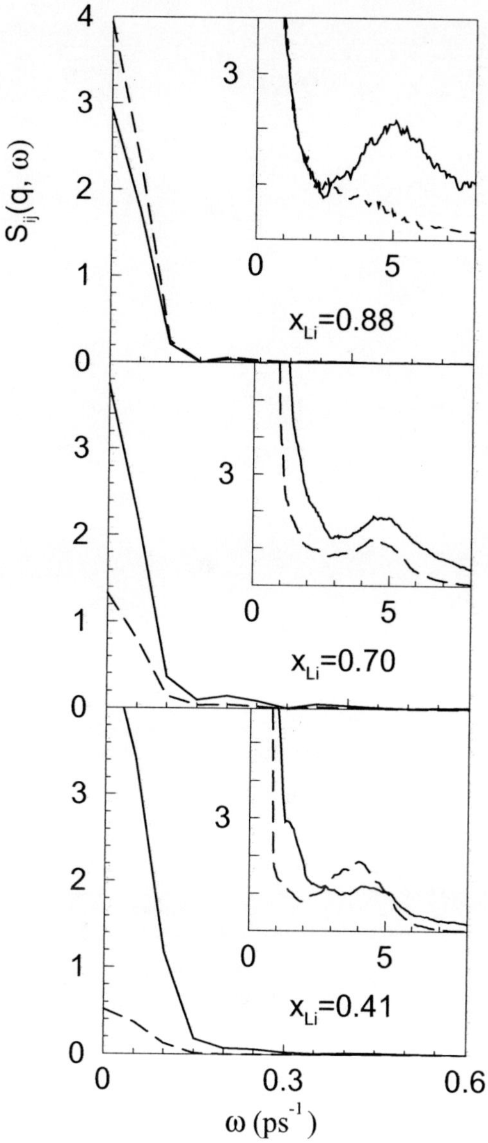

Figure 21: Partial dynamic structure factors, $S_{ij}(q,\omega)$, at $q = 0.19$ A^{-1} (x_{Li}=0.88, T=575 K), $q = 0.23$ A^{-1} (x_{Li}=0.70, T=575 K) and $q = 0.20$ A^{-1} (x_{Li}=0.41, T=775 K), for the liquid Li-Ba alloy. The full and dashed lines represent the $S_{\mathrm{LiLi}}(q,\omega)$ and $S_{\mathrm{BaBa}}(q,\omega)$ respectively. The inset stands for $10^3 S_{ij}(q,\omega)$.

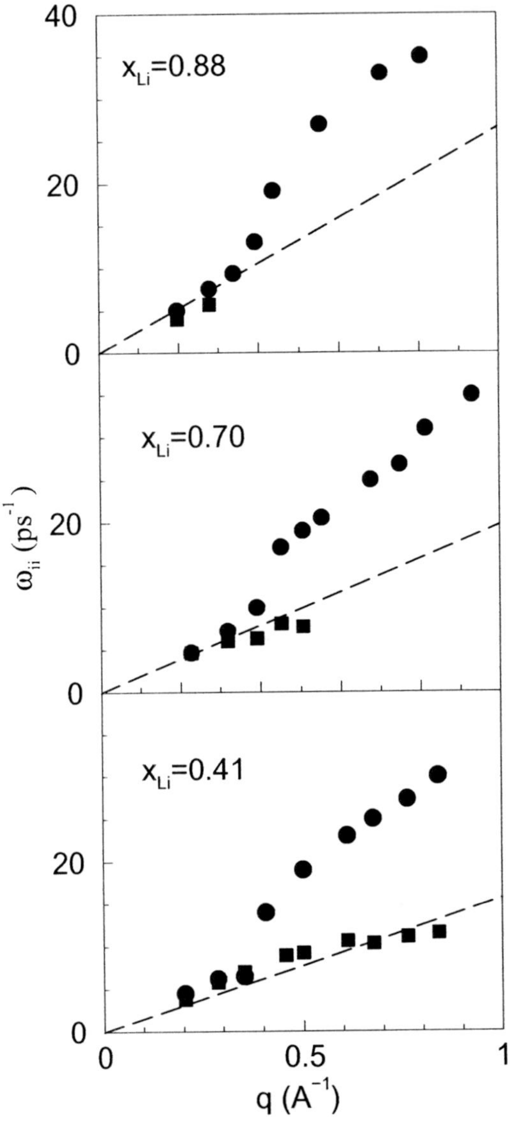

Figure 22: Dispersion relation of the collective modes in the partials, $S_{\mathrm{LiLi}}(q,\omega)$ and $S_{\mathrm{BaBa}}(q,\omega)$ (full circles and full squares respectively) for the liquid Li-Ba alloy at several concentrations. The slope of the dashed lines shows the respective experimental adiabatic sound velocities.

3.3 OFMD Studies of the Liquid-Vapor Interface in Simple Liquid Metals and Alloys

The study of the structure of the free liquid surface has attracted much theoretical and experimental work, [155, 156] with emphasis on the possible existence of liquid surface layers. Although it is now well established that ionic [157, 158, 159, 160] and dielectric [155, 161] liquids exhibit just a smooth monotonic decay from the bulk liquid density to the bulk vapor density, liquid surface layering appears in liquid crystals [162] and at the interface between a simple uid and a hard wall [163]. Special attention has been devoted to liquid metals since the early experiments suggested a liquid surface layered structure in Hg [164]; moreover computer simulation [165, 166, 167, 168, 169, 170] and theoretical studies [171] have predicted a significant structure, extending several atomic layers into the bulk liquid. These results have been recently corroborated by X-ray re ectivity experiments on several liquid metals (Hg, Ga, In, K) and alloys (Na-K, In-Ga, Bi-Ga and Sn-Ga). [172]

It not yet clear whether surface layering is a feature of all liquid metals or only characteristic of some. Rice and coworkers [165, 166, 167] have performed Monte Carlo (MC) simulations based on density-dependent pair potentials obtained from pseudopotentials; their results suggest that surface layering is due to the coupling between ionic and electronic density profiles (DP) and that the abrupt decay of the electron DP gives rise to an effective wall against which the ions, behaving like hard-spheres, stack. Other workers [173] have suggested that the many body forces arising from the delocalized electrons, tend to increase the ionic surface density so that its coordination approaches that of the bulk. Recently, Chacón *et al.* [174] proposed that surface layering may be a generic property of uids at low temperature, so that the only requirement for an oscillatory DP is a low melting temperature relative to the critical temperature so as to avoid crystallization.

In this section we present some recent OFMD results on the liquid-vapor interface of pure liquid Li, Mg, Al and the Li-Na liquid alloy at three concentrations [169, 170].

3.3.1 Liquid Li, Mg and Al

The OFMD simulations have been performed by considering slabs consisting of 2000 ions in a supercell with two free surfaces normal to the *z*-axis. The dimensions of the slab were $L_0 \cdot L_0 \cdot \alpha L_0$, ($\alpha$=2 for Li, Mg and α=1.57 for Al) and where L_0 is chosen so that the average number density of the slab coincides with the experimental bulk ionic number density of the system at the same temperature, i.e. ρ_0(Li) = 0.04451 A^{-3}, ρ_0(Mg) = 0.03829 A^{-3} and ρ_0(Al) = 0.0529 A^{-3}, leading to L_0(Li) =28.44 A and L_0(Mg) = 29.67 A. and L_0(Al) =28.97 A . A further 8 A of vacuum were added both above and below the slab. Although the periodic boundary conditions require that a particle moving out of the cell in the *z*-direction reappears on the other side of the slab, we have not observed this event during the present simulations. For all systems equilibration lasted for 5 ps and the cut-off energy for the plane-wave expansion was 9.5 (Li), 8.5 (Mg) and 11.25 (Al) Ryd. The ionic time-step was 0.0060 ps (Li) and 0.001 ps (Mg, Al), and the calculation of properties was made averaging over another 20 ps. During the simulations each slab contracted and the average density rose in response to the condition of zero external pressure so that the density in the central region of slab increased by $\approx 6\%$ (Li, Mg) and $\approx 7\%$ for Al. The longitudinal ionic DP have been computed from a histogram of particle positions relative to the slab's center of mass, with the profiles from both halves of the slab being averaged. The results are shown in figures 23-24. All the systems show a stratification for about five layers into the bulk liquid, a structural feature that has already been observed experimentally for the liquid-vapor interface of Hg, Ga, In, K and $Na_{33}K_{67}$ [172]. An interesting feature is the relative amplitude of the first and second oscillations from the surface. According to Figures 23-24 the amplitude of the first oscillation in Li is smaller than that of the second, the amplitudes are comparable in Mg, and for Al the first oscillation has the larger amplitude. An earlier OFMD simulation for Na [169] showed oscillations with a similar behavior as Li; in fact, the relative amplitude of the first oscillation was even smaller than in Li. The OFMD results are in close agreement with previous MC results of Rice *et al.* [165, 166, 167] for Na, K, Rb and Cs which produced a first oscillation with a smaller amplitude than the second. Their MC results for Al, Ga, In and Tl [167] showed a first oscillation with a larger amplitude than the second, which we find for Al, and which is in agreement with the reported experimental DP for Hg, Ga and In [172].

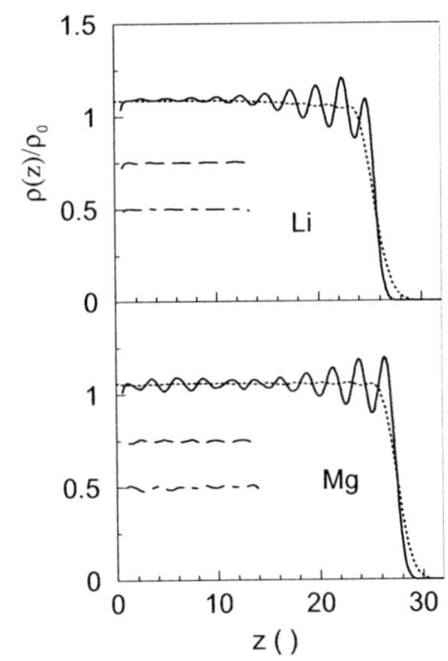

Figure 23: Electronic (dotted line) and ionic (full line) density profiles normal to the Li and Mg liquid-vapor interface. The densities are plotted relative to the bulk densities. The dashed and dot-dashed lines are the *x*-transverse (displaced by -0.5) and *y*-transverse (displaced by -0.75) ionic density profiles.

In order to check that the surface stratification is not an artifact of the finite size of the simulation box, we have computed the transverse ionic DP which, as shown in Figures 23-24, are rather uniform, but with some noise which is substantially smaller in amplitude than the oscillations in the longitudinal DP. As a further check on the reliability of these

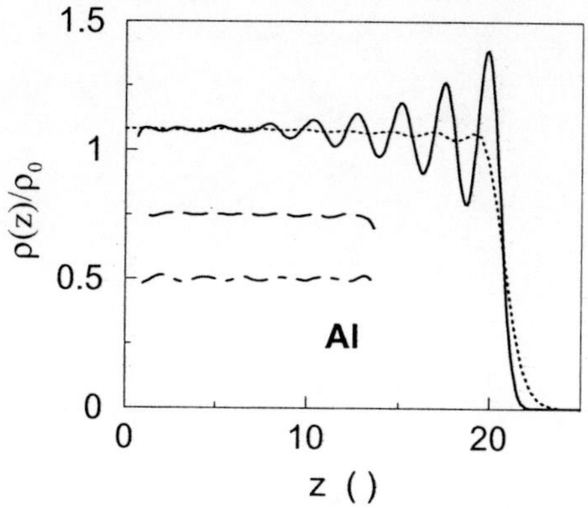

Figure 24: Same as the previous figure but for the Al liquid-vapor interface.

calculations, we have calculated the *bulk* pair distribution functions, $g(r)$, for the atoms within a 30.0 A wide central section of a slab. The results shown in figures 25-26 are in excellent agreement with their experimental counterparts [118].

Another interesting point is the in uence that the stratification of the liquid-vapor interface might have on the in-plane structure of the liquid. This would be seen in a variation of the transverse pair correlation function across the interface which we have investigated by dividing the outer region of the slab into slices located between consecutive minima of the oscillations, with the outermost slice between the outermost minima and the in ection point in the decaying ionic DP. For each metal, all the slices had the same width, except the outermost one which was slightly narrower, and the widths were: 2.5 A for Li, 2.6 A for Mg and 2.35 A for Al (2.15, 2.3 and 2.1 A respectively for the outermost slice). The calculated mean ionic number density in a slice deviated significantly from the bulk value for the two outermost two slices. At the outermost slice the relative changes in the ionic number density were $\approx$ -9 % (Li), -4 % (Mg) and 3 % (Al), and for the next slice the corresponding values were $\approx$ -3 % (Li), -2 % (Mg) and -2 % (Al). These changes were mirrored in deviations of the associated transverse pair correlation function from its bulk shape. Figures 25-26 show the transverse pair correlation functions $g(r)$ at the two outer slices. For both Li and Mg the outermost $g(r)$ is displaced towards larger r-values, with the main peak position shifted by $\approx$ 6% and $\approx$ 3 % respectively, pointing to an increased nearest neighbor distance. The outermost $g(r)$ for Al preserves the main peak at its position in the bulk while slightly increasing its height.

Further insight into the local structure is provided by the z-dependent coordination number $n(z)$, defined as the average number of neighbors within a distance r_m which is the position of the first minimum of the bulk $g(r)$. The results for Li, Mg and Al, give an average value $n(z)$ = 11.8, 11.5 and 11.2 respectively in most of their slabs, and only very close to the surfaces, starting around the second outer maximum, does $n(z)$ begin to decrease. Specifically, at the outermost maximum $n(z)$= 8.0 for Mg and $n(z) = 7.9$ for both Li and Al whereas at the preceding maximum $n(z) = 11.3$ for Li and 11.0 for both Mg and Al.

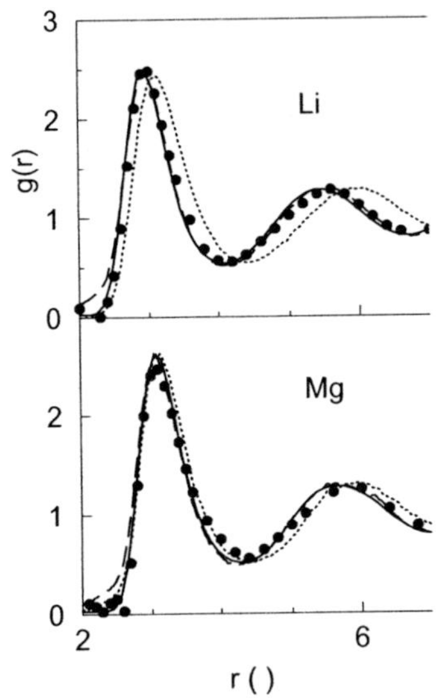

Figure 25: Transverse pair correlation functions for selected layers of the Li and Mg slabs, namely from the bulk (full line), a slice from outermost minimum outwards (dotted line) and from the region between the outermost minimum and the previous one (dashed line). The full circles are the bulk experimental data [118].

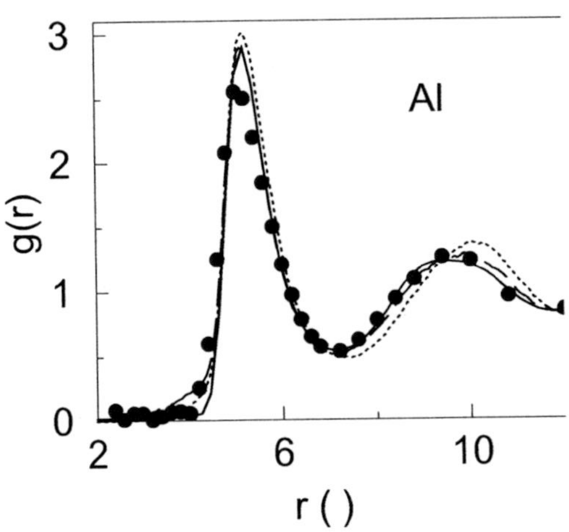

Figure 26: Same as the previous figure but for the Al slab.

The average electronic DP are also included in figures 23-24 and show some oscillations although with amplitudes much smaller than those of their respective ionic density profiles. The phase of the electronic oscillations is noteworthy. Whereas for Li the electronic and ionic DP are 180° out of phase, figure 23-24 show that for liquid Al the maxima and minima of the electronic oscillations are midway between the maxima and minima of the ionic oscillations, and Mg is intermediate.

The liquid-vapor interface exerts a rather weak in uence on the electronic density distribution for these metals. From the OFMD generated ionic positions we have constructed a configurational average of the displaced electronic density used in the construction of the ionic pseudopotential. This construction yields electronic DP, shown in figures 27-28, which are rather similar to the computed self-consistent OFMD electronic DP, although the latter ones are slightly more spilled out at the interface.

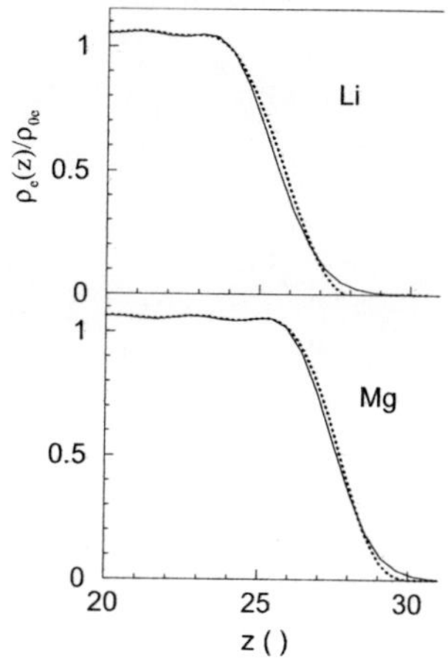

Figure 27: Electronic density profiles (relative to the bulk average value) for the Li and Mg liquid-vapor interfaces. The continuous line is the OFMD result and dashed line represents a linear superposition of displaced densities.

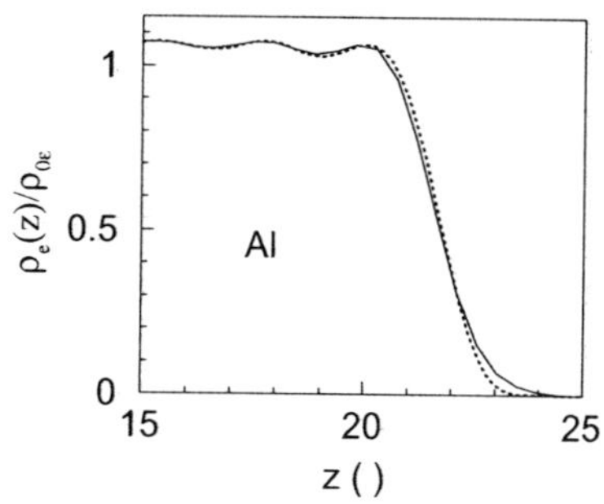

Figure 28: Same as before, but for Al.

Analysis of the ionic density oscillations in figs. 23-24 gives for their wavelengths

$\approx$ 2.5, 2.6 and 2.35 A for Li, Mg and Al respectively. These values, along with some preliminary results for Na, K, Cs and Ba slabs, at thermodynamic conditions near their respective triple points, show a rather good linear scaling of the wavelengths with the radii of their corresponding Wigner-Seitz spheres which suggests that the oscillations are due to atomic stacking against the surface. Indeed, some authors have linked the existence of ionic density oscillations to the metallic character of the interactions in the uid. According to Rice [165, 166] the abrupt decay of the delocalized electronic density at the interface generates an effective wall-like potential against which the ions lay orderly, similarly to a hard sphere uid close to a hard wall. Therefore, some relationship should be expected between the ionic oscillations and the radii of their corresponding Wigner-Seitz spheres. In addition the increasing mean valence electron density values from Li to Mg and Al, lead to a sharper decay of the density at the interface and an increasingly steep wall-like potential and therefore to a larger amplitude of the first oscillation.

Also of interest are the possible structural rearrangements induced by the interface. It has been suggested [168] to quantify such changes by comparison with an ideally terminated surface obtained by cutting abruptly the slab in the central region, i.e. at z=0. Then, the value of $n(z)$ at the outermost maximum should be compared with the values of $n(z_d)$, where z_d is the distance between the outermost maximum and the in ection point in the decaying ionic DP. The present OFMD results for Li, Mg and Al give z_d = 1.1, 1.1 and 0.9 A with the respective values for the ideal step-like surface being $n(z_d)$= 7.8, 7.7 and 7.2. Whereas the corresponding values of $n(z)$ at the utmost maximum are 7.9, 8.0 and 7.9. These values show that the surface structural rearrangement slightly increases the coordination number of an ideally terminated surface.

3.3.2 Liquid LiNa

The OFMD simulations for the liquid-vapor interface in the $Li_{1-x}Na_x$ alloy at T=590 K have been performed at the concentrations x_{Li}=0.80, 0.61 and 0.40. The slabs compounded 2000 ions in a supercell with two free surfaces normal to the z-axis. The dimensions of the slab were initially $L_0 \cdot L_0 \cdot \alpha L_0$, ($\alpha$=1.73) and L_0 was chosen so that the average number density of the slab coincides with the experimental bulk ionic number density of the system at the same temperature [142], leading to L_0=31.56, 33.01 and 34.48 A respectively. A further 8 A of vacuum were added above and below the slab. Although the periodic boundary conditions require that a particle moving out of the cell in the z-direction reappears on the other side of the slab, we have not observed this event during the present simulations. For all systems equilibration lasted for 20 ps and the cut-off energy for the plane-wave expansion was 7.5 Ryd. The ionic time-step was 0.0025 ps and the calculation of properties was made averaging over another 120 ps.

During the simulations each slab contracted and the average total ionic density rose in response to the condition of zero external pressure, so that the ionic density in the central region of slab increased by approximately 10%, 13% and 16% for x_{Na}=0.20, 0.39 and 0.60 respectively. The longitudinal partial and total ionic DPs were computed from a histogram of particle positions relative to the slab's center of mass, and the results are depicted in figure 29. The total longitudinal ionic DP shows a structureless appearance and as the interface region is approached we observe three weak oscillations which are superposed on a monotonic decay. Although weak, those oscillations are real features of the DP and not an artifact begot by the finite size of the simulation box. This idea is underpinned by the calculated total (x and y) transverse ionic DPs which, as observed in figure 29, are rather uniform albeit with some noise which is commensurable to the structure of the longitudinal ionic DP in the central region of the slab, but substantially smaller than the amplitudes of the first three oscillations in the interface region. Obviously, the shape of the total longitudinal ionic DP is induced by the partial DPs associated to the Li and Na ions. A common feature for the three concentrations is that the Na longitudinal DP is shifted relative to the Li one, with the consequence that the surface of the slab is virtually composed of Na ions only. This is because the component with the lower surface energy segregates to the surface, leading to an enrichment of Na (at their respective melting points, Na has a 50 % smaller surface tension than Li).

We have also examined how the stratification of the liquid-vapor interface in uences the in-plane structure of the liquid. Here we present results for the $Li_{0.61}Na_{0.39}$ system although analogous conclusions were attained for the other compositions. Specifically, we have analyzed the variation of the transverse partial pair correlation functions, $g_{ij}(r)$, across the interface. Therefore, we have divided the outer region of the slab into slices located between consecutive minima of the oscillations, with the outer slice comprising from the outermost minima to the in ection point in the decaying total ionic DP. In the $Li_{0.61}Na_{0.39}$ alloy, the outer slice was 2.3 A wide whereas the two inner slices were 3.0 A wide each. The total ionic number density in those slices departed significantly from the bulk value. At the outermost slice the relative change in the total ionic number density was $\approx$ -40 % whereas for the adjacent inner two slices the respective values were $\approx$ -18 % and -7 %. These changes were also matched by variations in the alloy concentration along the slab, as shown in figure 31. Specifically, we obtained for the outer slice $x_{Na} \approx 0.98$, whereas for the adjacent inner two slices the respective values were $x_{Na} \approx 0.70$ and 0.44. Those variations in the total ionic number density and concentration have consequences on the associated $g_{ij}(r)$ as compared with their bulk shapes. Figure 30 shows the corresponding $g_{ij}(r)$ at the two outer slices. At the outermost slice the $g_{NaNa}(r)$ is displaced towards larger r-values (we have not evaluated $g_{LiLi}(r)$ and $g_{LiNa}(r)$ because the Li concentration is negligible) due to the large drop in the total ionic number density which implies an increased nearest neighbor distance. However, at the adjoining inner slice, the concomitant $g_{ij}(r)$ become more akin to their bulk values.

Further insight into the local structure is provided by the z-dependent coordination numbers $n_{ij}(z)$, defined as the number of j-type ions around an i-type ion and within a distance r_{ij} (identified as the position of the first minimum of the bulk $g_{ij}(r)$). The $n_{ij}(z)$ undergo slight changes for most of the slab, although when the interface region is approached those changes become more substantial with an important increase (decrease) in the $n_{NaNa}(z)$ and $n_{LiNa}(z)$ ($n_{LiLi}(z)$ and $n_{NaLi}(z)$). We have also analyzed the ordering tendencies along the

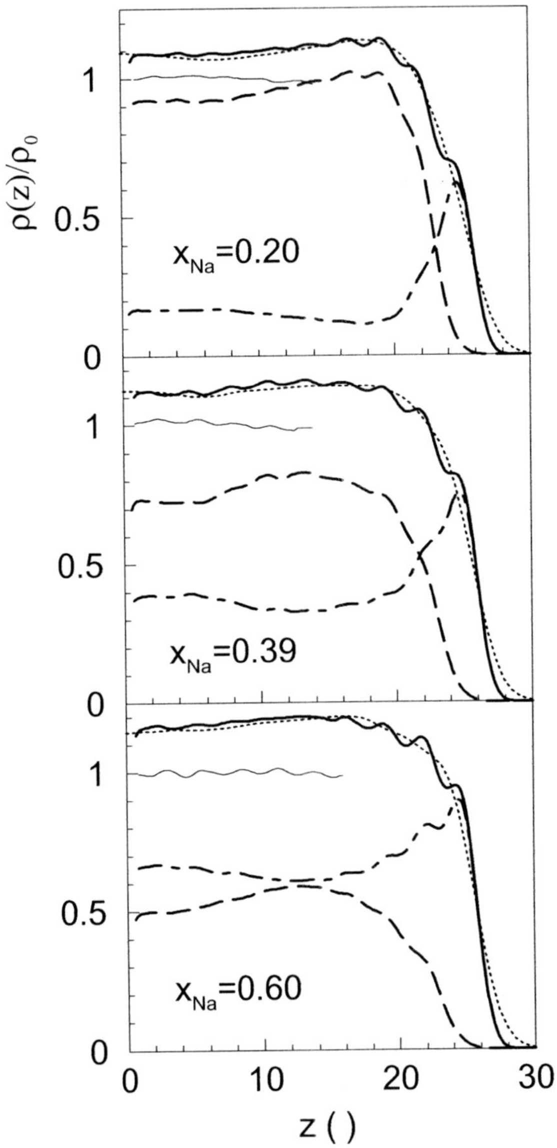

Figure 29: Electronic valence (dotted line), total ionic (full line), partial Li ionic (dashed line) and partial Na ionic (dash-dotted line) density profiles normal to the LiNa liquid-vapor interface at three concentrations. The densities are plotted relative to the total bulk densities. The continuous horizontal line located around one, represents the x-transverse total ionic density profile.

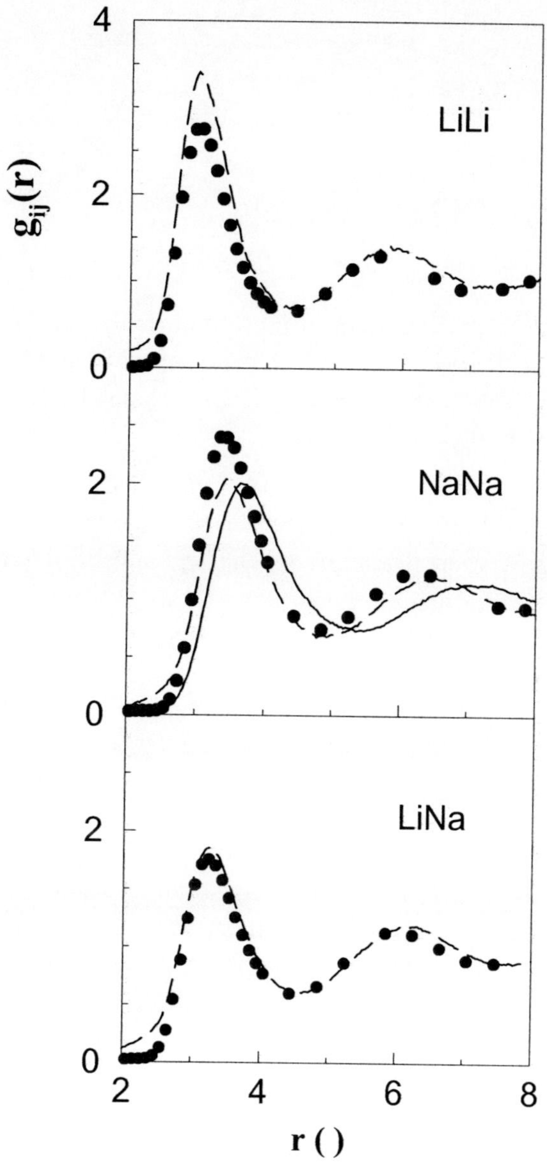

Figure 30: Transverse partial pair correlation functions, $g_{ij}(r)$, for selected layers of the $Li_{0.61}Na_{0.39}$ slab, namely from the bulk (full circles), a slice from outermost minimum outwards (continuous line) and from the region between the outermost minimum and the precedent one (dashed line).

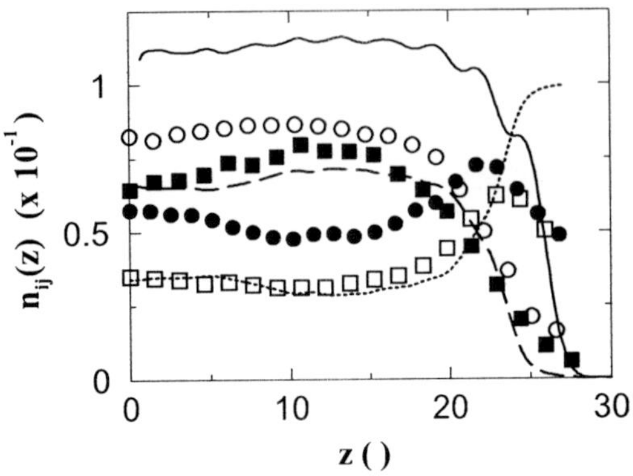

Figure 31: Calculated z-dependent coordination numbers, $n_{ij}(z)$, for the $Li_{0.61}Na_{0.39}$ liquid alloy. n_{LiLi} (open circles), n_{NaNa} (full circles), n_{LiNa} (open squares) and n_{NaLi} (full squares). The full line stands for the relative total ionic density profile, the broken (dotted) line is the Li (Na) local concentration.

slab by means of the local mole fraction method [175] and the results predict a homocoordinating tendency whose strength remains practically unchanged throughout the slab.

The partial longitudinal ionic DPs in figure 29 display a slowly varying monotonic evolution which approaches its bulk value rather deep into the slab. This is completely at variance with previous MC results for the partial ionic DPs in the $Na_{0.5}Cs_{0.5}$, $In_{0.25}Ga_{0.75}$ and $Ga_{0.86}Pb_{0.14}$ [176] liquid alloys, where the partial ionic DPs were oscillating functions around their respective bulk value. The reason underlying this pattern of the obtained partial ionic DPs in the $Li_{1-x}Na_x$ liquid alloy rests on the homocoordinating tendencies exhibited by this alloy which favor the surrounding of an ion by others of the same species and induce large concentration uctuations.

The average valence electronic DP, shown in figure 29, shows a structureless shape with a monotonic decreasing behavior at the liquid-vapor interface. On the other hand, it must be noted that the experimental study of the structure of the liquid-vapor interface is usually performed by XR and/or grazing incidence X-ray diffraction and therefore both techniques measure the total electronic density distribution in the interface. Based on the obtained longitudinal ionic DPs, we have calculated the corresponding total electronic (core and/or valence) DPs which are depicted in figure 32. Now, the longitudinal total electronic DP shows a different behavior at the interface because of the prominence of the contribution from the partial Na electronic DP (the Na ion has 10 core electrons and the Li ion has 2 core electrons); moreover the amplitude of the oscillations is slightly enhanced.

4 Conclusion

In this chapter we have described the orbital-free molecular dynamics method, which allows to perform large-scale simulations of metallic systems. The computational advantages

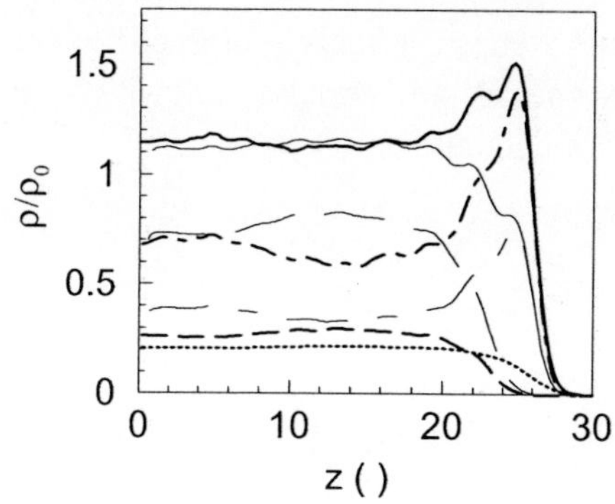

Figure 32: Longitudinal electronic density profiles for the Li $_{0.61}$Na$_{0.30}$ liquid alloy. Full thick line: total electronic (valence plus core) density profile. Dashed thick line: electronic core contribution from Li ions. Dash-dotted thick line: electronic core contribution from Na ions. Dotted line: electronic valence contribution from both Li and Na. The thin lines stand for total (full), partial Li (dashed) and partial Na (dash-dotted) longitudinal ionic density profiles.

(mostly linear scaling) and disadvantages (use of local pseudopotentials as well as an approximate expression for the electronic kinetic energy functional) of such a methodology have been highlighted. The method has then been applied to the efficient simulation of specific problems (thermal properties and melting-like transition in large metal clusters, static and dynamic properties of bulk simple metals and their alloys, and structural and electronic properties of the liquid-vapor interface of metallic systems) which nowadays are very difficult to obtain routinely from traditional (orbital-based) *ab initio* methods.

Let us now put the OFMD method in a more general perspective. With the presently developed approximations for the electronic kinetic energy functional and local pseudopotentials, the OFMD technique provides reliable results (competing in accuracy with KS methods indeed) for simple metals such as the alkalis, alkaline-earths, some trivalent metals like aluminum, and their alloys. Further progress in our understanding of the $T_s[n]$ functional, and/or in the methods for generating local pseudopotentials, is needed in order to extend the applicability of the OFMD method to other metals such as Ga or Ge, for which only qualitatively accurate results can be obtained at present. In general terms, it can be concluded that present $T_s[n]$ functionals work better the closer is the simulated system to a nearly-free electron metal. Strongly inhomogeneous electron densities, such as those of ionic (MgO, for example) or van der Waals crystals (Ne, Ar, etc) are much more difficult to represent with the available $T_s[n]$ functionals, partly because all these functionals have been generated by trying to reproduce known exact limits, most of which are related with small-amplitude departures from a homogeneous electron density.

We have good prospects for future work on metallic systems employing the OFMD methodology. On one hand, we plan to check in depth the possibility of extending the range of applicability of OFMD to more complicated metals. On the other hand, we will study other problems which can not yet be routinely addressed by orbital-based *ab initio* methods, such as the solid-liquid interface of bulk metals (from which we can extract melting curves) and the thermal properties of nanoalloys.

Acknowledgments

Thanks are due to Prof. Malcolm J. Stott, who was involved in specific parts of the research presented in this chapter. This work was supported by Junta de Castilla y León (Project VA073/02) and DGES (Project MAT2002-04393-C02-01). A. Aguado also acknowledges financial support under "Ramón y Cajal" program.

References

[1] Hohenberg, P.; Kohn, W. *Phys. Rev.* **1964**, *136*, 864B.

[2] Kohn, W.; Sham, L. J. *Phys. Rev.* **1965**, *140*, 1133A.

[3] Car, R.; Parrinello, M. *Phys. Rev. Lett.* **1985**, *55*, 2471.

[4] Verlet, L. *Phys. Rev.* **1967**, *159*, 98; Swope, W. C.; Andersen, H. C. *J. Chem. Phys.* **1982**, *76*, 637.

[5] Soler, J. M.; Artacho, E.; Gale, J. D.; Garc´a, A.; Junquera, J.; Ordejón, P.; Sánchez-Portal, D. *J. Phys.: Condens. Matter* **2002**, *14*, 2745.

[6] Bachelet, G. B.; Hamann, D. R.; Schlüter, M. *Phys. Rev. B* **1982**, *26*, 4199.

[7] Perdew, J. P.; Zunger, A. *Phys. Rev. B* **1981**, *23*, 5048.

[8] Ceperley, D.; Alder, B. *Phys. Rev. Lett.* **1980**, *45*, 566.

[9] Alonso, J. A.; Girifalco, L. A. *Phys. Rev. B* **1978**, *17*, 3735.

[10] Acharya, P. K.; Bartolotti, L. J.; Sears, S. B.; Parr, R. G. *Proc. Natl. Acad. Sci. U.S.A.* **1980**, *77*, 6978.

[11] Yang, W. *Phys. Rev. A* **1986**, *34*, 4575.

[12] DePristo, A. E.; Kress, J. D. *Phys. Rev. A* **1987**, *35*, 438.

[13] Lee C.; Parr, R. G. *Phys. Rev. A* **1987**, *35*, 2377.

[14] Lee, H.; Lee, C.; Parr, R. G. *Phys. Rev. A* **1991**, *44*, 768.

[15] Thakkar, A. J. *Phys. Rev. A* **1992**, *46*, 6920.

[16] Perdew, J. P. *Phys. Lett. A* **1992**, *165*, 79.

[17] Lacks, D. J. Gordon, R. G. *J. Chem. Phys.* **1994**, *100*, 4446.

[18] Vitos, L.; Skriver, H. L.; Kollár, J. *Phys. Rev. B 1998*, *57*, 12611.

[19] Herring, C. *Phys. Rev. A* **1996**, *34*, 2614; Herring, C.; Chopra, M. *ibid.* **1988**, *37*, 31.

[20] Chacón, E.; Alvarellos, J. E.; Tarazona, P. *Phys. Rev. B* **1985**, *32*, 7868; Tarazona, P.; Chacón, E. *Phys. Rev. B* **1989**, *39*, 10366; Garc´a-González, P.; Alvarellos, J. E.; Chacón, E. *Phys. Rev. A* **1996**, *54*, 1897; *Phys. Rev. B* **1996**, *53*, 9509; *Phys. Rev. A* **1998**, *57*, 4192; *Phys. Rev. B* **1998**, *57*, 4857; *Phys. Rev. A* **2000**, *62*, 014501.

[21] Wang, L. W.; Teter, M. P. *Phys. Rev. B* **1992**, *45*, 13196.

[22] Pearson, M.; Smargiassi, E.; Madden, P. A. *J. Phys.: Condens. Matter 1993*, *5*, 3221; Smargiassi, E.; Madden, P. A. *Phys. Rev. B* **1994**, *49*, 5220; Foley, M.; Madden, P. A. *Phys. Rev. B* **1996**, *53*, 10589.

[23] Perrot, F. *J. Phys.: Condens. Matter* **1994**, *6*, 431.

[24] Wang, Y. A.; Govind, N.; Carter, E. A. *Phys. Rev. B* **1998**, *58*, 13465; *ibid.* **1999**, *60*, 16350.

[25] Karasiev, V. V.; Ludeña, E. V.; Artemyev, A. N. *Phys. Rev. A* **2000** *62*, 062510; Ludeña, E. V.; Karasiev, V. V.; Nieto, P. *Theor. Chem. Acc.* **2003**, *110*, 395.

[26] Wang, B.; Stott, M. J.; von Barth, U. *Phys. Rev. A* **2001**, *63*, 052501.

[27] Perdew, J. P.; Burke, K.; Ernzerhof, M. *Phys. Rev. Lett. 1996*, *77*, 3865; *ibid.* **1997**, *78*, 1396(E).

[28] Perdew, J. P.; Tao, J.; Staroverov, V. N.; Scuseria, G. E. *J. Chem. Phys.* **2004**, *120*, 6898.

[29] Aguado, A.; López, J. M. *Phys. Rev. Lett.* **2005**, *94*, 233401.

[30] Aguado, A. *J. Phys. Chem. B* **2005**, *109*, 13043.

[31] Zhou, B. J.; Wang, Y. A.; Carter, E. A. *Phys. Rev. B* **2004**, *69*, 125109.

[32] Fiolhais, C.; Perdew, J. P.; Armster, S. Q.; McLaren, J. M.; Brajczewska, H. *Phys. Rev. B* **1995**, *51*, 14001; *ibid.* **1996**, *53*, 13193.

[33] Nogueira, F.; Fiolhais, C.; He, J.; Perdew, J. P.; Rubio, A. *J. Phys.: Condens. Matter* **1996**, *8*, 287.

[34] Schmidt, M.; Kusche, R.; Kronmüller, W.; von Issendorff, B.; Haberland, H. *Phys. Rev. Lett.* **1997**, *79*, 99.

[35] Schmidt, M.; Kusche, R.; von Issendorff, B.; Haberland, H. *Nature* **1998**, *393*, 238.

[36] Kusche, R.; Hippler, Th.; Schmidt, M.; von Issendorff, B.; Haberland, H. *Eur. Phys. J. D* **1999**, *9*, 1.

[37] Schmidt, M.; Ellert, C.; Kronmüller, W.; Haberland, H. *Phys. Rev. B* **1999**, *59*, 10970.

[38] Schmidt, M.; Kusche, R.; Hippler, Th.; Donges, J.; Kronmüller, W.; von Issendorff, B.; Haberland, H. *Phys. Rev. Lett.* **2001**, *86*, 1191.

[39] Schmidt, M.; Haberland, H. *Compt. Rend. Physique* **2002**, *3*, 327.

[40] Schmidt, M.; Donges, J.; Hippler, Th.; Haberland, H. *Phys. Rev. Lett.* **2003**, *90*, 103401.

[41] Haberland, H.; Hippler, Th.; Donges, J.; Kostko, O.; Schmidt, M.; von Issendorff, B. *Phys. Rev. Lett.* **2005**, *94*, 035701.

[42] Manninen, K.; Rytkönen, A.; Manninen, M. *Eur. Phys. J. D* **2004**, *29*, 39.

[43] Calvo, F.; Spiegelmann, F. *J. Chem. Phys.* **2004**, *120*, 9684.

[44] Aguado, A.; López, J. M.; Alonso, J. A.; Stott, M. J. *J. Chem. Phys.* **1999**, *111*, 6026.

[45] Aguado, A.; López, J. M.; Alonso, J. A.; Stott, M. J. *J. Phys. Chem. B* **2001**, *105*, 2386.

[46] Aguado, A.; Molina, L. M.; López, J. M.; Alonso, J. A. *Eur. Phys. J. D* **2001**, *15*, 221.

[47] Aguado, A. *Phys. Rev. B* **2001**, *63*, 115404.

[48] Aguado, A.; López, J. M. *Phys. Rev. B* **2005**, in press.

[49] Röthlisberger, U.; Andreoni, W. *J. Chem. Phys.* **1991**, *94*, 8129.

[50] Bulgac, A.; Kusnezov, D. *Phys. Rev. Lett.* **1992**, *68*, 1335; *Phys. Rev. B* **1992**, *45*, 1988; Ju, N.; Bulgac, A. *Phys. Rev. B* **1995**, *48*, 2721; Fosmire, M.; Bulgac, A. *Phys. Rev. B* **1995**, *52*, 17509; Bulgac, A. *Z. Phys. D* **1997**, *40*, 454; Fosmire, M.; Bulgac, A. *Z. Phys. D* **1997**, *40*, 458; Thompson, J. M.; Bulgac, A. *Z. Phys. D* **1997**, *40*, 462.

[51] Poteau, R.; Spiegelmann, F.; Labastie, P. *Z. Phys. D* **1994**, *30*, 57.

[52] Pacheco, J. M.; Schöne, W. D. *Phys. Rev. Lett.* **1997**, *79*, 4986.

[53] Bonacić-Koutecký, V.; Jellinek, J.; Wiechert, M.; Fantucci, P. *J. Chem. Phys.* **1997**, *107*, 6321; Reichart, D.; Bonacić-Koutecký, V.; Fantucci, P.; Jellinek, J. *Chem. Phys. Lett.* **1997**, *279*, 129.

[54] Rytkönen, A.; Häkkinen, H.; Manninen, M. *Phys. Rev. Lett.* **1998**, *80*, 3940.

[55] Calvo, F.; Spiegelmann, F. *Phys. Rev. Lett.* **1999**, *82*, 2270; *J. Chem. Phys.* **2000**, *112*, 2888.

[56] Calvo, F.; Spiegelmann, F. *Phys. Rev. Lett.* **2003**, *90*, 103401.

[57] Reyes-Nava, J. A.; Garzón, I. L.; Michaelian, K. *Phys. Rev. B* **2003**, *67*, 165401.

[58] Shah, V.; Nehete, D.; Kanhere, D. G. *J. Phys.: Condens. Matter* **1994**, *6*, 10773; Nehete, D.; Shah, V.; Kanhere, D. G. *Phys. Rev. B* **1996**, *53*, 2126; Shah, V.; Kanhere, D. G. *J. Phys.: Condens. Matter* **1996**, *8*, L253; Shah, V.; Kanhere, D. G.; Majumber, C.; Das, G. P. *ibid.* **1997**, *9*, 2165; Vichare, A.; Kanhere, D. G. *Eur. Phys. J. D* **1998**, *4*, 89; *J. Phys.: Condens. Matter* **1998**, *10*, 3309; Dhavale, A.; Shah, V.; Kanhere, D. G. *Phys. Rev. A* **1998**, *57*, 4522.

[59] Govind, N.; Mozos, J. L.; Guo, H. *Phys. Rev. B* **1995**, *51*, 7101; Wang, Y. A.; Govind, N.; Carter, E. A. *Phys. Rev. B* **1998**, *58*, 13465.

[60] Blaise, P.; Blundell, S. A.; Guet, C. *Phys. Rev. B* **1997**, *55*, 15856.

[61] Blaise, P.; Blundell, S. A. *Phys. Rev. B* **2001**, *63*, 235409.

[62] Vichare, A.; Kanhere, D. G.; Blundell, S. A. *Phys. Rev. B* **2001**, *64*, 045408.

[63] Chushak, Y. G.; Bartell, L. S. *J. Phys. Chem. B* **2003**, *107*, 3747.

[64] López, M. J.; Marcos, P. A.; Alonso, J. A. *J. Chem. Phys.* **1996**, *104*, 1056.

[65] Huang, S.; Balbuena, P. P. *J. Phys. Chem. B* **2002**, *106*, 7225.

[66] Joshi, K.; Kanhere, D. G. *Phys. Rev. A* **2002**, *65*, 043203.

[67] Joshi, K.; Kanhere, D. G. *J. Chem. Phys.* **2003**, *119*, 12301.

[68] Mainardi, D. S.; Balbuena, P. P. *Int. J. Quantum Chem.* **2001**, *85*, 580.

[69] Aguado, A.; González, L. E.; López, J. M. *J. Phys. Chem. B* **2004**, *108*, 11722.

[70] Aguado, A.; Núñez, A.; López, J. M. *Comput. Phys. Commun.* **2005**, in press.

[71] Aguado, A.; López, J. M. *Phys. Rev. B* **2005**, *71*, 075415.

[72] Aguado, A.; López, J. M. *J. Chem. Theory Comput.* **2005**, *1*, 299.

[73] Darby, S.; Mortimer-Jones, T. V.; Johnston, R. L.; Roberts, C. *J. Chem. Phys.* **2002** *116*, 1536.

[74] Wilson, N. T.; Johnston, R. L. *J. Mater. Chem.* **2002**, *12*, 2913.

[75] Bailey, M. S.; Wilson, N. T.; Roberts, C.; Johnston, R. L. *Eur. Phys. J. B* **2003**, *25*, 41.

[76] Baletto, F.; Mottet, C.; Ferrando, R. *Phys. Rev. B* **2002**, *66*, 155420.

[77] Baletto, F.; Mottet, C.; Ferrando, R. *Phys. Rev. Lett* **2003**, *90*, 135504.

[78] Baletto, F. ; Mottet, C.; Rapallo, A.; Rossi, G.; Ferrando, R. *Surf. Sci.* **2004**, *566-568*, 192.

[79] Rossi, G.; Rapallo, A.; Mottet, C.; Fortunelli, A.; Baletto, F.; Ferrando, R. *Phys. Rev. Lett.* **2004**, *93*, 105503.

[80] Rapallo, A.; Rossi, G.; Ferrando, R.; Fortunelli, A.; Curley, B. C.; Lloyd, L. D.; Tarbuck, G. M.; Johnston, R. L. *J. Chem. Phys.* **2005**, *122*, 194308.

[81] Rossi, G.; Ferrando, R.; Rapallo, A.; Fortunelli, A.; Curley, B. C.; Lloyd, L. D.; Johnston, R. L. *J. Chem. Phys.* **2005**, *122*, 194309.

[82] Ferrando, R.; Fortunelli, A.; Rossi, G. *Phys. Rev. B* **2005**, *72*, 085449.

[83] Mottet, C.; Rossi, G.; Baletto, F.; Ferrando, R. *Phys. Rev. Lett.* **2005**, *95*, 035501.

[84] Calvo, F.; Yurtsever, E. *Phys. Rev. B* **2004**, *70*, 045423.

[85] Doye, J. P. K.; Meyer, L. *Phys. Rev. Lett.* **2005**, *95*, 063401.

[86] Wales, D. J.; Scheraga, H. A. *Science* **1999**, *285*, 1368.

[87] Johnston, R. L. *Dalton Trans.* **2003**, *22*, 4193.

[88] Hamann, D. R.; Schlüter, M.; Chiang, C. *Phys. Rev. Lett.* **1979**, *43*, 1494.

[89] Kleinman, L.; Bylander, D. M. *Phys. Rev. Lett.* **1982**, *48*, 1425 (1982).

[90] Anglada, E.; Soler, J. M.; Junquera, J.; Artacho, E. *Phys. Rev. B* **2002**, *66*, 205101.

[91] Jellinek, J.; Beck, T. L.; Berry, R. S. *J. Chem. Phys.* **1986**, *84*, 2783; Davis, H. L.; Jellinek, J.; Berry, R. S. *ibid.* **1987**, *86*, 6456; Beck, T. L.; Jellinek, J.; Berry, R. S. *ibid.* **1987**, em 87, 545.

[92] Sugano, S. *Microcluster Physics*, Springer-Verlag, Berlin (1991).

[93] Shvartsburg, A. A.; Jarrold, M. F. *Phys. Rev. Lett.* **2000**, *85*, 2530.

[94] Balucani, U; Zoppi, M. *Dynamics of the Liquid State*, Clarendon, Oxford 1994; Hansen, J. P.; McDonald, I. R. *Theory of Simple Liquids*, (London: Academic, 1986); Boon, J. P.; Yip, S. *Molecular Hydrodynamics*, (New York: McGraw-Hill, 1980).

[95] Sjögren, L.; Sjölander, A. *J. Phys. C* **1979**, *12*,4369; Sjögren, L. *ibid.* **1980**, *13*, 705; *Phys. Rev. A* **1980**, *22*, 2866; *ibid.* **1980**, *22*, 2883.

[96] Casas, J.; González, D. J.; González, L. E.; Alemany, M. M. G.; Gallego, L. J. *Phys. Rev. B* **2000**, *62*, 12095.

[97] Jacucci, G.; McDonald, I. R. *J. Phys. F: Metal Phys. 1980*, *10*, L15.

[98] Jacucci, G.; Ronchetti, M.; Schirmacher, W. *J. Physique Coll.* **1984**, *8* C8, 385.

[99] Campa, A.; Cohen, E. G. D. *Phys. Rev. A* **1990**, *41*, 5451.

[100] Westerhuijs, P.; Montfrooij, W.; de Graaf, L. A.; de Schepper, I. M. *Phys. Rev. A* **1992**, *45*, 3749.

[101] Bryk, T.; Mryglog, I.; Kahl, G. *Phys. Rev. E* **1997**, *56*, 2903; Bryk, T.; Mryglod, I. *Phys. Lett. A* **1999**, *261*, 349; *J. Phys.: Cond. Matter* **2000**, *12*, 6063.

[102] Enciso, E.; Almarza, N. G.; Dominguez, P.; Bermejo, F. J. *Phys. Rev. Lett.* **1995**, *74*, 4233.

[103] Crevecoeur, R. M.; Smorenburg, H. E. de Schepper, I. M. *J. Low Temp Phys.* **1996**, *105*, 149.

[104] Fernández-Perea, R.; Alvarez, M.; Bermejo, F. J.; Verkerk, P.; Roessli, B.; Enciso, E. *Phys. Rev. E* **1998**, *58*, 4568.

[105] Montfrooij, W.; Westerhuijs, P.; de Haan, V. O.; de Schepper, I. M. *Phys. Rev. Lett.* **1989**, *63*, 544.

[106] de Jong, P. H. K.; Verkerk, P.; de Vroege, C. F.; de Graaf, L. A.; Howells, W. S.; Bennington, S. M. *J. Phys.: Condens. Matter* **1994**, *6*, L681.

[107] Alvarez, M.; Bermejo, F. J.; Verkerk, P.; Roessli, B. *Phys. Rev. Lett.* **1998**, *80*, 2141.

[108] Allen, M. P.; Tildesley, D. J. *Computer Simulation of Liquids* (Oxford: Clarendon, 1987).

[109] de Jong, P. H. K.; Verkerk, P.; de Graaf, L. A. *J. Phys.: Condens. Matter* **1994**, *6*, 8391; *J. Non-Cryst. Solids* **1993**, *156*, 48.

[110] Sinn, H.; Sette, F.; Bergmann, U.; Halcoussis, Ch.; Krisch, M. Verbeni, R.; Burkel, E. *Phys. Rev. Lett.* **1997**, *78*, 1715.

[111] Scopigno, T.; Balucani, U.; Ruocco, G.; Sette, F. *J. Phys.: Condens. Matter* **2000**, *12*, 8009.

[112] González, L. E.; González, D. J.; Silbert, M.; Alonso, J. A. *J. Phys: Condens. Matter* **1993**, *5*, 4283.

[113] González, D. J.; González, L. E.; Hoshino, K. *J. Phys.: Condens. Matter* **1994**, *6*, 3849.

[114] Canales, M.; Padró, J. A.; González, L. E.; Giró, A. *J. Phys: Condens. Matter* **1993**, *5*, 3095; Canales, M.; González, L. E.; Padró, J. A. *Phys. Rev. E* **1994**, *50*, 3656.

[115] Kresse, G. *J. Non-Cryst. Solids* **1996**, *205-207*, 833.

[116] Vanderbilt, D. *Phys. Rev. B* **1990**, *41*, 7892.

[117] Anta, J. A.; Madden, P. A. *J. Phys.: Cond. Matter* **1999**, *11*, 6099; Anta, J. A.; Jesson, B. J.; Madden, P. A. *Phys. Rev. B* **1998**, *58*, 6124.

[118] Waseda, Y. *The Structure of Non-Crystalline Materials*, (New York: McGraw-Hill, 1980); IAMP database of SCM-LIQ, Tohoku University, URL: http://www.iamp.tohoku.ac.jp/database/scm/LIQ.

[119] Olbrich, H.; Ruppersberg, H.; Steeb, S. *Z. Naturforsch. A 1983*, *38*, 1328.

[120] Hornung, K. in *Handbook of Thermodynamic and Transport Properties of Alkali Metals*, edited by R.W. Ohse (Oxford: Blackwell, 1985) Chap. 6.4.

[121] Murday J. S.; Cotts, R. M. *Z. Naturforsch. A* **1971**, *26*, 85; Gerl, M.; Bruson, A. in *Handbook of Thermodynamic and Transport Properties of Alkali Metals* , edited by Ohse, R. W. (Oxford: Blackwell, 1985) Chap. 7.5.

[122] Pilgrim, W. -C; Hosokawa, S.; Saggau, H.; Sinn, H.; Burkel, E. *J. Non-Cryst. Solids* **1999**, *250-252*, 96.

[123] Pasqualini, D.; Vallauri, R.; Demmel, F.; Morkel, C.; Balucani, U. *J. Non-Cryst. Solids* **1999**, *250-252*, 76.

[124] Bodensteiner, T.; Morkel, C.; Glaser, W.; Dorner, B. *Phys. Rev. A* **1992**, *45*, 7509.

[125] Alemany, M. M. G.; Casas, J.; Rey, C.; González, L. E.; Gallego, L. J. *Phys. Rev. E* **1997**, *56*, 6818.

[126] Alemany, M. M. G. *private communication*.

[127] de Wijs, G. A.; Pastore, G.; Selloni, G.; van der Lugt, W. *Phys. Rev. Lett.* **1995**, *75*, 4480.

[128] Kawakita, Y.; Hosokawa, S.; Enosaki, T.; Ohshima, K.; Takeda, S.; Pilgrim, W.; Tsutsui, S.; Tanaka, Y.; Baron, A. Q. R. *J. Phys. Soc. Jpn.* **2003**, *72*, 1603.

[129] Jacucci, G.; Taylor, R.; Tenenbaum, A.; van Doan, N. *J. Phys. F: Metal Phys.* **1981**, *11*, 783.

[130] Hafner, J.; Jank, W. *Phys. Rev. A* **1990**, *42*, 11530; Ebbsjo, I.; Kinell, T.; Waller, J. *J. Phys. C: Solid State Physics* **1980**, *13*, 1865.

[131] Blochl, P. E.; Parrinello, M. *Phys. Rev. B* **1992**, *45*, 9413.

[132] Takeda, S.; Harada, S.; Tamaki, S.; Waseda, Y. *J. Phys. Soc. Japan* **1991**, *60*, 2241.

[133] Seemann, H. J.; Klein, F. K. *Z. Angew. (Math.) Phys.* **1965**, *19*, 368.

[134] Alfe, D.; Gillan, M. J. *Phys. Rev. Lett.* **1998**, *81*, 5161.

[135] Scopigno, T.; Balucani, U.; Ruocco, G.; Sette, F. *Phys. Rev. E* **2000**, *63*, 011210.

[136] Webber, G. M. B.; Stephens, R. W. B. *Physical Acoustics*, Vol. IVB, edited by Mason, W. P. (New York: Academic, 1968).

[137] Shimoji, M.; Itami, I. *Atomic Transport in Liquid Metals*, (Trans. Tech. Publications, Switzerland, 1986)

[138] Ruppersberg, H.; Knoll, W. *Z. Naturforsch. A* **1977**, *32*, 1374.

[139] Gartrell-Mills, P. R.; McGreevy, R. L; van der Lugt, W. *Physica B* **1988**, *154*, 1.

[140] Senda, Y.; Shimojo, F.; Hoshino, K. *J. Phys. Soc. Jpn.* **1998**, *67*, 2753.

[141] Costa Cabral, B. J.; Martins, J. L. *J. Chem. Phys.* **1999**, *111*, 5067.

[142] Jost, J.; Heydt, D.; Spehr, J.; Ruppersberg, H. *J. Phys: Condens. Matter 1994*, *6*, 321.

[143] González, D. J.; González, L. E.; López, J. M.; Stott, M. J. *Phys. Rev. B* **2004**, *69*, 031205.

[144] Beyer, R. T. *Handbook of Thermodynamic and Transport Properties of Alkali Metals* edited by Ohse, R. W. (Oxford: Blackwell, 1985) p. 525.

[145] Sampoli, M.; Bafile, U.; Guarini, E.; Barocchi, F. *Phys. Rev. Lett.* **2002**, *88*, 085502.

[146] Bafile, U.; Verkerk, P.; Guarini, E.; Barocchi, F. *Phys. Rev. Lett.* **2001**, *86*, 1019.

[147] Shpil'rain, E. E.; Yakimovich, K. A.; Fomin, V. A.; Skovorodjko, S. N.; Mozgovoi, A. G. *Handbook of Thermodynamic and Transport Properties of Alkali Metals* edited by Ohse, R. W. (Oxford: Blackwell, 1985) p. 753.

[148] Keller, D. V.; Kanda, F. A.; King, A. J. *J. Phys. Chem* **1958**, *62*, 732.

[149] Haase, H. W. Dissertation, Hamburg (1960).

[150] Saar, J.; Ruppersberg, H. *Phys. Chem. Liquids* **1987**, *17*, 45.

[151] Ruppersberg, H. *Phys. Chem. Liquids* **1987**, *17*, 73.

[152] González, L. E.; Dalgic, S.; González, D. J.; Silbert, M. *J. Non-Cryst. Solids* **1996**, *205-207*, 901.

[153] Mori, H.; Hoshino, K.; Watabe, M. *J. Phys.: Condens. Matter* **1991**, *3*, 9791.

[154] González, D. J.; González, L. E.; López, J. M.; Stott, M. J. *J. Phys.: Condens. Matter* **2005**, *17*, 1429.

[155] Rowlinson, J. S.; Widom, B. in *Molecular Theory of Capillarity* (Clarendon, Oxford 1982); Beaglehole, D. in *Fluid Interfacial Phenomena* edited by Croxton, C. A. (Wiley, New York 1986).

[156] Penfold, J. *Rep. Prog. Phys.* **2001**, *64*, 777.

[157] Aguado, A.; Wilson, M.; Madden, P. A. *J. Chem. Phys.* **2001**, *115*, 8603.

[158] Aguado, A.; Scott, W.; Madden, P. A. *J. Chem. Phys.* **2001**, *115*, 8612.

[159] Aguado, A.; Madden, P. A. *J. Chem. Phys.* **2002**, *117*, 7659.

[160] Groh, B.; Evans, R.; Dietrich, S. *Phys. Rev. E* **1998**, *57*, 6944; González-Melchor, M.; Alejandre, J.; Bresme, F. *Phys. Rev. Lett.* **2003**, *90*, 135506.

[161] Beaglehole, D. *Phys. Rev. Lett.* **1979**, *43*, 2016; *Physica B* **1980**, *100*, 163.

[162] Ocko, B. M.; Breslau, A.; Pershan, P. S.; Als-Nielsen, J.; Deutsch, M. *Phys. Rev. Lett.* **1986**, *57*, 94; Martinez-Raton, Y.; Somoza, A.; Mederos, L.; Sullivan, D. *Faraday Discuss.* **1996**, *104*, 111.

[163] Evans, R. *Fundamentals of Inhomogeneous Fluis*, edited by Henderson, D. (Dekker, New York 1992).

[164] Barton, S. W. *et al*, *Nature (London)* **1986**, *321*, 685.

[165] D'Evelyn, M. P.; Rice, S. A. *Phys. Rev. Lett.* **1981**, *47*, 1844; *J. Chem. Phys.* **1983**, *78*, 5225; Harris, J. G.; Gryko, J.; Rice, S. A. *ibid.* **1987**, *87*, 3069.

[166] Chekmarev, D.; Zhao, M.; Rice, S. A. *J. Chem. Phys.* **1998**, *109*, 768; *Phys. Rev. E* **1999**, *59*, 479.

[167] Chekmarev, D.; Zhao, M.; Rice, S. A. *J. Chem. Phys.* **1998**, *109*, 1959.

[168] Fabricius, G.; Artacho, E.; Sánchez-Portal, D.; Ordejón, P.; Drabold, D. A; Soler, J. M. *Phys. Rev. B* **1999**, *60*, R16283.

[169] González, D. J.; González, L. E.; Stott, M. J. *Phys. Rev. Lett.* **2004**, *92*, 085501.

[170] González, D. J.; González, L. E.; Stott, M. J. *Phys. Rev. Lett.* **2005**, *94*, 077801.

[171] Gómez, M. A.; Chacón, E. *Phys. Rev. B* **1992**, *46*, 723; *ibid.* **1994**, *49*, 11405.

[172] Magnussen, O. M.; Ocko, B. M.; Regan, M. J.; Penanen, K.; Pershan, P. S.; Deutsch, M. *Phys. Rev. Lett.* **1995**, *74*, 4444; Regan, M.; Kawamoto, E. H.; Lee, S.; Pershan, P. S.; Deutsch, M.; Magnussen, O. M.; Ocko, B. M.; Berman, L. E. *ibid.* **1995**, *75*, 2498; DiMasi, E.; Tostmann, H.; Ocko, B. M.; Pershan, P. S.; Deutsch, M. *Phys. Rev. B* **1998**, *58*, R13419; *ibid.* **1999**, *59*, 783; Tostmann, H.; DiMasi, E.; Pershan, P. S.;

Ocko, B. M.; Shpyrko, O. G.; Deutsch, M. *ibid.* **2000**, *61*, 7284; Shpyrko, O.; Huber, P.; Grigoriev, A.; Pershan, P. S.; Ocko, B. M.; Tostmann, H.; Deutsch, M. *ibid.* **2003**, *67*, 115405.

[173] Iarlori, S.; Carnevali, P.; Ercolessi, F.; Tosatti, E. *Surf. Sci.* **1989**, *211/212*, 55.

[174] Chacón, E.; Reinaldo-Falagan, M.; Velasco, E.; Tarazona, P. *Phys. Rev. Lett.* **2001**, *87*, 166101; Velasco, E.; Tarazona, P.; Reinaldo-Falagan, M.; Chacón, E. *J. Chem. Phys.* **2002**, *117*, 10777.

[175] Blanco, J.; González, D. J.; González, L. E.; López, J. M.; Stott, M. J. *Phys. Rev. E* **2003**, *67*, 041024.

[176] Gryko, J.; Rice, S. A. *J. Chem. Phys.* **1984**, *80*, 6318; Lei, N.; Huang, Z.; Rice, S. A. *ibid.* **1997**, *107*, 4051; Rice, S. A.; Zhao, M. *Phys. Rev. B* **1998**, *57*, 13501; *ibid.* **2001**, *63*, 085409.

In: Trends in Chemical Physics Research
Editor: A.N. Linke, pp. 271-294
ISBN 1-59454-483-2

Chapter 11

OBSERVATION AND CONTROL OF MOLECULAR MOTION USING ULTRAFAST LASER PULSES

S.A. Malinovskaya*
Michigan Center for Theoretical Physics
and Department of Physics, University of Michigan
Ann Arbor, MI 48109

Abstract

Generation of femtosecond and subfemtosecond pulses is a key to the direct observation of electron and nuclear rearrangements in molecules in real time. The feasibility of investigations relies on the concept of coherence. Using ultrafast laser pulses in time-resolved inner-shell spectroscopy allows one to probe dynamics of core-excited atoms and molecules.

Understanding of dynamical changes in molecules, as well as advances in laser technology and molecular spectroscopy stimulate the study of laser control of molecular motion. Modern adaptive control techniques use feed-back to correct the shape of the pulse. Selective excitation of molecular vibrations is achieved in stimulated Raman scattering experiments using adaptive control based on genetic type algorithm. To build up a connection between the pulse shapes and mode selective excitation, a semi-classical theory is developed that investigates the possibility of control of excitations of closely spaced Raman transitions in molecules using broadband pulse shaping.

A method is proposed for achieving either enhancement or suppression of particular vibrational excitations using broadband pulses. It implies pulse shapes possessing a broad spectral dip at the frequency to be suppressed and a suitably chosen intensity of the field at the frequency to be efficiently excited [32]. The role of coupling between vibrational modes is elucidated in the controllability of selective excitation in weak and strong fields. Phase control of the initially populated vibrational states gives new possibilities for the coherence manipulation in molecules [33].

Keywords: ultrafast molecular dynamics, coherent control, broadband pulse shaping, Raman spectroscopy

---

*E-mail address: smalinov@umich.edu

1 Introduction

The chapter is devoted to the studies of interaction of molecules with ultrafast laser pulses that gives insight into a time-dependent picture of ultrafast molecular motion and stimulates ideas for quantum control over it. Femtosecond wave packet spectroscopy was pioneered in works by Zewail *et.al.* [1, 2], where the potential and structural information about bound isolated systems has been extracted. Nowadays ultrafast time-resolved spectroscopy is broadly used to investigate reactive and non-reactive molecular dynamics [3]. Fundamental progress in generation of energetic ultrafast pulses using Kerr-lens lasers gave rise to sub-femtosecond x-ray pulses, that may be used to observe electronic processes [4] and to achieve control over them [5]. The goal of stirring an ensemble of atoms or molecules to a specific quantum state by means of ultrafast pulse techniques may be successfully achieved in the near future by acting on both nuclei and electrons. With respect to control of specific nuclear motion in molecules using ultrafast lasers, modern techniques based on adaptive learning algorithms [6] serve as an excellent tool. Series of successful experiments implementing these algorithms offer proof that it is possible to exercise quantum coherent control over a range of processes. Two such experiments employ stimulated Raman scattering in liquid methanol [7] and gas phase carbon dioxide [8]. The controlling optical field may be constructed using coherent quantum control theory, see, e.g. [9, 10]. An alternative approach involves a search for an analytical pulse function. For example, in [11, 12] an antisymmetric phase function was proposed to observe two energetically close Raman levels of pyridine in CARS signals [13].

In this chapter we specify implementations of ultrafast pulses for observation of electron and nuclear motion in time domain and for control over nuclear dynamics using pulse shaping based on adaptive learning techniques such as genetic type algorithm. Two examples of exciting experiments on quantum control in stimulated Raman spectroscopy are given. Significant part of the chapter is devoted to the theory of stimulated Raman scattering and control aspects by means of pulse amplitude shaping. A method to achieve selective excitation of desired Raman transitions is proposed using broadband pulse amplitude shaping. The dynamics of two, two-level systems representing normal vibrational modes in a molecule is considered in the weak and strong fields. Part of discussions is assigned to mechanisms of interaction of femtosecond laser pulses with molecules in the stimulated Raman scattering and, particularly, to the coupling between active vibrational modes as a factor affecting controllability of quantum yield. When investigating the ways of implementing the coupling mechanism, it was found that selective, high level coherence can be achieved based on the choice of the relative phase of the initially populated vibrational states in a molecule. For a particular phase, a molecular dark state was formed with zero eigenvalue of the interaction Hamiltonian.

A general purpose of the chapter is to show the uniqueness of femto- and subfemtosecond pulse discovery with respect to a longstanding problem of penetration into microscopic dynamics within the time domain far beyond the speed of everyday occurrences, besides, to give one an idea about a resent, detailed insight into the problem of quantum control of molecular vibrational motion using femtosecond pulse shaping in stimulated Raman spectroscopy.

2 Observation of Electron and Nuclear Motion

Molecules involved in chemical, physical and biological processes undergo dynamical changes associated with rearrangements of electrons and nuclei. Nuclear motion within molecules is on a fundamental time scale of molecular vibrations with the rate on the order of angstroms per femtosecond. To observe this motion in real time a femtosecond time resolution is required. Innovative ultrafast pulsed laser techniques [14, 15, 16] provide the necessary resolution and make possible the direct observation of nuclear dynamical changes. Time-resolved spectroscopy based on the pump-probe approach is commonly used requiring a pump pulse to initiate the dynamics and a probe pulse to analyze the time evolution of the system. This methodology is now widely applied to investigate nuclear wave packet dynamics in the time domain. It was initiated in works on preparation and probing transition state dynamics of chemical reactions with ultrafast pulses [17, 18]. Broad range of experiments on time-resolved observations of elementary microscopic motions using various spectroscopies are carried out, including collective vibrations of crystal lattices [19], molecular vibrations [20] and rotations [21], breaking and formation of chemical bonds [22], etc.

The feasibility of investigations relies on the concept of molecular and ensemble coherence. Molecular coherent wave packet is prepared on the electron potential energy surface of interest by irradiation of molecules with an ultrafast laser pulse, Fig.1. Femtosecond time scale of exciting pulse characterizes wave packet in terms of coherent superposition of vibrational levels included. Prepared by femtosecond laser pulse molecular coherent wave packet is spatially localized on the atomic scale of length and undergoes, in general, an unharmonic motion. For longer times, beyond the *molecular coherence time*, the wave packet dephases and looses coherence. Propagating molecular coherent wave packet is probed by a second ultrafast pulse or an accurately timed sequence of pulses. Spectroscopy of the probe pulse interaction with molecules carries explicit information about the vibrational and/or rotational motion in molecular electronic state.

Initiated by a pulse much shorter than a typical active vibrational period, excitation of an ensemble of molecules occurs instantaneously, giving rise to a synchronous vibrational motion. A signal from millions of molecules, measurable within the *ensemble coherence time*, is as if a signal from a single molecule. This picture is valid until molecular coherence and ensemble coherence are destroyed by intra- and intermolecular perturbations. The unique harmony of femtosecond pulse duration and coherence time of microscopic molecular motion makes possible observation of nuclear rearrangements in real time.

Electrons rearrange within molecules much faster than nuclei, and observation of electron dynamics in real time requires much higher resolution - that is where attosecond time scale comes in. Electron relaxation following core excitation in atoms and molecules is ultrafast, life time of the core excited states ranges from a few attoseconds to a few fem-

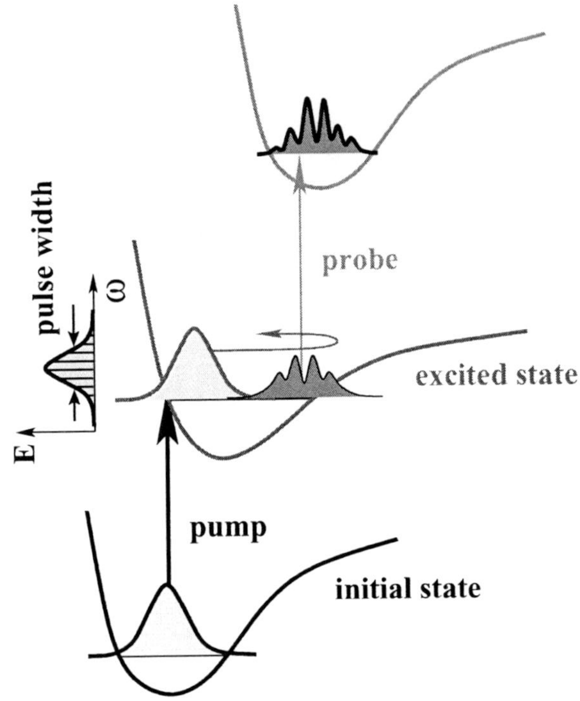

Figure 1: Schematic picture of a pump-probe experiment. Coherent molecular wave packet is created on a potential energy surface of interest with the bandwidth determined by the duration of a femtosecond pump laser pulse. A delayed probe pulse transfers an instant wave packet to a control state, from which the signal is detected.

toseconds. The information about the dynamics of short-living core excited states could be gained so far only indirectly from the energy-domain measurements. However, due to the specifics of multi-electron dynamics, the energy spectra of emitted photons or electrons often unable to provide a detailed insight into the process. Recently demonstrated time-resolved technique implementing sub-femtosecond pumping and attosecond sampling of electron wave packets [4, 16] suggests new possibilities for exploration of electron rearrangements of core-excited atomic and molecular systems in real time.

Electron promotion from the core shell into valence or continuum taking place upon molecular interaction with an incident sub-femtosecond pulse, gives rise to a short-living core-excited molecular state, the life-time of which depends on the energy of the core hole and the strength of electronic coupling. The scheme of the process is depicted in Fig.2. An excited system with excess internal energy tends to relax by filling the core vacancy with an electron from an outer shell. This process is followed by extreme ultraviolet or x-ray photon emission, or Auger electron emission depending on the magnitude of binding energies. The possibility of recording of electron rearrangements relies on the fact that the exciting pump pulse is much shorter than the life-time of core excited state. Then attosecond sampling

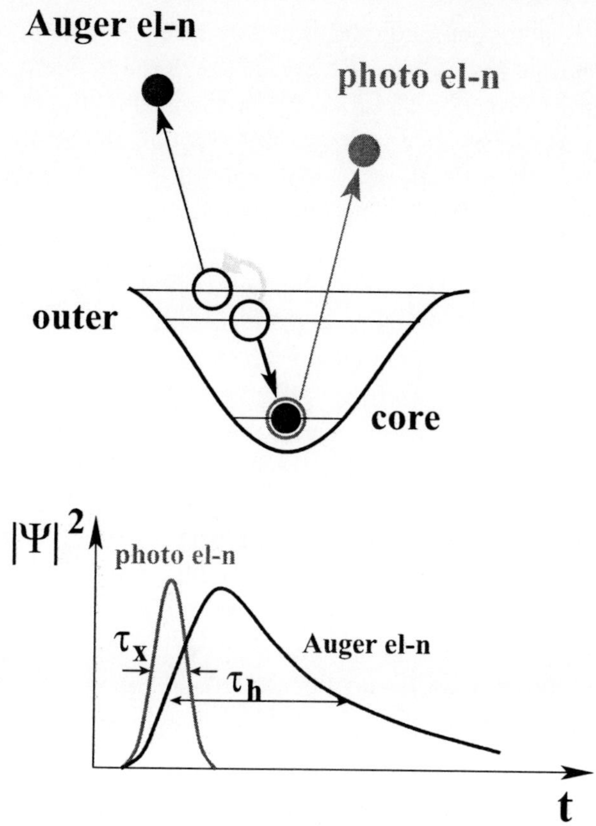

Figure 2: Top figure illustrates core electron ionization and relaxation via Auger electron emission. Bottom figure shows time-dependence of photo- and Auger electron wave packet probability. $|\Psi|^2$ of photo electrons follows intensity profile of the exciting sub-femtosecond x-ray pulse and shows the rate of the creation of core holes; τ_x is the duration of the pulse. $|\Psi|^2$ of Auger electrons has a trailing edge, that shows the decay rate of core holes; τ_h is the decay time of core holes.

of wave packet dynamics of photo- and Auger electrons reproduces electron dynamics in real time [23, 4]. Dynamic symmetry breaking accompanying core electron excitation or ionization in highly symmetrical molecules [24] may be observed experimentally using power gaining time-resolved attosecond spectroscopies.

3 Quantum Control of Nuclear Dynamics

3.1 Overview

Advances in laser technology, molecular spectroscopy as well as understanding of molecular dynamics stimulate studies in a challenging field of quantum control of molecular motion using laser pulse phase-amplitude shaping. Laser control methods began to develop about twenty years ago and have advanced from the simplest to modern sophisticated methods of coherent control devoted to various goals in chemistry, physics, biology and interdisciplinary sciences. Books by Rice and Zhio [25], Shapiro and Brumer [26] comprehensively expose contemporary methods of quantum control and their applications. Method of adaptive control, pioneered by Rabitz and Judson [6], is nowadays most commonly used in experimental groups pursuing goal of quantum control of the dynamics in systems of varying complexity, from diatomic molecules to polymers [27] and biological systems [28]. The adaptive method uses feed-back to optimize the phase and amplitude of the field to stir a molecular system to a desired quantum yield. The method is based on global optimization algorithm, e.g., the genetic type algorithm. The idea about this algorithm was inspired by evolution theory, where species develop subject to variability, inheritance and natural selection. Laser pulses, possessing a particular field amplitude and phase, are coded as strings of numbers - genes [7]. A set of genes represents a generation with a single gene being an individual. Each individual is tested through the interaction with a molecular medium and evaluated based upon fitness of an outcome signal to an 'ideal' function. Adaptive operators make selection among 'variability' operators by assigning each of them weighting, which allows operators that produce the most fit pulse shape to get high reproductive rates. Such a 'natural' selection based on evaluation of generated individual pulses provides the fittest pulses features to be inherited to future generations and to form the best pulse shape.

3.2 Experimental Realization of Coherent Control in Stimulated Raman Spectroscopy

Experiments on coherent control of molecular dynamics in stimulated Raman scattering using ultrafast shaped laser pulses are of a special demand and attraction owing to their fundamental goal of control of vibrational degrees of freedom. Adaptive learning technique based on the genetic type algorithm is efficiently used in stimulated Raman spectroscopy to design pulses to provide maximum gain into a predetermined Raman field. Experiment in molecular gas CO_2 [8] implements a pump-probe technique in impulsive regime of interaction of laser pulse with molecular media. Impulsive regime implies a broadband pump pulse possessing frequency components of both pump and Stokes transitions. Duration of such pulse is shorter or about an active vibrational period of molecules. In molecular gas CO_2 selective excitation of two Raman active vibrational modes is achieved with an intense pump pulse shaping. These modes are totally symmetrical mode at 38.6 THz and two quanta of bending mode at 41.6 THz. Selective excitation is analyzed by a delayed probe pulse, applied within the 'coherence time'. A weak probe pulse, when propagates through the selectively excited molecular medium, induces the Raman fields at the frequency of one of two transitions. Prob pulse spectra, shown in Fig.3, are obtained for the pump pulse optimized to excite symmetric stretch (a) and bending mode (b). The difference in intensity of two side bands in prob-pulse spectrum is used as a feedback in an evolutionary algorithm resulting in excellent selectivity of Raman mode excitation. Pump pulses applied in the ex-

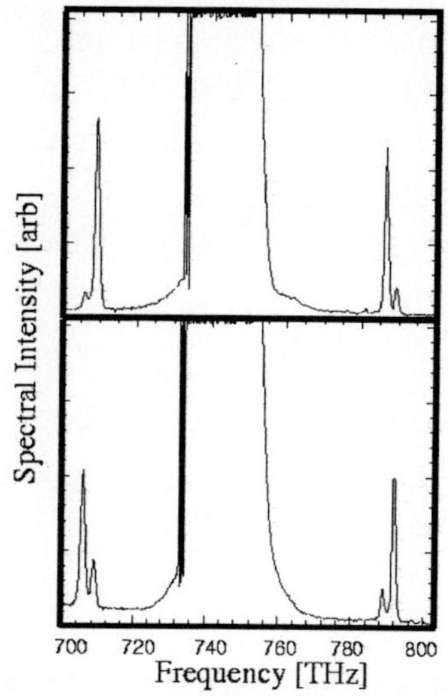

Figure 3: Probe spectra for pulses optimized to drive totally symmetrical mode (top panel) and two quanta of bending mode (bottom panel). (From Chem. Phys. Lett. 2001, **344**, 333, reprinted by permission.)

periment had frequency 375 THz and were generated as short as 15 fs. Temporarily delayed probe pulses with a frequency of 750 THz and a bandwidth 1.5 THz were injected collinear with the driving pulse in order to probe the achieved vibrational excitation. Temporal delay was about 100 fs, which was longer than a duration of a pump pulse, but shorter than a coherence time. Notably, described above impulsive regime was firstly discussed by Weiner *et. al.* in [29]. In this work accurately timed sequences of femtosecond pulses were used to initiate coherent excitation of several phonon modes in organic molecular crystal. For an impulse force to be exerted, the pulse duration had to be short compared to vibrational oscillation period.

Selective excitation of vibrational modes was also achieved in an experiment performed in liquid methanol [7]. In this experiment a pump pulse duration was longer than a vibrational period of two Raman active vibrational modes, which implies non-impulsive regime of interaction. The bandwidth of the pulse was about 100 cm^{-1}, while the magnitude of Raman active symmetric and asymmetric stretch modes was 2990 and 3000 cm^{-1}, respectively. Because of a large difference between transitional frequencies and the pulse bandwidth, the component of the field, stimulating Stokes transition in methanol molecules, was seeded by the noise signal. The feed-back goal for the pump pulse to excite symmetric and asymmetric modes alone resulted in generation of spectra shown in Fig.4. Panel a shows the forward scattered spectrum for an incident unshaped laser pulse. Panel b shows the spectrum after the learning algorithm optimized excitation of both modes while minimizing peak broadening due to other nonlinear effects. Panel c shows spectra for optimization of each mode independently. Panel d shows the spectrum for a pulse that minimized Raman scattering from both modes.

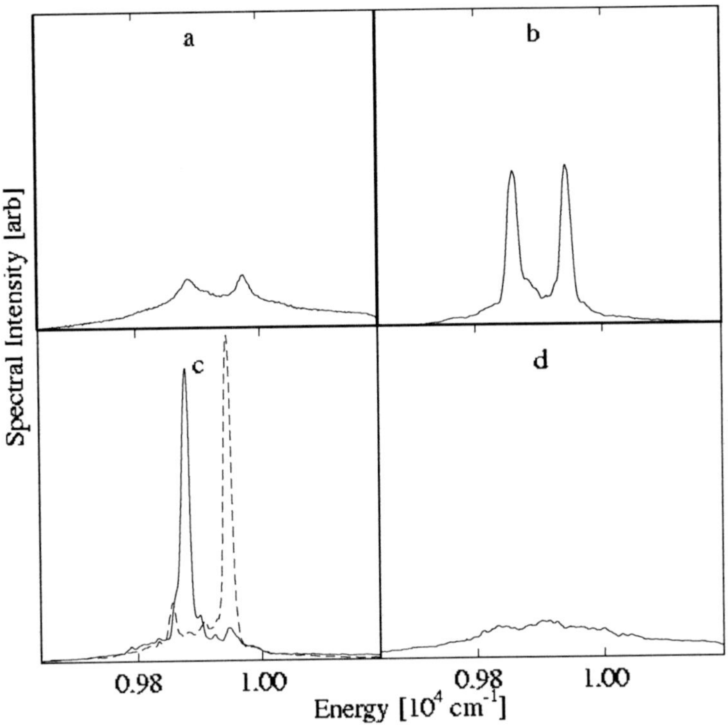

Figure 4: Control of Raman scattering in methanol. Panel a shows the forward scattered spectrum for an incident unshaped laser pulse. Panel b shows the spectrum after the learning algorithm optimized excitation of both modes while minimizing peak broadening due to other nonlinear effects. Panel c shows spectra for optimization of each mode independently. Panel d shows the spectrum for a pulse that minimized Raman scattering from both modes. (From Phys. Rev. A 2001, **63**, 063412, reprinted by permission.)

To reveal the role of the coupling between vibrational modes in the mode selective excitation, the methanol was replaced with a mixture of benzene and deuterated benzene. The ring breathing Raman mode of benzene is at 992 cm^{-1} frequency while in deuterated benzene a frequency of this mode is shifted by 47 cm^{-1}, similar to the mode splitting in methanol. Pulse shapes were found that drove transitions at ring breathing mode frequency for pure benzene and deuterated benzene, shown in Fig.5 (top panel). But when the mixture

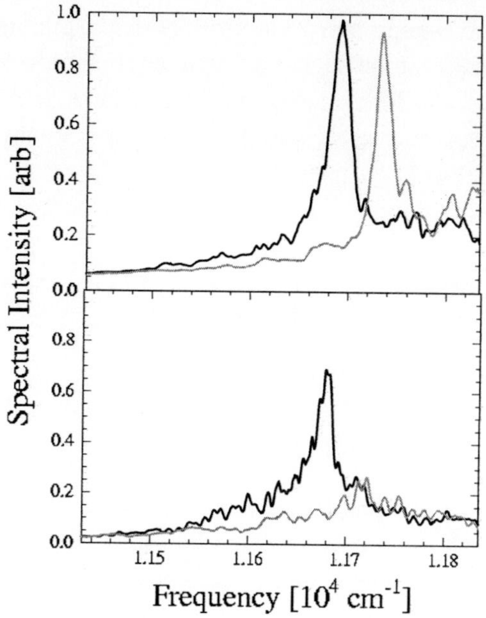

Figure 5: Stimulated Raman scattering in C_6H_6 and C_6D_6. Top panel shows spectra for C_6H_6 (left curve) and C_6D_6 (right curve) separately after optimization of the pulse shape to excite the breathing mode of each molecule. Bottom panel shows the results after using the learning algorithm to excite each molecule separately in a 50-50 mixture of the two. (From Phys. Rev. A 2001, **63**, 063412, reprinted by permission.)

of benzene and deuterated benzene was exposed to laser radiation with the goal of selective excitation of Raman transitions in each of two species, the learning algorithm could have created pulse shapes that selected the breathing mode in benzene but never in deuterated benzene, Fig.5 (bottom panel).

The experiments on vibrational mode selective excitation in Raman spectroscopy give rise to fundamental questions about pulse shape properties that induce a specific response in molecular systems, as well as about mechanisms of molecular interaction with the laser field that reveal the possibilities for manipulation of molecular motion in a designed way.

4 Semiclassical Theory of Stimulated Raman Scattering

4.1 Overview

Experiments on quantum control of molecular vibrations in stimulated Raman spectroscopy initiated a theoretical work having complementary goals, including the pulse shape design that results in mode selective excitation in polyatomic molecules. Polyatomic molecules usually have a number of energetically close vibrational modes that are visible in Raman spectroscopy. Intense femtosecond pulse instantly exciting molecular medium, is broad enough to drive more than one Raman transition. The Raman excitation of a single vibrational mode within bandwidth of the pulse, having frequency ω normally requires a pulse with $\omega\tau \leq 1$, where τ is the pulse duration. Optimal excitation may occur if the intensity is modulated at a frequency that corresponds to the transition frequency. For any two vibrational modes ω_1, ω_2 within the pulse bandwidth, $(\omega_1 - \omega_2)\tau \ll 1$, in order to induce a particular transition among them a pulse with a designed shape must be applied. In the next section we propose a method for achieving selective excitation using broadband pulses. The method is stated for two Raman active vibrational modes, and its applicability is discussed for the case of many transitions within the pulse bandwidth.

We start with developing a general theory that describes stimulated Raman scattering in a molecular medium having two Raman active vibrational modes. Vibrational modes are represented by two-level systems with the lower and upper levels of each system being the ground and excited vibrational states of a single mode. Duration of a femtosecond pump pulse with respect to a typical active vibrational period in molecules determines regime of interaction. Larger pulse duration corresponds to a non-impulsive regime of interaction, smaller duration leads to an impulsive interaction. In non-impulsive regime, molecules interact with different pulse intensity and phase at different time moments. This interaction is projected onto a time dependent picture of light scattering. Propagation effects in this case are significant and should be taken into account in the theoretical description of the scattering process. In impulsive regime of interaction, when a pump-probe technique is applied, a Raman spectrum is formed upon a weak probe pulse propagation through the excited medium. Probe pulse does not excite additionally the molecular medium and propagates subject to prepared molecular state coherence. Thus the outcome spectrum is fully determined by the pump pulse excitation. We describe stimulated Raman scattering in non-impulsive and impulsive regime of interaction using Maxwell-Bloch equations.

As shown in Fig.6, model consists of two, two-level systems, having transition frequencies ω_{21} and ω_{43}. The molecular medium is considered as an ensemble of such two-level systems with no relaxation processes taken into account. Each two-level system interacts with an intense off-resonant femtosecond pulse that initiates stimulated Raman scattering via an off-resonant interaction with a virtual state. Green curve shows the pulse spectral profile corresponding to impulsive interaction, having transitional frequencies of both two-level systems within the bandwidth. Blue curve shows the pulse corresponding to non-impulsive interaction, with the bandwidth broader than the frequency difference of the transitions. A general view of interaction Hamiltonian written in the basis of five states is

$$H = \frac{\hbar}{2} \begin{pmatrix} \alpha_1\omega & 0 & 0 & 0 & 2\Omega_1 cos(\omega_p t) \\ 0 & \alpha_2\omega & 0 & 0 & 2\Omega_2 cos(\omega_s t) \\ 0 & 0 & \alpha_3\omega & 0 & 2\Omega_3 cos(\omega_p t) \\ 0 & 0 & 0 & \alpha_4\omega & 2\Omega_4 cos(\omega_s t) \\ 2\Omega_1^* cos(\omega_p t) & 2\Omega_2^* cos(\omega_s t) & 2\Omega_3^* cos(\omega_p t) & 2\Omega_4^* cos(\omega_s t) & E_b \end{pmatrix}, \tag{1}$$

where $\alpha_j\omega$ is the frequency of a single level, such that, e.g., $(\alpha_2 - \alpha_1)\omega = \omega_{21}$, Ω_j is a

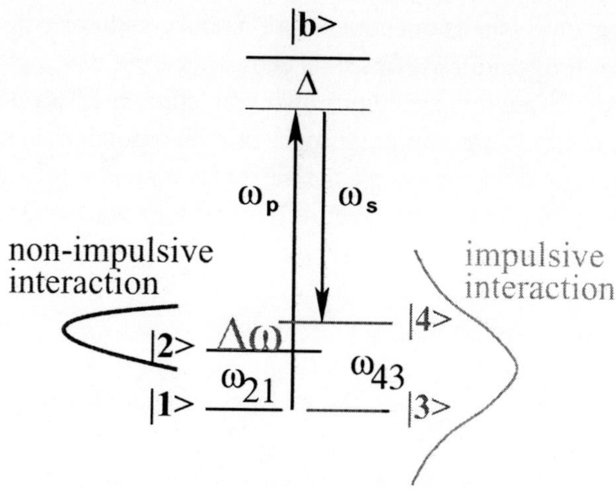

Figure 6: Schematic picture of a model system consisting of two, two-level systems having frequencies ω_{21} and ω_{43}. Initially, the lower levels are equally populated. The Raman transitions may be driven by an off-resonant femtosecond pulse in impulsive (green curve) and non-impulsive (blue curve) regime of interaction.

Rabi frequency, μ_{jb} is a dipole moment matrix element. For non-impulsive scattering

$$\Omega_{1,3} = -\frac{\mu_{jb} E_{p0}(t)}{\hbar}, \qquad \Omega_{2,4} = -\frac{\mu_{jb} E_{s0}(t)}{\hbar}, \tag{2}$$

where $E_{p0}(t)$, $E_{s0}(t)$ are the pump and Stokes pulse envelopes. For impulsive regime of interaction $\omega_p = \omega_s$ and $\Omega_{2,4} = \Omega_{1,3}$ because the same field provides transitions between four states. In this case the Rabi frequencies may differ only owing to different dipole moment matrix elements.

Initially lower levels $|1>$ and $|3>$ of both two-level systems are equally populated. The time evolution of two, two-level systems is described in terms of probability amplitudes, derived from time-dependent Schrödinger equation with the Hamiltonian defined in Eq.(1). In the interaction representation first order differential equations for probability amplitudes are

$$\dot{a}_j = i\frac{\Omega_j}{4\Delta}\sum_{j'=1}^{4}\Omega_{j'}^{*} e^{-(\alpha_j - \alpha_{j'})\omega t} a_{j'}, \tag{3}$$

Δ is the detuning of the frequency of the pulse from the frequency of the virtual state $|b>$. Eqs. (3) are obtained by adiabatically eliminating state $|b>$ within the rotating wave approximation.

Eqs. (3) are used to study the pulse interaction with two uncoupled two-level systems having zero probability for population transfer between them, and with two coupled two-level systems having all four states coupled via the external field. For the former, Eqs.(3)

cast into two independent systems of coupled differential equations with two variables. For the latter, a system of four coupled differential equations is valid.

Pulse propagation through the medium involving stimulated Raman scattering is described as follows. Initially, a pump pulse induces a vibrational coherence of states $|\rho_{12}|$ and/or $|\rho_{34}|$. Then, e.g., the coherence $|\rho_{12}|$ serves as a source for the generation of the Raman field on the 2-b transition. The evolution of quantum states and propagating fields is calculated by means of the coupled system of quantum equations for the states and Maxwell equations for the fields.

Here we present the solution for the 1-2 two-level system. Equations for the $|3>$ and $|4>$ quantum states may be written by analogy. We assume here that the time evolution of two, two-level systems is uncoupled. We chose the density matrix formulation because we need to calculate the expectation value of the induced dipole moment, $<\mu>= tr(\widehat{\rho\mu})$. This quantity is the source for the radiation fields. Differential equations for coherence and populations of the 1-2 two-level system is derived from Liouville equation of motion for the density matrix [30] with Hamiltonian

$$H = \frac{\hbar}{2}\begin{pmatrix} \omega_1 & 0 & 2\Omega_1 cos(\omega_p t) \\ 0 & \omega_2 & 2\Omega_2 cos(\omega_R t) \\ 2\Omega_1^* cos(\omega_p t) & 2\Omega_2^* cos(\omega_R t) & \omega_b \end{pmatrix}, \tag{4}$$

where $\Omega_1 = -\frac{\mu_{1b}E_{p0}(t)}{\hbar}$, $\Omega_2 = -\frac{\mu_{2b}E_{R0}(t)}{\hbar}$ are the Rabi frequencies related to the pump and Raman fields.

First order differential equations for coherence ρ_{12} and populations ρ_{11} and ρ_{22} are

$$\begin{aligned} \dot{\rho}_{12} &= -i\Omega_{12}^{eff}(\rho_{11}-\rho_{22}) - i(\Delta\omega_{12}-\delta)\rho_{12} \\ \dot{\rho}_{22} &= -i\Omega_{12}^{eff} 2Im[\rho_{12}], \qquad \dot{\rho}_{22} = -\dot{\rho}_{11} \end{aligned} \tag{5}$$

where

$$\begin{aligned} \delta &= \omega_{12} - (\omega_p - \omega_R), \\ \Omega_{12}^{eff} &= \mu_{1b}\mu_{b2}\frac{E_R^*(t)E_p(t)}{4\hbar^2\Delta}, \\ \Delta\omega_{12} &= -\mu_{1b}\mu_{b1}\frac{|E_p(t)|^2}{4\hbar^2\Delta} + \mu_{2b}\mu_{b2}\frac{|E_R(t)|^2}{4\hbar^2\Delta}. \end{aligned}$$

The Raman fields are driven by the second order time derivative of the polarization $P_R = 2Re[\rho_{2b}\mu_{b2}]$, which is

$$\begin{aligned} \frac{\partial^2 P_R}{\partial t^2} &= -\omega_R^2 Re[\mu_{b2}\rho_{2b}\exp\{i(\omega_R t - k_R z)\}], \\ \rho_{2b} &= -\frac{\mu_{1b}E_p(t)}{2\hbar\Delta}\rho_{21} - \frac{\mu_{2b}E_R(t)}{2\hbar\Delta}\rho_{22}. \end{aligned} \tag{6}$$

The incident pulse evolution depends on the polarization $P_p = 2Re[\rho_{2b}\mu_{b2}]$ with the second order derivative equal to

$$\frac{\partial^2 P_p}{\partial t^2} = -\omega_p^2 Re[\mu_{b1}\rho_{1b}\exp\{i(\omega_p t - k_p z)\}], \quad (7)$$
$$\rho_{1b} = -\frac{\mu_{2b}E_R(t)}{2\hbar\Delta}\rho_{12} - \frac{\mu_{1b}E_p(t)}{2\hbar\Delta}\rho_{11},$$

where ρ_{1b}, ρ_{2b} are slowly varying complex polarizations.

Evolution of the fields, propagating through molecular medium, is determined by Maxwell equations for slowly varying field functions:

$$\frac{dE_p(z,t)}{dz} + \frac{1}{c}\frac{dE_p(z,t)}{dt} = -\frac{k_p}{2\epsilon_0}\mu_{b1}\mu_{2b}\frac{E_R(z,t)}{2\hbar\Delta}Im[\rho_{12}] \quad (8)$$
$$\frac{dE_R(z,t)}{dz} + \frac{1}{c}\frac{dE_R(z,t)}{dt} = -\frac{k_R}{2\epsilon_0}\mu_{b2}\mu_{1b}\frac{E_p(z,t)}{2\hbar\Delta}Im[\rho_{21}]$$

A set of Eqs.(5) and Eqs.(8) describes an evolution of a quantum system in stimulated Raman scattering. The theory is valid for the conditions of intensive laser fields and the pulse duration longer than the inverse of the transition frequency. Example of application is a non-impulsive stimulated Raman scattering in the gas and liquid molecular media with coherence time longer than the duration of the pulse. The model is also valid for the description of a weak probe pulse propagation in the pump-prob experiments in an impulsive regime of interaction, the change in field indexes from p to pr in Eq.(8) should be made.

5 Control of Raman Transitions Using Ultrafast Pulse Shaping

5.1 Method for Selective Excitation of Raman Vibrational Modes Using Broadband Pulses

In impulsive interaction of laser field with molecular media in pump-probe spectroscopy the Raman field amplitudes generated upon propagation of the probe pulse are determined by the coherence of two-level systems, see the model in Fig.6 and Eqs.(8). Propagating weak probe pulse does not change notably this coherence, hence the intensity of the Raman fields depends purely on the coherence formed by pumping. Control of vibrational excitations manifesting itself in the ratio of Raman field intensities at transitional frequencies of two, two-level systems is entirely determined by pump pulse shaping. In order to get a significant ratio between peak intensities, pump pulse has to provide maximum coherence of one two-level system and zero coherence of another. Therefore, it is of prime interest to calculate the coherence of two-level systems resulting from the interaction with the broadband shaped pump pulse. We assume now that the coupling between two-level systems is zero, e.g., the coupling between Raman vibrational modes in a molecule via an external field is negligibly small. We put the transition dipole moment matrix elements all equal to unity in Eqs.(3). This allows us to focus on the effects of pulse shaping. In the next subsection we will analyze the effects caused by the coupling between two-level systems and different oscillator strength of the transitions. The goal now is to choose a pump pulse shape that, e.g., efficiently suppresses coherence $|\rho_{12}|$ and enhances coherence $|\rho_{34}|$. We present here a method for achieving selective excitation by broadband pulse amplitude shaping, (see also [31, 32]), investigations on phase shaping will be presented elsewhere. In impulsive interaction, the time derivative of the probability amplitudes in Eqs.(3) is proportional to the second order in the field amplitude, e.g., the intensity envelope. The idea about an analytical function for the shaping of the intensity envelope came from the solution in a perturbative regime. Eqs.(3) for the upper levels of two, two-level systems without Stark shift are written as

$$\dot{a}_4 = i\frac{\mu_{4b}\mu_{3b}^*}{4\Delta\hbar^2}I(t)e^{i\omega_{43}t}a_3, \qquad \dot{a}_2 = i\frac{\mu_{2b}\mu_{1b}^*}{4\Delta\hbar^2}I(t)e^{i\omega_{21}t}a_1 \tag{9}$$

Stark shift has the same sign for all four states and does not change probability amplitude dynamics, that is why it is omitted in present consideration.

In the perturbative limit the solution for a_2 and a_4 is

$$a_4 = i\frac{\mu_{4b}\mu_{3b}^*}{4\Delta\hbar^2}\int_{-\infty}^{\infty} I(t)e^{i\omega_{43}t}dt, \qquad a_2 = i\frac{\mu_{2b}\mu_{1b}^*}{4\Delta\hbar^2}\int_{-\infty}^{\infty} I(t)e^{i\omega_{21}t}dt. \tag{10}$$

It means that in weak fields complex probability amplitudes of upper levels of two uncoupled two-level systems are proportional to the Fourier transform of the intensity of the field at corresponding transitional frequencies:

$$a_4 \sim i\tilde{I}(\omega_{43}), \qquad a_2 \sim i\tilde{I}(\omega_{21}) \tag{11}$$

In order to suppress excitation at frequency ω_{21}, an intensity envelope has to vanish for a modulation frequency of the pulse equal to that frequency. From the other hand, the intensity at frequency ω_{43} has to be optimally chosen to provide probability amplitude a_4 that results in maximum coherence ρ_{34}.

We make an initial guess for an analytical function for intensity envelope in the frequency domain. To suppress excitation at frequency ω_{21} and enhance excitation at frequency ω_{43}, we choose

$$\tilde{I}(\omega) = I_0 e^{-(\omega-\omega_{43})^2T^2}\left(1 - e^{-(\omega-\omega_{21})^2{T_1}^2}\right), \tag{12}$$

where T and T_1 are free parameters. When the modulation frequency of the pulse ω is equal to ω_{43} the intensity approaches its maximum I_0 for a sufficiently large parameter T_1,

$$\tilde{I}(\omega_{43}) = I_0(1 - e^{-\Delta\omega^2{T_1}^2}), \qquad \Delta\omega = \omega_{43} - \omega_{21}. \tag{13}$$

For ω equal to ω_{21} the intensity is zero, $\tilde{I}(\omega_{21}) = 0$. The intensity envelope as a function of frequency is drawn in Fig.7a. The frequencies of the two-level systems are ω_{21}=1 and ω_{43}=1.1 in frequency units of ω_{21}, parameter $T = 3[\omega_{21}^{-1}]$. The intensity of the field at ω_{43} and frequency region near ω_{21} over which $\tilde{I}(\omega)$ is approximately zero depend on T_1. The larger T_1, the greater is the selectivity for suppressing the ω_{21} frequency.

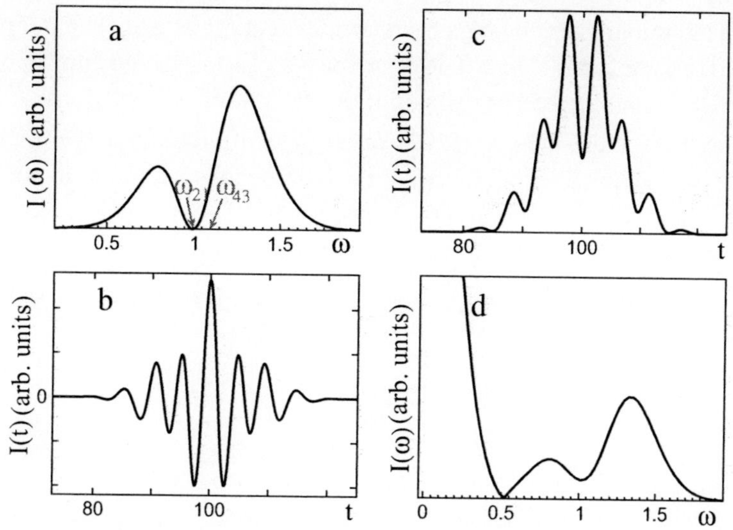

Figure 7: (a) Frequency dependence of the initial guess pulse intensity; (b) Fourier transform of the curve in (a); pump pulse intensity (c) temporal and (d) spectral profile; parameters used in calculations are $T = 3, T_1 = 3$ in frequency units of ω_{21}^{-1}, $\omega_{21} = 1$, and $\omega_{43} = 1.1$.

The inverse Fourier transform of the spectral density in Eq.(12) is a complex function. To arrive at a physically acceptable temporal pulse function, as a first step we take the real part of the inverse Fourier transform, given by

$$I(t) = I_0\left((\sqrt{2}T)^{-1}e^{-\frac{t^2}{4T^2}}cos(\omega_{43}t) - (\sqrt{2}\tau)^{-1}e^{-\Delta\omega^2T^2(1-\frac{T^2}{\tau^2})-\frac{t^2}{4\tau^2}}cos((\omega_{21} - \Delta\omega\frac{T^2}{\tau^2})t)\right), \tag{14}$$

where $\tau^2 = T^2 + T_1^2$.

The dependence of this function on time is plotted in Fig.7b. This function is not positive definite for all times. To constrain the pulse intensity to be non-negative for all times we make a shift into positive region by adding a 'unit' gaussian $C = (\sqrt{2}\tau)^{-1}e^{-\frac{t^2}{4\tau^2}}$. Then the expression for the field intensity is modified to

$$I(t) = I_0C\left(1 + A/Ccos(\omega_{43}t) - B/Ccos((\omega_{21} - \Delta\omega\tfrac{T^2}{\tau^2})t)\right), \tag{15}$$

$$A = (\sqrt{2}T)^{-1}e^{-\frac{t^2}{4T^2}}, B = (\sqrt{2}\tau)^{-1}e^{-\Delta\omega^2T^2(1-\frac{T^2}{\tau^2})-\frac{t^2}{4\tau^2}}, C = (\sqrt{2}\tau)^{-1}e^{-\frac{t^2}{4\tau^2}}.$$

The temporal profile of the intensity of the field is shown in Fig.7c. Eq.(15), it is used for the calculation of the probability amplitudes.

The Fourier transform of Eq.(15) reads

$$\tilde{I}(\omega) = I_0\left[e^{-(T^2+T_1^2)\omega^2} + e^{-(\omega-\omega_{43})^2T^2}\left(1 - e^{-(\omega-\omega_{21})^2T_1{}^2}\right)\right]. \tag{16}$$

It differs from the initial guess by a gaussian centered at the origin, which reproduces a dc component of the field. In Fig.7d a part of the curve $\tilde{I}(\omega)$ is shown emphasizing common features with Fig.7a and manifesting a part of the dc component.

Polyatomic molecules often possess several or many Raman active modes with frequencies close enough to be within the bandwidth of the pulse. In order to enhance a single vibration and suppress other vibrations the function for the pulse may be constructed as a product of several terms

$$\tilde{I}(\omega) = I_0 \left[e^{-(T^2+T_1^2)\omega^2} + e^{-(\omega-\omega_{43})^2T^2} \Pi_j \left(1 - e^{-(\omega-\omega_j)^2{T_1}^2}\right)\right]. \tag{17}$$

Numerical results

Analytical function for the intensity of the field described by Eq.(15) is used in Eqs.(3) to calculate probability amplitudes and coherence of two uncoupled two-level systems. A system of differential equations is solved numerically using Runge-Kutta method. With the proposed function for the intensity envelope, the behavior of two-level systems is investigated in weak and strong fields. The solution in a perturbative limit leads to stronger excitation of coherence $|\rho_{34}|$ than coherence $|\rho_{12}|$, with the magnitude proportional to $\tilde{I}(\omega_{43})$ and $\tilde{I}(\omega_{21})$ respectively. In the strong fields, the field amplitude becomes a sensitive variable and plays a crucial role in selective excitation.

Parameters for the system, used in numerical calculations, are taken from the experimental data on impulsive excitation of vibrational modes in the molecular gas CO_2 [7]. In CO_2 the frequencies of two selectively excited Raman modes are 38.6 and 41.6 THz. The FWHM of the applied intense pulse is taken equal to 18 THz. In our calculations the frequency ω_{21} is set equal to unity; in these units the frequency ω_{43} is equal to 1.1. From experimental data, we estimate that the parameter T is about equal to 3, in frequency units of ω_{21}^{-1}. The intensity of the field is determined by the parameter I_0. The parameter T_1 is related to the width of the spectral dip in $\tilde{I}(\omega)$, see Eq.(16), centered at frequency ω_{21}. Although a value of $T_1 \gg T$ would provide the best selectivity, the choice for the parameter T_1 is strongly restricted by the requirement that the duration of the applied pulse be within a typical molecular vibrational period and in calculations the parameter $T_1 = 3[\omega_{21}^{-1}]$ is used.

In Fig.8 the coherence is plotted as a function of the intensity of the field for parameters T=3 and $T_1 = 3$. Coherence $|\rho_{34}|$ of the 3-4 system is represented by red line and coherence $|\rho_{12}|$ of the 1-2 system by black line; blue and green lines show populations of the upper levels of both two-level systems.

For the weak fields coherence $|\rho_{34}|$ prevails over coherence $|\rho_{12}|$ in accordance to the solution in a perturbative limit, shown in Eqs. (10-11). For the intense fields, coherence of the excited and suppressed systems possess somewhat chaotic structure. Several values of the intensity, e.g., $I_0 = 1.8\pi$ and $I_0 = 2.55\pi$ give rather low coherence of the 3-4 system but maximum coherence of the 1-2 system. A desired solution for maximum coherence of the 3-4 system is achieved for the intensity coefficient $I_0 = 2\pi$. This is the result of redistribution of population within that two-level system: half of population is transferred

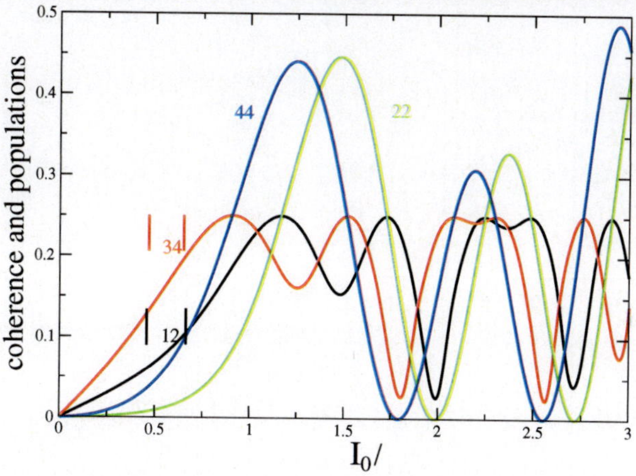

Figure 8: Intensity dependence of the coherences $|\rho_{12}|$ and $|\rho_{34}|$ and upper states populations ρ_{22} and ρ_{44}. Maximum coherence of the 3-4 system and negligibly small coherence of the 1-2 system are observed for $I_0 = 2\pi$ in the intensity region shown with $T = 3, T_1 = 3$ in frequency units of ω_{21}^{-1}.

to the upper level providing maximum coherence. The corresponding coherence of the 1-2 system at this intensity is nearly zero, where most population remains in the lower level.

The goal of control of the coherence of two uncoupled two-level systems is achieved with a pulse shape possessing a broad spectral dip at the suppressed frequency and a suitably chosen intensity of the field at the frequency to be efficiently excited. This technique allows one to use pulses of duration T to selectively excite transitions having frequency separations less than the bandwidth of the pulse, e.g., when $\Delta\omega < 1/T$. Our results should not be taken to imply that one can spectroscopically determine frequencies to better than the inverse temporal width of the pulse. On the other hand, if the frequencies are known from previous measurements, it is possible to suppress one transition and enhance the other by the method outlined above.

Had we taken a frequency profile centered at ω_{43} with $T = 3$, the curves for the coherence and populations of the excited and suppressed systems would differ qualitatively from those show in Fig.8. The desired selectivity could not be achieved.

The time-dependence of the coherence, populations and the field is shown in Fig.9 for $T = 3, T_1 = 3$, $I_0 = 1.4\pi$. The pulse duration is about 50 fs (thin solid line). It induces oscillations in the population distribution (thin lines) which lead to oscillations of the coherence of the two-level systems (bold lines). At long times the coherence and populations of levels achieve stationary values.

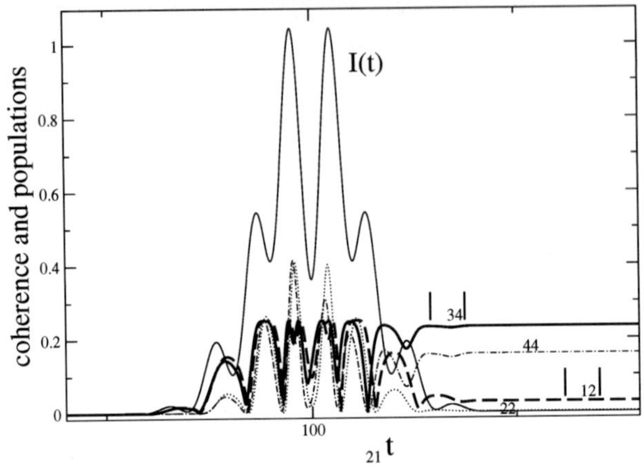

Figure 9: Time evolution of $|\rho_{12}|$, $|\rho_{34}|$, ρ_{22}, and ρ_{44} for T=3, $T_1 = 3$, and $I_0 = 1.4\pi$. The pulse intensity I(t), in arbitrary units, is also shown in the figure.

5.2 Raman Mode Coupling as a Factor of Control

One of an important steps needed for efficient control of molecular motion is understanding of the factors that govern the system's time evolution. In this section the coupling between normal vibrational modes is investigated as a mechanism contributing to coherent control in stimulated Raman spectroscopy [33]. Specific questions about the mechanism of interaction of molecular media with an external field have been addressed: How the coupling via an external field in uences the controllability of excitation of transitions, how much the achieved control depends on the coupling strength as well as on the strength of the field.

In [7] a comparative analysis of Raman spectra of liquid methanol and a mixture of benzene and deuterated benzene showed experimental evidence of the dependence of the selective excitation of Raman transitions on an intramolecular coupling of normal vibrational modes. A semiclassical model proposed in Section 4 is used to study the coupling between Raman active modes as a factor governing the system's time evolution. The effects caused by the presence of coupling between two, two-level systems are considered as are the effects caused by different strengths of the coupling constants. We investigate the effects of coupling as a function of field strength and the relative phase of the amplitudes of the initially populated states.

Notably, within our model selective, high level coherence can be achieved in two coupled two-level systems, depending on the relative phase of the initially populated states.

Coupling strength of states is determined by coupling constants which are transition dipole moment matrix elements of the states with a virtual intermediate state $|b>$, equal to μ_{ib}. Generally they may be different. Transition matrix elements of the 3-4 two-level system are assumed to be equal, $\mu_{3b} = \mu_{4b}$, and transition matrix elements of the 1-2 two-level system satisfy the following conditions $\frac{\mu_{1b}}{\mu_{3b}} = \frac{\mu_{2b}}{\mu_{3b}} = r$. Originated from the Hamiltonian in Eq.(1), the equations of motion for the probability amplitudes of two coupled two-level systems written in the interaction representation are:

$$i\frac{d}{dt}\begin{pmatrix} a_1 \\ a_2 \\ a_3 \\ a_4 \end{pmatrix} = -\chi \begin{pmatrix} r^2 & r^2 & r & r \\ r^2 & r^2 - \frac{\omega_{21}}{\chi} & r & r \\ r & r & 1 & 1 \\ r & r & 1 & 1 - \frac{\omega_{43}}{\chi} \end{pmatrix} \begin{pmatrix} a_1 \\ a_2 \\ a_3 \\ a_4 \end{pmatrix}, \tag{18}$$

where χ is effective Rabi frequency $\frac{|\mu_{3b}|^2}{4\hbar^2\Delta}I(t)$, $I(t)$ is the pulse intensity envelope and Δ is the detuning of the frequency of the pulse from the frequency of the virtual state $|b>$. The chosen form of the Hamiltonian allows for an adequate description of various coupling schemes. Note, that the pulse intensity envelope $I(t)$ is the same for all transitions, since the impulsive scattering is considered. The specific form for the pulse shape used in calculations is that described by Eq.(15). In the weak field regime the pulse selectively excites transitions of predetermined frequencies, while in strong fields the result of selective excitation depends on the coupling strength.

Numerical results

A numerical solution of the differential equations for probability amplitudes in Eq.(18) is obtained for two values of r, equal to $\frac{1}{2}$ and 1. The results reveal the importance of the relative phase between the initial state amplitudes a_1 and a_3. This relative phase could be established by optical pumping into state $|1>$ and using a Raman pulse to create the $|1> - |3>$ state coherence. An equilibrium ensemble consists of molecules with various relative phases of populated states related to different vibrational modes. Following this statement we first calculated coherence $|\rho_{12}|$ and $|\rho_{34}|$ as an average over relative phases between initially populated states $|1>$ and $|3>$. For $r = 1$, $|\rho_{12}|$ and $|\rho_{34}|$ are shown in Fig.10a, (bold black and bold red lines, respectively), as a function of the dimensionless intensity of the ultrafast laser pulse.

Phase averaging is equivalent to calculating $|\rho_{34}|$ and $|\rho_{12}|$ as a sum of two contributions resulting from two initial population distributions: $(\rho_{11} = \frac{1}{2}, \rho_{33} = 0)$ and $(\rho_{33} = \frac{1}{2}, \rho_{11} = 0)$. This approach eliminates the role of the phase between initially populated states. Coherences as a function of the intensity of the field calculated in such a way are identical to those in Fig.10a.

Also shown in Fig.10a, are values of $|\rho_{34}|$ and $|\rho_{12}|$, (represented by thin red and thin black lines, respectively), when there is no coupling between the two, two-level systems (obtained formally by setting $r^2 = 1, r = 0$ in Eq.(18)). From a comparison of the two sets of curves it is seen that phase averaged solution for two, coupled two-level systems gives much lower values of $|\rho_{12}|$ and $|\rho_{34}|$ than that for uncoupled systems. In weak fields the coherence $|\rho_{34}|$ prevails over $|\rho_{12}|$ regardless to the strength of coupling. However, the coherence $|\rho_{12}|$ increases faster than that for the case of two uncoupled two-level systems. In strong fields $|\rho_{12}|$ and $|\rho_{34}|$ for coupled two-level systems oscillate synchronously due to strong interaction with the dc component of the field. The coherence $|\rho_{12}|$ is almost always larger than $|\rho_{34}|$ owing to its slightly smaller transition frequency. In the weak

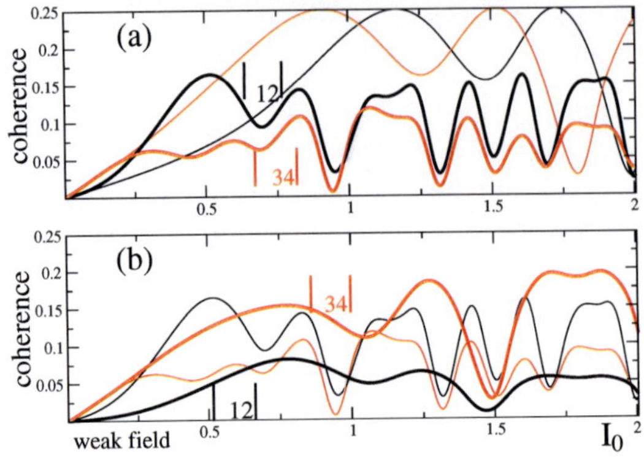

Figure 10: Intensity dependence of the coherences of two, two-level systems; $|\rho_{12}|$ is shown by black lines and $|\rho_{34}|$ by red lines. In (a) bold curves depict the case for $r = 1$ corresponding to two coupled two-level systems with equal coupling constants; thin curves depict coherences for two uncoupled two-level systems. The phase averaged solution gives much lower values of coherences than that for zero phase and coupling. In (b) bold curves show coherences for $r = 1/2$, and thin curves for $r = 1$. Weak coupling constants of the 1-2 system result in efficient population transfer to the 3-4 system, strongly coupled to the field.

fields, the proposed laser pulse efficiently excites vibrations in a molecule at frequency ω_{43}, while the excitation of vibrations at ω_{21} is essentially suppressed regardless to the strength of coupling between two vibrational modes. In the strong fields, an efficient population transfer is induced between all four levels in a molecule having nonzero mode coupling, causing excitation of both vibrations. Consequently, the amplitude of the induced electric dipole moment is modulated at both vibrational frequencies giving rise to two Raman fields.

Strong laser pulses shaped in accordance to Eq.(15) may result in significant differences between $|\rho_{12}|$ and $|\rho_{34}|$ in a system with different coupling constants μ_i. For $r = \frac{1}{2}$ the dependence of $|\rho_{34}|$ and $|\rho_{12}|$ on the intensity of the field is shown in Fig.10b by bold red and bold black lines, respectively. For such coupling constants, the probability of population transfer between two-level systems is equal to 1/2, between states $|1>$ and $|2>$ it is equal to 1/4, and between states $|3>$ and $|4>$ to 1. As the result, population transfers from the 1-2 to the 3-4 two-level system, maximizing coherence $|\rho_{34}|$.

Now we consider the effect of the relative phase between initially populated states. Notably, it shows the dependence on the field amplitude. For various values of intensity of the field I_0 the dependence of $|\rho_{12}|$ and $|\rho_{34}|$ on the relative phase between initially populated states $|1>$ and $|3>$ has been calculated. In Fig.11a the case for $I_0 = 2.625\pi$, $r = 1$ is represented (which gives almost equal values of $|\rho_{12}|$ and $|\rho_{34}|$ for the phase averaged solution). For phases from zero to π, $|\rho_{34}|/|\rho_{12}| > 1$, and a vibrational mode having frequency ω_{43} is excited stronger in a molecule than that having frequency ω_{21}. For phases from π to 2π, $|\rho_{34}|/|\rho_{12}| < 1$ indicating the reverse picture of the strength of excitation of vibrational modes. Using phase control of the initially populated states allows one to

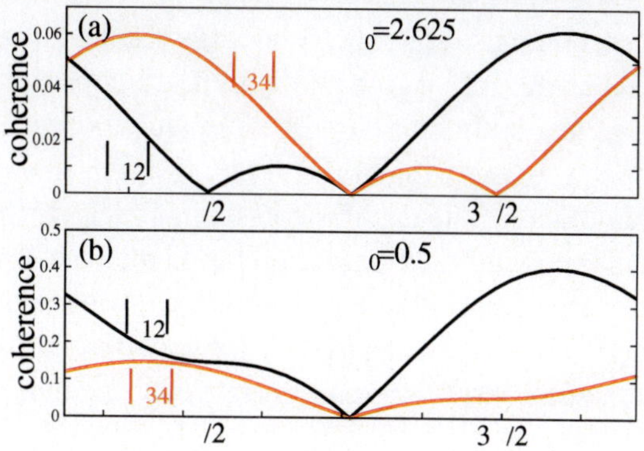

Figure 11: Relative phase dependence of $|\rho_{12}|$, $|\rho_{34}|$, ρ_{22}, and ρ_{44} for T=3, $T_1 = 3$, and a) $I_0 = 2.625\pi$ b) $I_0 = 0.5\pi$.

enhance the coherence between desired vibrational levels in a molecule and consequently enhance the intensity of a respective side band in the stimulated Raman spectrum. The behavior of $|\rho_{34}|/|\rho_{12}|$ as a function of the relative phase is sensitive to the intensity of field. For example, when $I_0 = 0.5\pi$, (which gives a phase averaged solution $|\rho_{12}| = 0.163$ and $|\rho_{34}| = 0.059$ in Fig.10a), the phase dependent calculation shows that $|\rho_{12}|/|\rho_{34}| > 1$ for all phases except for $\phi = \pi$, see Fig.11b. The maximum value of coherence $|\rho_{12}^{max}| = 0.4$ at $\phi = 7\pi/4$ is much higher than that for the phase averaged solution. This result demonstrates that in a molecular media with prepared relative phase between initially populated states and accurately chosen intensity of the field an abnormally high polarization may be induced. When the initial relative phase ϕ is equal to π, coherences $|\rho_{12}|$ and $|\rho_{34}|$ are equal to zero. This is the case for any external field. Populations of all states exhibit no time evolution. This result indicates an existence of a dark state as can be deduced directly from the Hamiltonian in Eq.(1). For an arbitrary value r, the necessary conditions for a dark state are $\frac{n_3}{n_1} = r^2$ and $\phi = \pi$.

6 Conclusion

Availability of femtosecond pulses makes it possible to observe direct nuclear wave packet dynamics in time domain. Applying an ultrafast intensive pump pulse creates a coherent superposition of vibrational states, time evolution of which is probed by a femtosecond snapshot. The uniqueness of the procedure is that it takes course within the time scale natural for ultrafast nuclear motion, which allows one to make use of the phenomena of the molecular and ensemble coherence.

Perspectives in investigation of electron wave packet dynamics in real time are uncovered due to recent demonstration of generation of isolated sub-femtosecond pulses and light-field-controlled photoemission technique. Measured Auger electron emission of krypton atoms using attosecond sampling demonstrates successful steps in this direction [4].

Fast evolving femtosecond pulse shaping technology resulted in a pool of new ideas for optical control of molecular motion. Most generally used approach is an adaptive learning technique based on global optimization procedure, such as genetic type algorithm. It gave rise to a series of exciting experiments with two of them in stimulated Raman spectroscopy discussed in this chapter. The experiments demonstrate quantum control of molecular vibrations in molecular gas carbon dioxide and liquid methanol, manifested itself in the Raman spectra. In both cases pulse shaping is optimized using adaptive control with implementation of the genetic algorithm. Inspired by the experiments, a semi-classical theory of stimulated Raman scattering involving the selective excitation of closely spaced Raman transitions is developed to uncover the physical reason for the pulse shapes giving optimal Raman spectra, to analyze the mechanisms of a pulse-molecule interactions and to reveal factors contributing to molecular time evolution. A semiclassical theory describes stimulated Raman scattering in a molecular medium with two Raman active vibrational modes represented by two-level systems. The dynamics of two-level systems is demonstrated in terms of probability amplitudes and depends on the shape of a femtosecond laser pulse, that has a duration shorter than a typical molecular vibrational period. An analytical function for the shape of the broadband pulse intensity envelop is proposed, that allows for the selective excitation of a predetermined vibrational motion. The pulse shape leads to maximum coherence of a desired vibrational transition and consequently to maximum gain into Raman side bands when the probe pulse is applied. It has a broad spectral deep at frequency of unfavorable transition and an optimally chosen intensity of the field at frequency to be efficiently excited. Practically, it may be used as an initial guess for the control of bond excitation in chemistry as well as a fit function within an adaptive learning algorithm.

The effects of the coupling between Raman active vibrational modes via a laser field are analyzed. The coupling is shown to cause an efficient population transfer between two-level systems which may result in desired enhancement and suppression of coherences $|\rho_{12}|$ and $|\rho_{34}|$, given inequivalent transition dipole moment matrix elements of two-level systems. The coherences determine the induced polarization in a molecular medium and the amplitudes of Raman fields generated from these coherences using a probe field. Effects of the relative phase between the initially populated vibrational states in a single molecule are revealed. Depending on the intensity of the field, the initial phase may cause predominance of the coherence of one two-level system with respect to another. When the relative phase between initially populated levels $|1>$ and $|3>$ is equal to π, two, two-level systems encounter stationary conditions. Regardless to the external field, all levels stay uncoupled exposing no time evolution, which implies the existence of molecular dark state. These results may be useful for investigation of decoherence processes caused by the driving field.

Acknowledgement

The author acknowledges support from and many valuable discussions with P. H. Bucksbaum, P. R. Berman, B. J. Pearson, T. C. Weinacht and V. S. Malinovsky. This work was supported by the National Science Foundation (No. PHY-9987916) through the Center for Frontiers in Optical Coherent and Ultrafast Science (FOCUS).

References

[1] Dantus, M.; Bowman, R.M.; Zewail, A.H. *Nature* 1990, 343, 737.

[2] Gruebele, M.; Zewail, A.H. *J. Chem. Phys.* 1993, 98, 883.

[3] *Femtochemistry and femtobiology: ultrafast events in molecular science*; Martin, M.M., Hynes J.T.;Ed.; Elsevier Science 2004.

[4] Drescher, M.; Hentschel, M.; Klenberger, R.; Ulberacker, M.; Yakovlev, V.; Scrinzl, A.; Westerwalbesloh, Th.; Kleineberg, U.; Heinzmann, U.; Krausz, F. *Nature* 2002, 419, 803.

[5] Baltuska, A.; Udem, Th.; Uiberacker, M.; Hentschel, M.; Goulielmakis, E.; Gohle, Ch.; Holzwarth, R.; Yakovlev., V.S.; Scrinzi, A.; Hänsch, T.W.; Krausz, F. *Nature* 2003, 421, 611-615.

[6] Judson, R.S.; Rabitz, H. *Phys. Rev. Lett.* 1992, 68, 1500.

[7] Pearson, B.J.; White, J.L.; Weinacht, T.C.; Bucksbaum, P.H. *Phys. Rev. A* 2001, 63, 063412.

[8] Weinacht, T.C.; Bartels, R.; Backus, S.; Backsbaum, P.H.; Pearson, B.; Geremia, J.M.; Rabitz, H.; Kapteyn, H.C.; Murnane, M.M. *Chem. Phys. Lett.* 2001, 344, 333.

[9] Shi, S.; Woody, A.; Rabitz, H. *J. Chem. Phys.* 1988, 88, 6870.

[10] Shi, S.; Rabitz, H. *Chem. Phys.* 1989, 139, 185.

[11] Dudovich, N.; Oron, D.; Silberberg, Y. *Phys. Rev. Lett* 2002, 88, 123004.

[12] Oron, D.; Dudovich, N.; Yelin, D.; Silberberg, Y. *Phys. Rev. Lett.* 2002, 88, 063004.

[13] D. Oron, N. Dudovich, D. Yelin, and Y. Silberberg, *Phys. Rev. A* 2002, 65, 043408.

[14] Sartania, S.; Cheng, Z.; Lenzner, M.; Tempea, G.; Spielmann, Ch.; Krausz, F.; Ferencz, K. *Opt. Lett.* 1997, 22, 1562.

[15] Shirakawa, A.; Sakane, I.; Takasaka, M.; Kobayashi, T. *Appl. Phys. Lett.* 1999, 74, 2268.

[16] Brabec, T.; Krausz, F. *Rev. Mod. Phys.* 2000, 72, 545.

[17] Zewail, A.H. *J. Phys. Chem.* 2000, 104, 5660.

[18] Dantus, M.; Rosker, M.J.; Zewail, A.H. *J. Chem. Phys.* 1987, 87, 2395.

[19] Misochko, O.V.; Hase, M.; Isioka, K.; Kitajima, M. *Phys. Rev. Lett.* 2004, 92, 197401; Hase, M.; Kitajima, M.; Constantinescu, A.M.; Petek, H. *Nature* 2003, 426, 51-54.

[20] Blanchet, V.; Zgierski, M.Z.; Seideman, T.; Stolow, A. *Nature* 1999, 401, 52-54.

[21] Cicerone, M.T.; Blackburn, F.R.; Ediger, M.D. *J. Chem. Phys.* 1995, 102, 471-479.

[22] Rice, S.A. *Nature* 2004, 429, 255-257.

[23] Hentchel, M.; Klenberger, R.; Spielmann, Ch.; Reider, G.A.; Milosevic, N.; Brabec, T.; Corcum, P.; Heinzmann, U.; Drescher, M.; Krausz, F. *Nature* 2001, 414, 509.

[24] Malinovskaya S.A.; Cederbaum L.S. *Phys. Rev. A* 2000, 61, 42706.

[25] Rice, S. A.; Zhao, M. *Optical Control of Molecular Dynamics*, John Wiley&Sons, Inc. 2000.

[26] Brumer, P. W.; Shapiro, M. *Principles of the Quantum Control of Molecular Processes*, John Wiley&Sons, Inc. 2003.

[27] Zeidler, D.; Frey, S.; Wohlleben, W.; Motzkus, M. *J. Chem. Phys.* 2002, 116, 5231.

[28] Herek, J.L.; Wohlleben, W.; Cogdell, R.J.; Zeidler, D.; Motzkus, M. *Nature* 2002, 417, 533.

[29] Weiner, A.M.; Leaird, D.E.; Wiederrecht, G.P.; Nelson, K.A. *Science* 1990, 247, 1317.

[30] Scully, M.O.; Zubairy, M.S. *Quantum optics*, Cambridge University Press 1997.

[31] Malinovskaya, S. *Int. J. Quant. Chem.* 2005, 102, 313.

[32] Malinovskaya, S.; Bucksbaum, P.; Berman, P. *Phys. Rev. A* 2004, 69, 013801.

[33] Malinovskaya, S.; Bucksbaum, P.; Berman, P. *J. Chem. Phys.* 2004, 121, 3434.

In: Trends in Chemical Physics Research
Editor: A.N. Linke, pp. 295-310

ISBN 1-59454-483-2

Chapter 12

POSSIBLE ROLE OF TELOMERES IN DNA IN AGING PROCESS

Md. Ashrafuzzaman* ***and Ahmed Shafee***[1]
Institute of Physics, University of Neuchatel, Switzerland
[1]Department of Physics, University of Dhaka, Bangladesh

Abstract

The interaction energies between the different types of bases of a single strand of DNA molecule have been calculated. Using these original values of energies the harmonic behavior of a number of base patterns of DNA has been studied. In view of the great interest aroused by the discovery of the role of the telomere segment of the DNA in the replication process and its possible link with the aging process, we have investigated, with simple models, the harmonic behavior of the telomeric pattern of bases as well as the thermodynamic response in the biological system. With these results a conclusion on the probable role of the telomeric pattern on aging has also been drawn. Here the calculated values of harmonic frequencies of the telomeric pattern of bases and of other possible patterns show that the telomeric pattern is associated with the highest vibrational frequency among all patterns of base combinations at the replication end of DNA. This seems to suggest that due to the existing telomeric pattern being closest to the frequencies of the electromagnetic radiation coming from sunlight, resonance of the telomeric frequency with such radiation may be responsible for damage to the reproductive ability of the cells and consequent aging and other problems. On the other hand in the last part of this work we have calculated the thermal vibrational amplitudes of the telomeric pattern and other possible patterns which show that the amplitude for the telomeric pattern is the least, and this suggests that the telomeric pattern is more mechanically and thermally stable than other possible patterns in the biological environment against damage from ordinary heat and mechanical effects.

PACS: 87.14.G,82.39.P,87.15.K

---

*E-mail address: md.ashrafuzzaman@unine.ch

1 Introduction

Most of the activities of the different parts of a living system are controlled in nucleic acids DNA (deoxyribonucleic acids) and RNA (ribonucleic acids). The nucleic acids are made of some combinations of purine and pyrimidine bases called genetic codons. The behavior and functioning of a living system are all controlled by different combinations of genetic codons. The possibility of the control of genetic characteristics is the most important area of the latest genome research. People are trying to find out how the genetic codes arrange themselves to make DNA and how they have evolved with time. Since the central dogma of biology is that genetic changes are brought about by changes in the macromolecules, it is of great importance also to investigate the stability of the molecules and their patterns responsible for the genotypical variations. It can of course be expected that some of the patterns of bases are more stable and they may be more persistent in organisms.

Attempts are being made to gain a full understanding of the role of the different molecules involved in the passing of genetic information to succeeding generations and of the pitfalls which lie in the path. That structure and function of macromolecules are interrelated is fully established at present. Hence any effort to understand biological processes at the molecular level requires a reasonable understanding of their structure, which in turn, can only be achieved by a thorough study of stability and energy considerations. Molecular biophysics, therefore, is an integral part of man's quest for complete control of his inner world.

For a long time biochemists and biologists have been trying to gather more information on structural and functional behavior of biomolecules in terms of enzymatic activities. They met some success in certain areas but the complete ideas of the behavior of the molecules are to be established. The nature of biochemical or enzymatic approach to know the structural and functional behavior of biomolecules is biological aspects and the limitations of these methods force the scientists to find other ways to explain the cause of stability or instability and some functional behavior of biomolecules. Therefore scientists are trying to impose the physical methods to explain the structural and functional behavior of biomolecules. In this new process to learn the behavioral aspects of biomolecules, the established methods or sometimes some modifications of the existing methods according to the problem of real or imaginary structures of biomolecules and their functioning are used.

For more than two decades physicists have been studying the structure-function relationship of the proteins. People now have a fairly good idea about how muscle filaments work or how some of the simpler enzymes perform their catalytic activities. Protein crystallography is an established science and using this method coordinates of the atoms of the smaller proteins, such as dihydrofolate reductase (DHFR), are known accurately. Many research groups all over the world are active in the ambitious task of designing molecules which will accelerate or inhibit the action of enzymes by attaching themselves to the enzyme at the active site. While this approach does remain a promising field, more recently it is becoming attractive to try to do the same with nucleic acids, i.e., to suppress or promote specific genes. Genes that cause diseases should be repaired, and if that is not possible, at least those causing production of harmful proteins should be inhibited. Some antibiotics prevent bacterial growth by stopping nucleic acid (NA) synthesis, others attempt to arrest the growth of cancer cells. However present methods do not allow choice of specific targets, e.g., bacterial NA or the NA of cancer cells. More knowledge is required about the conformational structures of the NAs, about their exibility and their binding to different ligands. Here too energy considerations are of vital importance.

Recently scientists have become aware of the existence of the link between aging and NAs. It is well known that cells can not replicate indefinitely in vivo. It is possible that despite the high accuracy of the process of replication, with proofreading arrangements to correct errors, it is impossible to eliminate errors altogether. The biological process continues independently and man can not control any biological activity completely. Any disorder initiated inside the biological system therefore sometimes grows multi-dimensionally. The random noise is accumulated and eventually replication stops.

Very recently scientists have begun to look into another possible source of loss of replicability of nucleic acids. The DNA molecule contains a tail called telomere, a certain pattern of bases which do not code for any gene, but on which the DNA polymerase has to stand when performing replication. It has been suspected that with each new cell generation in an organism, i.e., in the replication of somatic cells, parts of the telomere gets lost. As the telomere keeps shrinking with age, eventually a time comes when they can no longer replicate. As dying cells are not replaced by new cells, the organism decays or 'gets old' and finally faces death. This biological activity of DNA [1,2] is known to scientists but they don't know why and how this activity occurs.

M. Olovnikov formulated the problem of terminal under replication of linear DNA molecules in 1971[3]; this phenomenon is caused by the inability of DNA polymerases to replicate several nucleotides at 3 ends of DNA templates. Olovnikov also suggested that a specific biological mechanism should normally prevent this phenomenon. This mechanism was expected to be active in gametes, cancer cells, as well as in cells of vegetatively reproducing organisms. In most other cases, e.g., in many human somatic cells, this mechanism is suppressed.

Further studies revealed the enzyme telomerase [4,5] (whose existence had been predicted by Olovnikov) that compensates for DNA shortening in the mentioned cell types. The function of telomerase is to add a repeated sequence (the hexamer TTAGGG in humans), which forms the so-called telomere, to ends of nuclear DNA. Here A is for adenine, G for guanine and T for thymine. After this, underreplication of the linear DNA molecule only shortens this nontranscribed sequence of the telomeric fragment of the chromosome without damaging the genetic information or the mechanism that reads it. At certain stages of development in early embryogenesis, the gene encoding telomerase in the majority of human somatic cells is switched off, thereby making the genome susceptible to shortening. The telomere shortens at a low but appreciable rate which impairs the functioning of the chromosome. This impairment begins long before the disappearance of the whole telomere, which removes protection from genetic information contained in transcribed regions. This role is still poorly understood.

There is close correlation between shortening of telomeric DNA regions and Hay ick's limit [1,2]. To surpass this limit and continue reproduction, the cell should activate its telomerase gene.

The most remarkable feature is that switching off the telomerase gene is an ontogenetic stage that occurs at a distinct time point in the life of an organism and involves only some of its cell types. This event seems to perform a specific function and cannot be regarded as a disorder in the living system or a kind of unpredictable age-related defect, although it clearly promotes aging. In this context, we are reminded of an observation made to investigate the association of human telomerase activity with immortal cells and cancer [6]. During the development the telomere suddenly loses 50 kb (kilo bases). It loses an additional 20 kb during growth of the spike. The mechanism responsible for this event remains unknown. If the arrest of telomerase synthesis is considered as an accidental fault, telomere shortening in cells containing no functional telomerase appears to be an act of deliberate damage to the organism. An alternative interpretation is that the inhibition of the telomerase gene and the telomere shortening are biologically important events. To understand the meaning of this process, one should remember that telomerase genes of somatic cells are switched off only in sexually reproducing organisms, but not in vegetatively reproducing organisms. It is exactly in the former case that the appearance of a new trait, which can result from a combination of parental genomes, becomes the most probable event. However, in vegetatively reproducing organisms, the appearance of a new trait results from random mutations occurring in the same cell.

The above mentioned properties of the telomere make understanding the cause of its instability is an exciting problem. Again, it is possible that evolution has chosen the most stable pattern. This may be investigated by studying the energy of the telomere portion of the DNA. We can try to interpret this stability with respect to radiation, vibration etc i.e., by using physical rather than biological methods. If the energies involved between the bases of DNAs can be determined and if using these harmonic behaviors of the telomere can be calculated (i.e., the harmonic frequencies and thermal vibrational amplitudes of bases) it can probably be possible to get some ideas about the stability of this all important tail which needs to be retained for good health and to continue normal living behavior. Application of quantum mechanics can give a very good solution of energies and other necessary parameters to explain the states of a single body or in some cases two or three body problems. The absolutely correct way of carrying out such investigations would be to solve the Schrodinger equation for the entire DNA, which is, of course, quite impossible. Even small molecules can not be solved quantum mechanically, exactly or even by approximation methods. However, one can, as a zeroth order approximation, completely ignore the details of the interactions and study only the symmetry aspects of the problem by taking square well potentials or harmonic behavior, as is often done in condensed matter physics. The harmonic attitude has been adopted in this work and the telomere has been studied as a specific pattern of bases represented by harmonic oscillators. For this work the bases are considered as simple atoms connected to the nearest atom (base) by a force constant 'k' i.e., between the bases linear restoring forces are involved. We believe that even with such a drastic approximation a meaningful beginning can be made in trying to understand the stability problem of the telomere and the response of the pattern with the thermal, electromagnetic and other agents involved in environment. We have also introduced the thermodynamic effects on the telomere molecules in this work which would represent the dependence of structure and function of biological molecules, especially of telomere on thermal condition of the biological system.

In sec.II, the energies between the different bases in DNA have been calculated. For this some established methods of energy calculation have been used and the values of some parameters needed for this work of calculating the energies involved in DNA have also been calculated. In sec.III, a method of calculating the harmonic frequencies of telomere and other possible patterns taking four purines and two pyrimidines and also for few other imaginary patterns of bases has been developed. For this a matrix has been developed and using the energies involved between the bases in DNA the harmonic frequency of telomere has been calculated and a conclusion on the stability of telomere pattern in the replication end of DNA has been drawn. In sec.IV, the thermodynamic behavior of DNA molecules especially the temperature effect on the stability of telomere has been discussed. Here it is tried to find the effect of the environment especially the biological environment where the DNA exists on the structure and function of telomere. Finally in sec.V, the results are discussed and the conclusions are presented.

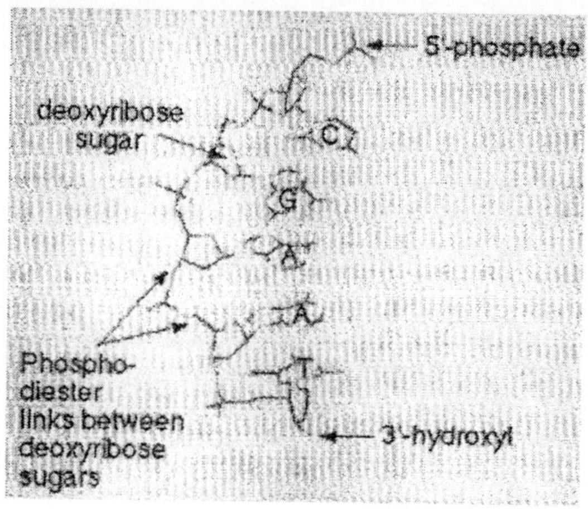

Figure 1: Sequence of bases in DNA single strand are shown. Here it is seen that the aromatic rings of the bases are parallel to each other.

2 Energy Calculation for Base Sequences in DNA

2.1 Introduction

In DNA the successive bases exist in energetically equilibrium state. FIG.1 shows a possible sequences of bases and their attachment with the sugar rings. The most important energies between the successive bases are stacking energy and van der Waals interaction. Stacking energy is due to interaction between the magnetic moments of the helical structure of bases and van der Waals interaction is the well known 6-12 potential that arises when the two bases come very near to each other. In this work these two types of interactions between the bases of DNA have been calculated using the well known formulae.

2.2 Energy Modeling

A computer model has been constructed to make DNA double helices according to whatever base pattern sequence we required. This was done by smoothly fitting experimental data for each base pair configuration with one another to form a straight double helical structure. However some exibility were also retained so as to vary the inter base separation to some extents. While lengthening the double helix by stretching it our method also changes the twist angles to some extends so that the interbase angles which are normally 36° for 10 base pairs per turn per piece can vary to some extends in keeping with the inter base separation h which is normally 0.36 nm [7-12]. It is well known that the most important energy in forming double helix is stacking energy. The model described here has been used to calculate this stacking energy for different base pair neighbors. The pyrimidine (Cytosine and Thymine) have single aromatic six side a ring pattern whereas for the ring in purine (Adenine and Guanine) each has adjoining six sided and a five sided aromatic rings. These rings content electrons that contribute magnetic moments. The energy between two rings i.e., between two such magnetic moments is given by

$$U(\mathbf{r}_1,\mathbf{r}_2) = \frac{\mu_o}{4\pi}[\frac{\mu_1 \cdot \mu_2}{|\mathbf{r}_1 - \mathbf{r}_2|^3} - 3\frac{\mu_1 \cdot (\mathbf{r}_1 - \mathbf{r}_2)\mu_2 \cdot (\mathbf{r}_1 - \mathbf{r}_2)}{|\mathbf{r}_1 - \mathbf{r}_2|^5}] \quad (1)$$

When taking into account the orthogonality of the s (i.e., the base planes) with the helical axis, this reduces to

$$U(|\mathbf{r}_1 - \mathbf{r}_2|) = \frac{\mu_o(lg)^2}{4\pi}[1 - 3\frac{d^2}{s^2}] \quad (2)$$

where $\mu_o = 4\pi \times 10^{-7}$ SI unit, is the magnetic permeability in free space, s is the actual distance between the centers of the two rings and d is the 2 dimensional projection of the separation in the plane of the base l is a no. and r's are coordinates. μ_1 and μ_2 are proportional to the angular momentum of electrons by the usual gyromagnetic ratio (g) and can be taken identical for other rings because always π electrons are involved.

It is to be mentioned the part of energy that can counter the DNA to be collapsed by the attractive stacking energy. This type of energy is the van der Waals interaction which is given by

$$U(i,j) = -\frac{A}{r_{ij}^6} + \frac{B}{r_{ij}^{12}} \quad (3)$$

where A and B are van der Waals constants. The van der Waals interactions becomes dominating when two atoms come within van der Waals limit distance. This type of interaction is also called the 6-12 potential.

Special feature for DNA double helix is that all the base rings are perpendicular to the axis of the double helix. This feature is included in the energy calculation. It has also been taken in the term the rotation of the bases as the helix goes down as is already mentioned. Combining the two energies mentioned above the equation for the calculation of the total interaction energy [13] between the two bases of DNA is being calculated which is as follows

$$V_{ij} = \frac{A}{(z^2+r^2)^6} + \frac{B}{(z^2+r^2)^{3/2}}(1-3\frac{z^2}{z^2+r^2}) \tag{4}$$

where z represents the variable separation between the i th and j th bases and r represents the radius.

In the real configuration of DNA the average separation of two successive bases is about 0.36 nm and in the calculation of the total energy here this separation has been used as the minimum energy state and thus this state is more stable. Using this minimum energy state at 0.36 nm separation the constants A and B have been calculated. In calculating the values of the force constants the value of energy has been taken from the deviated portion from this minimum level.

2.3 Results

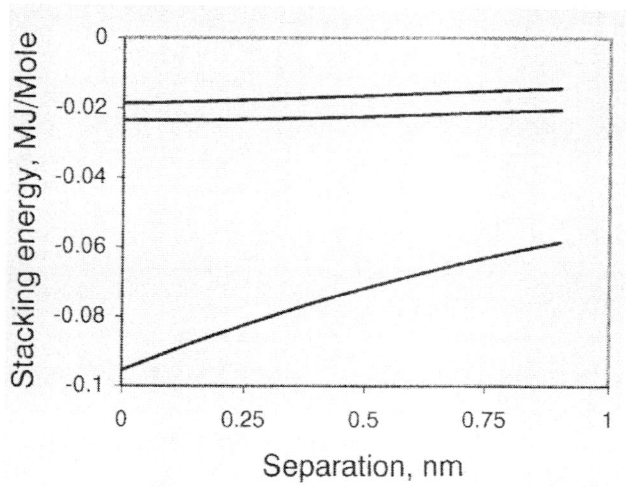

Figure 2: Stacking Energy versus Separation plot between bases in DNA. Different curves from upper to lower are for successive pyrimidine pyrimidine, purine pyrimidine, purine purine bases in DNA.

The results for the stacking energies and the total energies including the van der Waals energy with respect to the different distances between the bases in DNA are shown in Fig.2 and Fig.3 respectively. In these calculations the stacking energies between purine-purine bases at different separations are seen to be higher towards negative than those between

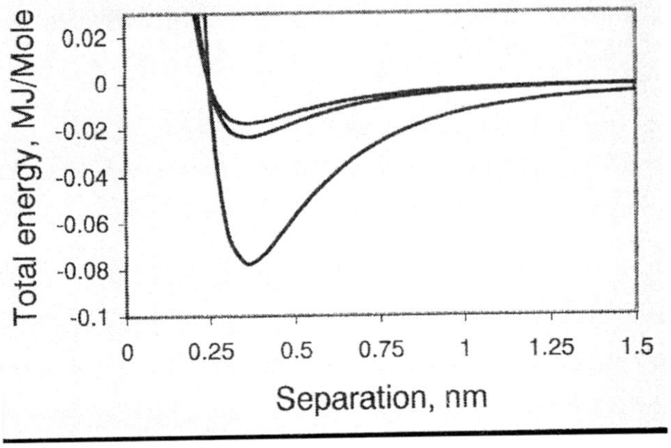

Figure 3: Total Energy (stacking+van der Waals) versus Separation plot between bases in DNA. Different curves from upper to lower are for successive pyrimidine pyrimidine, purine pyrimidine, purine purine bases in DNA.

purine-pyrimidine and pyrimidine-pyrimidine bases respectively. Similar sequence is also seen for the calculation of total interaction energies between the bases.

2.4 Discussion

The stacking energy calculation between the bases shows that the stacking energies for different order of separation between purine purine bases are higher than those between purine pyrimidine and pyrimidine pyrimidine bases. This is because pyrimidine (Cytosine and Thymine) have single aromatic six side a ring pattern whereas for the ring in purine (Adenine and Guanine) each has adjoining six sided and a five sided aromatic rings. So for purine purine interaction extra stacking for extra rings exists and therefore the stacking energy between purine purine bases is much higher than that between pyrimidine pyrimidine bases. When the two rings come close due to their attraction by stacking energy they should collapse as the stacking interaction is only attractive. So in lieu of being stable the whole DNA molecule should collapse. But this type of collapsing is encountered by van der Waals high repulsive 12 potential which becomes dominating when the charge distributions of two atoms overlap. So the attractive interaction of stacking energy is thus compensated by the high repulsive van der Waals interaction.

3 Frequency Calculation for Different Base Sequences in DNA

3.1 Introduction

DNA bases are considered as molecules those are attached to each other by intermolecular forces that are harmonic. So between two bases there exists force constants. The values of force constants k between different pairs e.g., purine-purine, pyrimidine-pyrimidine, purine-pyrimidine etc. are different. Due to this force DNA molecules follow the harmonic behavior and therefore they vibrate with some harmonic frequencies.

3.2 Derivation of Expression for Frequency

The values of k's are calculated from the van der Waals interaction and stacking interaction [7,11,12] between bases of different types e.g., purine-purine, pyrimidine-pyrimidine, purine-pyrimidine etc. Using these values the harmonic frequencies of the telomere in DNA are calculated. To illustrate the technique for obtaining the resonant frequencies and normal modes it is considered in detail a model based on a linear symmetrical six atomic molecule. All these six atoms are considered to be on one straight line, the equilibrium distances apart being denoted by b. For simplicity it is considered that only vibrations along the line of the molecule, and the actual complicated interatomic potential are approximated by two spring of two different or similar force constants on both sides of an atom. There are six obvious coordinates marking the position of the six atoms on the line. The atomic arrangement then repeats for the telomere in DNA like AGGGTT then again AGGGTT and so on upto a huge number of repeatation. If the mean positions of the atoms are considered as $x_1, x_2, x_3, x_4, x_5, x_6$ then the positions of the atoms repeat and thus the arrangement completes a cyclic arrangement. In telomere system it is assumed that there exists two types of atoms: four purine (one Adenine and three Guanine) and two pyrimidine (two Thymine). It is true that Adenine and Guanine have about the same mass which is considered as M and the mass of pyrimidine (Thymine and Cytocine) is different from that of purine and let this be m. Both these M and m have been calculated from the total atomic weights of their constituent atoms. In the arrangement there may exist the sequences like purine-purine, purine-pyrimidine, pyrimidine-pyrimidine and the force constants for these arrangements are considered k_1, k_2, k_3 respectively. If a force constant k in between two atoms of mass M each, the potential energy in this system is given by Potential energy

$$V_2 = \frac{1}{2}k(x_2 - x_1 - b)^2 \tag{5}$$

For a three atoms system this potential energy becomes

$$V_3 = \frac{1}{2}k(x_2 - x_1 - b)^2 + \frac{1}{2}k(x_3 - x_2 - b)^2 \tag{6}$$

with the atomic arrangement as follows
$m(x_1) - k - M(x_2) - k - m(x_3)$

Similarly if we increase the number of atoms according to our real problem of telomere (AGGGTT) in DNA with six atoms the arrangement becomes like this
$M(x_1) - k_1 - M(x_2) - k1 - M(x_3) - k1 - M(x_4) - k2 - m(x_5) - k3 - m(x_6) - k2 - M(x_1)$

For this system of atoms the total potential energy V is

$$V = +\frac{1}{2}k_1(x_2 - x_1 - b)^2 + \frac{1}{2}k_1(x_3 - x_2 - b)^2$$

$$+\frac{1}{2}k_1(x_4-x_3-b)^2+\frac{1}{2}k_2(x_5-x_4-b)^2$$
$$+\frac{1}{2}k_3(x_6-x_5-b)^2+\frac{1}{2}k_2(x_1-x_6-b)^2 \tag{7}$$

Now we introduce coordinates relative to the equilibrium positions:

$$y_i = x - x_{oi} \tag{8}$$

where

$$x_{o2} - x_{o1} = b = x_{o3} - x_{o2}, \quad etc. \tag{9}$$

The potential energy then reduces to

$$\begin{aligned} V &= +\frac{1}{2}k_1(y_2-y_1)^2+\frac{1}{2}k_1(y_3-y_2)^2 \\ &\quad +\frac{1}{2}k_1(y_4-y_3)^2+\frac{1}{2}k_2(y_5-y_4)^2 \\ &\quad +\frac{1}{2}k_3(y_6-y_5)^2+\frac{1}{2}k_2(y_1-y_6)^2 \end{aligned} \tag{10}$$

Hence the V matrix has the form,

$$V_m = \begin{pmatrix} k_1+k_2 & -k_1 & 0 & 0 & 0 & -k_2 \\ -k_1 & 2k_1 & -k_1 & 0 & 0 & 0 \\ 0 & -k_1 & 2k_1 & -k_1 & 0 & 0 \\ 0 & 0 & -k_1 & k_1+k_2 & -k_2 & 0 \\ 0 & 0 & 0 & -k_2 & k_2+k_3 & -k_3 \\ -k_2 & 0 & 0 & 0 & -k_3 & k_2+k_3 \end{pmatrix} \tag{11}$$

The kinetic energy T has the form

$$T = \frac{1}{2}M(\dot{y}_1^2+\dot{y}_2^2+\dot{y}_3^2+\dot{y}_4^2)+\frac{1}{2}m(\dot{y}_5^2+\dot{y}_6^2) \tag{12}$$

The T matrix is diagonal which is of the form,

$$T_m = \begin{pmatrix} M & 0 & 0 & 0 & 0 & 0 \\ 0 & M & 0 & 0 & 0 & 0 \\ 0 & 0 & M & 0 & 0 & 0 \\ 0 & 0 & 0 & M & 0 & 0 \\ 0 & 0 & 0 & 0 & m & 0 \\ 0 & 0 & 0 & 0 & 0 & m \end{pmatrix} \tag{13}$$

Combining these two matrices, the secular equation appears as

$$|V_m - \omega^2 T_m| = 0 \tag{14}$$

So the square of the frequencies ($i\omega$) is given by the eigen values of the following matrix

$$\begin{pmatrix} \frac{k_1+k_2}{M} & \frac{-k_1}{M} & 0 & 0 & 0 & \frac{-k_2}{M} \\ \frac{-k_1}{M} & \frac{2k_1}{M} & \frac{-k_1}{M} & 0 & 0 & 0 \\ 0 & \frac{-k_1}{M} & \frac{2k_1}{M} & \frac{-k_1}{M} & 0 & 0 \\ 0 & 0 & \frac{-k_1}{M} & \frac{k_1+k_2}{M} & \frac{-k_2}{M} & 0 \\ 0 & 0 & 0 & \frac{-k_2}{m} & \frac{k_2+k_3}{m} & \frac{-k_3}{m} \\ \frac{-k_2}{m} & 0 & 0 & 0 & \frac{-k_3}{m} & \frac{k_2+k_3}{m} \end{pmatrix} = 0 \qquad (15)$$

Here the values for k's (at the separation 0.39 nm between bases) are used from the calculated values from the total energies (stacking and van der Waals interaction) between the bases of DNA. The masses of purine and pyrimidine are the total masses of their constituent atoms.

3.3 Results

The eigenvalues of the derived matrix gives the square of the frequencies ($i\omega$). Hence negative values of the eigen values give the real parts of the frequencies and positive values give the imaginary parts of the frequencies. From the eigenvalues of the matrix the values of the frequencies of the harmonic oscillator of telomeric pattern have been calculated. Similarly rearranging the patterns of the matrices for other patterns of base combination of telomere and for some imaginary patterns taking all purine or all pyrimidines the eigenvalues and thus the frequencies have been calculated.

The highest values of the frequencies [13] of telomere and other different patterns of base combinations are as follows:

Telomere (AGGGTT) pattern:
ω =-3.104028110062502 $\times 10^{12}$ Hz, ω =+3.104028110062502 $\times 10^{12}$ Hz

AGTGGT pattern:
ω =-2.586737456962528 $\times 10^{1}2$ Hz, ω =+2.586737456962528 $\times 10^{12}$ Hz

AGTGTG pattern:
ω =-2.944491229994985 $\times 10^{12}$ Hz, ω =+2.944491229994985 $\times 10^{12}$ Hz

AAAAAA or GGGGGG pattern:
ω =-3.333518830239974 $\times 10^{12}$ Hz, ω =+3.333518830239974 $\times 10^{12}$ Hz

TTTTTT or CCCCCC pattern:
ω =-1.668814797430148 $\times 10^{12}$ Hz, ω =+1.668814797430148 $\times 10^{12}$ Hz

3.4 Discussion

In the calculation of the frequencies of the oscillators of the telomere pattern of bases and the other patterns it is seen that the value of the frequency for the telomere pattern is the highest among those patterns having four purine and two pyrimidine bases. The frequency of the telomere pattern is close to the frequencies available in the sunlight. So a resonance of this frequency with the frequencies of the infrared rays existing in the lower side of the band of light coming from the sun may occur. Most of the people stay under shade in most of their life time. The infrared light of the frequency of the order of the frequency of telomere are in the lower end of the band of light coming from the sunlight and staying in the shade (in most time) people experience this infrared lights. The interference of this infrared rays with the frequency of the telomere may be one of the possible causes of the aging i.e., the shrinking of the skin and decaying of the normal activities of the cells of human body and finally leading to death. The frequencies of the two artificial patterns of bases e.g., all purine and all pyrimidines are different from the original pattern with four purines and two pyrimidines. For all purines (GGGGGG or AAAAAA) the value of the frequency is higher than that for telomeric pattern which supports for being more stable than the telomeric pattern but there may arise some problem in replicability of DNA due to the heavy attraction of the pattern GGGGGG in the first strand with CCCCCC in the second strand because GC has very high attraction between two strands. On the other hand for all purines like AAAAAA the unstability also arises due to very weak TA or AT coupling between the bases from two strands.Similar compliments are valid for the base sequences like TTTTTT and CCCCCC where the frequencies take the least values. Hence to continue the normal living behavior of biological molecules telomere pattern should the the most stable pattern at the replication end of DNA.

4 Amplitude of Vibration in Thermal Equilibrium

4.1 Introduction

Human body exists at a particular temperature. This temperature is at an average about 37^oC. The temperature normally does not change with the change of the external or environmental temperature. DNA molecules exist inside this temperature state. The telomeres in DNA thus exist in the thermodynamic state of temperature $37\,^oC$. The telomere pattern AGGGTT has been considered as an oscillator that vibrates with some frequencies. This vibration should create some thermodynamic motion of these molecules due to the thermal state. It is known from the well known kinetic interpretation of temperature that an atom being in the thermal state of temperature T has an average translational kinetic energy equal to $\frac{3}{2}kT$. Here k is the Boltzmann constant that is equal to $1\,.38\times10^{-23}$ joule/molecule K and T is the absolute temperature of the environment i.e., $T=273+t$, where t is in ^oC and T is in ^oK. For one dimentional motion the average energy is $\frac{1}{2}kT$. Now any atom or molecule of mass M vibrating with a frequency ω should have the total energy $\frac{1}{2}Ma^2\omega^2$ where a is the amplitude of the vibration. In this problem it is considered for DNA telomere bases (M

for purine and m for pyrimidine)

$$\frac{1}{2}Ma^2\omega^2 = \frac{1}{2}kT \tag{16}$$

The calculated values of the frequencies for telomere pattern AGGGTT and other possible random patterns have been used here. The calculation gives us the following results for the amplitudes of vibration.

4.2 Results and Discussions

The following results show the amplitudes of thermal vibration for purine (Adenine and Guanine) and pyrimidine (Cytosine and Thymine) in DNA.

TELOMERE:
Pyrimidine: $a = +/-0.0311nm$
Purine: $a = +/-0.0292nm$

AGTGTG pattern:
Pyrimidine: $a = +/-0.0328nm$
Purine: $a = +/-0.0308nm$

AGTGGT pattern:
Pyrimidine: $a = +/-0.0374nm$
Purine: $a = +/-0.0351nm$

AAAAAA pattern:
Pyrimidine: $a = +/-0.0290nm$
Purine: $a = +/-0.0272nm$

TTTTTT pattern:
Pyrimidine: $a = +/-0.0579nm$
Purine: $a = +/-0.0543nm$

The least values for amplitudes for purine and pyrimidine are found in the telomere pattern (AGGGTT) among the patterns having two pyrimidine and four purine base sequences in DNA. The higher the values of the thermodynamic vibrational amplitudes from the mean position of rest the higher the possibility of the breaking of the biological structure. Hence the lowest values of the amplitudes of thermodynamic vibrational for telomere structure support for the stability of telomere pattern from the thermodynamic point of view. Here all purine and pyrimidine base arrangements, showing extreme low and high amplitudes, are also not physically favorable to exist to continue the biological processes.

5 Discussion and Conclusions

In this work some basic parameters involved in DNA, the biggest molecule, have been calculated. In doing this some approximations are considered but these approximations do not affect the reality of the problem. Though the method of calculation is fairly simple, using sometimes somewhat drastic approximations, it is remarkable that the results which are achieved here are either in agreement with experimental observations, or, when that is lacking, in agreement with the idea of molecular evolution, viz. the approach to states of stability or lower energy in molecular configurations that presumably are less susceptible to perturbations of the environment.

A possible cause of the stability of the telomeric pattern in the replication end of DNA has been suggested in this work. Before that the interaction energies involved between the successive bases in a DNA single helix have been calculated. With these energies the values of the force constants have been calculated for the different patterns of base combinations i.e., between the purine-purine, purine-pyrimidine and pyrimidine-pyrimidine sequence. It is seen that the minimum energy state for the sequence purine-purine is lower than all other sequences i.e., purine-pyrimidine and pyrimidine-pyrimidine. In reality too it should be so, because in purine there exists one additional ring than the single ring in pyrimidine base.

Using the values of force constants k's [13] from slightly deviated portions of the minimum energy positions, the frequencies for the telomeric pattern (AGGGTT) and other patterns have been calculated. Here the results show that the highest frequency exists in the telomeric pattern among the different possible patterns, taking four purines and two pyrimidines. The value of the frequency for telomeric pattern is close to the infrared band of the frequencies available in sunlight. The frequency of the telomeric pattern may have thus a resonance with the infrared rays existing below the lower end of visible spectrum of sunlight. Human beings can not avoid exposure to direct or indirect sunlight during most of their lifetime, and are therefore vulnerable to the effects of the infrared frequencies. The possible resonance of the telomeric frequency with the infrared frequencies may play an important role in the aging process and other maladies such as cancer in the lives of human beings.

On the other hand this frequency for the telomeric pattern may be the cause of the thermal and mechanical stability of the pattern AGGGTT in the replication end of DNA and therefore the pattern repeats such a large number of times. In this work the frequencies for some imaginary patterns with all purines and with all pyrimidines in a single helix are also calculated where the results show that the frequency of the pattern with all purines (Adenine or Guanine) is the highest among all the patterns and the frequency for the pattern with all pyrimidines (Thymine or Cytosine) is the lowest. Therefore the pattern AAAAAA or GGGGGG show most stable than other patterns. Here only a single strand of the DNA is considered. When both strands are considered together, of course there is an equality of the purines and pyrimidines. A problem may arise here due to the anomaly of attraction of these patterns with their counter strand in DNA as GC has higher attraction than AT. So if all Gs exist in the 1st coil then in the 2nd coil there will be all Cs. So highest binding between GC bonding may be a cause of stopping of the replication of DNA, but that would be impossible for continuing the living activity of human beings, or any other organism. So, for stability of the telomeric pattern (AGGGTT) there should be an optimization between the stability and replicability of DNA and that might be a reason why the specific telomeric pattern continues in human DNA at the replication end.

With the values of the frequencies of the telomeric pattern and other possible patterns of DNA bases with four purines and two pyrimidines, the thermodynamic vibrational amplitudes have been calculated in this work. Here too telomeric pattern shows the lowest value of the amplitude of vibration, which supports the stability of the configuration with this pattern (AGGGTT) in the thermal environment of human cells. For purines (A or G) the amplitudes are smaller than that of pyrimidine (T or C) because purines are more massive than pyrimidines due to the presence of extra ring in purines.

In this work some physical methods are used and some mathematical formulas have been developed for calculating the interaction energies and frequencies of DNA molecules. Calculations have been performed with these formulas and appropriate conclusions have been drawn. In future when more crystallographic data become available for base sequences of DNA for human genome or other living organs, the work can be extended to test this hypothesis about the association between lower energy and preferred molecular configurations with molecular arrangements. Thus perhaps the work can be extended to a level which can explain the most important causes of aging and other phenomena involved in living bodies, not only for human beings but also for other living systems.

Acknowledgment

We thank Gias uddin Ahmad for interesting discussions.

References

[1] Hay ick, L., and Moorhead, P. S. (1961) *Exp. Cell Res.*, **25**, 585-621.

[2] Hay ick, L. (1997) *Biochemistry* (Moscow), **62**, 1380-1393 (Russ.).

[3] Olovnikov, A. M. (1971) *Dokl. Akad. Nauk SSSR*, **201**, 1496-1498.

[4] Melek, M., Greene, E. C., and Skippen, D. E. (1996) *Mol. Cell Biol.*, **16**, 3437-3445.

[5] Harley, C. B. (1991) *Mutat. Res.*, **256**, 271-282.

[6] Kim, N. W., Piatyszek, M. A., Prowse, K. R., Harley, C. B., West, M. D., Ho, P. L. C., Coviello, G. M., Wright, W. E., Weinrich, S. L., and Shay, J. W. (1994) *Science*, **266**, 2011-2015.

[7] Fengelman, M. et al.: *Nature* London 175, 834 (1955)

[8] *Biophysics by Walter Hoppe*, Lohmann, H. Markl, H. Ziegler, ed. 2nd p25, t2.2

[9] G.N. Ramachandran and V. Sasisekharan, *Adv. Protein Chem*. **23** (1968) 283-437

[10] G.N. Ramachandran, C. Ramakrisanan and V. Sasisekharan, *J. Mol.*,**7** (1963) 95-99

[11] Volkenshtein, M.V.: *Molecules and Life*. New York-London, Plenum Press 1970

[12] *Biophysics by Walter Hoppe*, Lohmann, H. Markl, H. Ziegler, ed. 2nd p24, s2.2.4 Thesis, Md. Ashrafuzzaman, Department of Physics, BUET, Dhaka, July, 2000.

INDEX

A

B

C

D

E

J

K

L

M

N

O

P

Q

R

S

T

U

V

W

X

Y

Z